Computational Methodologies for Electrical and Electronics Engineers

Rajiv Singh
G.B. Pant University of Agriculture and Technology, Pantnagar, India

Ashutosh Kumar Singh
Indian Institute of Information Technology, Allahabad, India

Ajay Kumar Dwivedi
SIET, India & Dr. A.P.J. Abdul Kalam Technical University, Lucknow, India

P. Nagabhushan
Indian Institute of Information Technology, Allahabad, India

A volume in the Advances in Computer and Electrical Engineering (ACEE) Book Series

IGI Global
PUBLISHER of TIMELY KNOWLEDGE

Published in the United States of America by
IGI Global
Engineering Science Reference (an imprint of IGI Global)
701 E. Chocolate Avenue
Hershey PA, USA 17033
Tel: 717-533-8845
Fax: 717-533-8661
E-mail: cust@igi-global.com
Web site: http://www.igi-global.com

Library of Congress Cataloging-in-Publication Data

Names: Singh, Rajiv, editor. | Singh, Ashutosh Kumar, editor. | Dwivedi, Ajay Kumar, 1987- editor. | Nagabhushan, P., editor.
Title: Computational methodologies for electrical and electronics engineers / Rajiv Singh and Ashutosh Kumar Singh, and Ajay Kumar Dwivedi and P. Nagabhushan, editors.
Description: Hershey, PA : Engineering Science Reference, an imprint of IGI Global, [2021] | Includes bibliographical references and index. | Summary: "This book examines the application of intelligence computational techniques and methods in energy and power systems"-- Provided by publisher.
Identifiers: LCCN 2019051060 (print) | LCCN 2019051061 (ebook) | ISBN 9781799833277 (hardcover) | ISBN 9781799833284 (paperback) | ISBN 9781799833291 (ebook)
Subjects: LCSH: Electrical engineering--Data processing. | Electronics--Data processing.
Classification: LCC TK153 .C573 2021 (print) | LCC TK153 (ebook) | DDC 621.30285--dc23
LC record available at https://lccn.loc.gov/2019051060
LC ebook record available at https://lccn.loc.gov/2019051061

This book is published in the IGI Global book series Advances in Computer and Electrical Engineering (ACEE) (ISSN: 2327-039X; eISSN: 2327-0403)

British Cataloguing in Publication Data
A Cataloguing in Publication record for this book is available from the British Library.

Advances in Computer and Electrical Engineering (ACEE) Book Series

Srikanta Patnaik
SOA University, India

ISSN:2327-039X
EISSN:2327-0403

Mission

The fields of computer engineering and electrical engineering encompass a broad range of interdisciplinary topics allowing for expansive research developments across multiple fields. Research in these areas continues to develop and become increasingly important as computer and electrical systems have become an integral part of everyday life.

The **Advances in Computer and Electrical Engineering (ACEE) Book Series** aims to publish research on diverse topics pertaining to computer engineering and electrical engineering. **ACEE** encourages scholarly discourse on the latest applications, tools, and methodologies being implemented in the field for the design and development of computer and electrical systems.

Coverage

- VLSI Design
- Computer Architecture
- Computer Hardware
- Algorithms
- Sensor Technologies
- Programming
- Computer Science
- Optical Electronics
- Power Electronics
- Digital Electronics

IGI Global is currently accepting manuscripts for publication within this series. To submit a proposal for a volume in this series, please contact our Acquisition Editors at Acquisitions@igi-global.com or visit: http://www.igi-global.com/publish/.

The Advances in Computer and Electrical Engineering (ACEE) Book Series (ISSN 2327-039X) is published by IGI Global, 701 E. Chocolate Avenue, Hershey, PA 17033-1240, USA, www.igi-global.com. This series is composed of titles available for purchase individually; each title is edited to be contextually exclusive from any other title within the series. For pricing and ordering information please visit http://www.igi-global.com/book-series/advances-computer-electrical-engineering/73675. Postmaster: Send all address changes to above address. ©

Titles in this Series

For a list of additional titles in this series, please visit: www.igi-global.com/book-series

Innovations in the Industrial Internet of Things (IIoT) and Smart Factory
Sam Goundar (The University of the South Pacific, Fiji) J. Avanija (Sree Vidyanikethan Engineering College, India) Gurram Sunitha (Sree Vidyanikethan Engineering College, India) K. Reddy Madhavi (Sree Vidyanikethan Engineering College, India) and S. Bharath Bhushan (VIT Bhopal University, India)
Engineering Science Reference • © 2021 • 311pp • H/C (ISBN: 9781799833758) • US $225.00

Innovative Applications of Nanowires for Circuit Design
Balwinder Raj (National Institute of Technical Teachers Training and Research, Chandigarh, India)
Engineering Science Reference • © 2021 • 263pp • H/C (ISBN: 9781799864677) • US $225.00

Optimizing and Measuring Smart Grid Operation and Control
Abdelmadjid Recioui (Université M'hamed Bougara de Boumerdes, Algeria) and Hamid Bentarzi (Université M'hamed Bougara de Boumerdes, Algeria)
Engineering Science Reference • © 2021 • 400pp • H/C (ISBN: 9781799840275) • US $225.00

New Methods and Paradigms for Modeling Dynamic Processes Based on Cellular Automata
Stepan Mykolayovych Bilan (State University of Infrastructure and Technology, Ukraine) Mykola Mykolayovych Bilan (Mayakskaya Secondary School, Moldova) and Ruslan Leonidovich Motornyuk (Main Information and Computing Center, Ukraine)
Engineering Science Reference • © 2021 • 326pp • H/C (ISBN: 9781799826491) • US $195.00

Examining Quantum Algorithms for Quantum Image Processing
HaiSheng Li (Guangxi Normal University, China)
Engineering Science Reference • © 2021 • 336pp • H/C (ISBN: 9781799837992) • US $195.00

Design and Investment of High Voltage NanoDielectrics
Ahmed Thabet Mohamed (Aswan University, Egypt & Qassim University, KSA)
Engineering Science Reference • © 2021 • 363pp • H/C (ISBN: 9781799838296) • US $195.00

Research Advancements in Smart Technology, Optimization, and Renewable Energy
Pandian Vasant (University of Technology Petronas, Malaysia) Gerhard Weber (Poznan University of Technology, Poland) and Wonsiri Punurai (Mahidol University, Thailand)
Engineering Science Reference • © 2021 • 407pp • H/C (ISBN: 9781799839705) • US $225.00

IGI Global
PUBLISHER of TIMELY KNOWLEDGE

701 East Chocolate Avenue, Hershey, PA 17033, USA
Tel: 717-533-8845 x100 • Fax: 717-533-8661
E-Mail: cust@igi-global.com • www.igi-global.com

Table of Contents

Detailed Table of Contents

This chapter is about a basic power-efficient object dropping game named "Catch Me If You Can," which works on the Arduino platform. A dropping mechanism is developed using the servo motors interfaced to Arduino. The mechanism includes an object holder attached to the servomotor and a loading tube. The dropping of the objects is controlled by any wireless Bluetooth controlling device like a mobile phone or joystick and by the keypad interface installed in the design. The game is about catching the objects dropping randomly as a challenge which is controlled by the operator. The overall simulated design can be done in EasyEDA platform. The overall game can be controlled by an app which is designed by MIT App Inventor. This game can be implemented in Amusement parks, exhibitions, kid schools, and shopping malls. Besides this entertainment aspect, commercially it works as a small-scale product packaging system by involving DC motors, which are needed in moving packaging belts. This mechanism is efficient in packaging products in some specific count.

3D reconstruction is a long-standing complication when comes to testing happening from decades from machine learning, computer graphics, and computer perspective environments. Using CNN for the reconstruction of the 3D image has enchanted growing attentiveness and shown spectacular execution. Emerging in the new era of abrupt development, this chapter lays out an in-depth study of the latest developments in the field. Its focuses on activities that use in-depth learning strategies for measuring the 3D status of common things from one or more RGB images. It organizes based on literature in the layout presentations, network structures, and training methods they use. As the survey was conducted for methods of reconstructing common objects, this chapter also evaluates some of the latest efforts that emphasize particular categories of an object such as the shape of the human body and face. This provides an examination and correspondence of the execution of some important papers, summarizing some open-ended issues in the field, and exploring encouraging indications for subsequent research.

Machine learning-based intrusion detection system (IDS) is a research field of network security which depends on the effective and accurate training of models. The models of IDS must be trained with new attacks periodically; therefore, it can detect any security violations in the network. One of most frequent security violations that occurs in the network is denial of service (DoS) attack. Therefore, training of IDS models with latest DoS attack instances is required. The training of IDS models can be more effective when it is performed with the help of machine learning algorithms because the processing capabilities of machine learning algorithms are very fast. Therefore, the work presented in this chapter focuses on building a model of machine learning-based intrusion detection system for denial of service attack. Building a model of IDS requires sample dataset and tools. The sample dataset which is used in this research is NSL-KDD, while WEKA is used as a tool to perform all the experiments.

Smart grid can work effectively only when a reliable, fast communication network is available. The communication network is a prerequisite to connect different protection, control, and monitoring equipment within the substation. Ethernet fulfills all the requisites such as reliable, fast, secure, interoperable, LAN-based communication system for smart substations. Therefore, the main aspect is to improve the reliability of the network by prioritizing the critical components by using the knowledge of component importance measures (CIM). In this chapter, analysis of IEC 61850 ethernet-based substation communication network (SCN) architectures has been examined using various reliability importance measures (RIM). The importance measures namely Birnbaum, improvement potential, criticality importance, and reliability achievement worth have given their justified rankings of the various components of SCN architectures. The practice of these CIMs works towards the identification of the components that can be allocation of resources for the improvement of system reliability.

Organizations now have knowledge of big data significance, but new challenges stand up with new inventions. These challenges are not only limited to the three Vs of big data, but also to privacy and security. Attacks on big data system ranges from DDoS to information theft, ransomware to end user level security. So implementing security to big data system is a multiple phase-based ongoing process in which security is imposed from perimeter level to distributed file system security, cloud security to data security, storage to data mining security, and so on. In this chapter, the authors have identified some key vulnerable point related to big data and also proposed a security model.

In a human body, the heart is the second primary organ after the brain. It causes either a long-term impairment or death of a person if suffering from a cardiovascular disease. In medical science, a proper medical analysis and examination of a cardiovascular disease is very crucial, convincing, and sophisticated task for saving a human life. Data analytics rises because of the absence of sufficient practical tools for exploring the trends and unknown relationships in e-health records. It predicts and achieves information which can ease the diagnosis. This survey examines cardiovascular disease prediction systems developed by different researchers. It also reviews the trend of machine learning approaches used in the past decade with results. Related studies comprise the performance of various classifiers on distinct datasets.

Amputation, especially of the upper limbs, is a condition that exists in almost all parts of the world. There are more than 110 thousand amputees in India itself. It is extremely difficult for amputees to carry out their daily activities and to deal with daily life as normal people do. The purpose of the myoelectric prosthesis is to restore the basic functions of the lost organs in the joint using neural signals produced by the muscles. Unfortunately, the use of such myosignals is complicated. In addition, once detected, it usually requires a computational force strong enough to convert it into a user-controlled signal. Its modification to the actual function of the implant is limited by a number of factors, especially those associated with the fact that each amputee has a different muscle movement. Modified artificial intelligence systems designed for pattern recognition have the potential to improve the size of implants but still fail to provide a system in which artificial arms can be controlled by brain signals.

This chapter proposes a switched inductor configuration-based non-isolated DC-DC converter with high voltage gain. The proposed converter has two output capacitors instead of a single output capacitor for voltage boosting capabilities. Enhancement of the output voltage with the addition of more number of switched inductor cells is also possible in this configuration. The most advantageous factor of this proposed converter is the use of low-voltage semiconductor devices as they don't require large heat sinks. The converter operation in the steady state is fully analyzed. In addition to that, for the purpose of stability analysis, the small signal model for the proposed converter has also been developed. The frequency response using the small-scale transfer function of the converter has also been done by employing MATLAB. A suitable controller with suitable parameters has also been designed to improve the overall stability of the DC-DC converter in consideration. The results obtained after simulation verifies the feasibility of the converter.

Cardiovascular disease prediction is a research field of healthcare which depends on a large volume of data for making effective and accurate predictions. These predictions can be more effective and accurate when used with machine learning algorithms because it can disclose all the concealed facts which are helpful in making decisions. The processing capabilities of machine learning algorithms are also very fast which is almost infeasible for human beings. Therefore, the work presented in this research focuses on identifying the best machine learning algorithm by comparing their performances for predicting cardiovascular diseases in a reasonable time. The machine learning algorithms which have been used in the presented work are naïve Bayes, support vector machine, k-nearest neighbors, and random forest. The dataset which has been utilized for this comparison is taken from the University of California, Irvine (UCI) machine learning repository named "Heart Disease Data Set."

The world is becoming smart as IoT is now an integral part of individuals' routine lives. To control any devices at any place and at anytime from anywhere is now just a matter of access. The goal of this work is to provide simple, efficient, cost-effective, and reliable communication system for traffic management. Keeping in view the aim of smart city, after cleanliness, traffic is the major concern nowadays. A case study is presented through proposed model in this work that will help in improving traffic condition of the city. The available data is analyzed and processed through Raspberry-pi. This data is simultaneously being updated at the web server through cloud. Based on the data available in real time, the system enables controlling traffic system dynamically. This helps in reducing congestion and provides fast going way for heavy vehicular traffic. The system can be clubbed with existing centralized traffic control system in the Indore city to manage traffic conditions in a better way.

Security is one of the fundamental issues for both computer systems and computer networks. Intrusion detection system (IDS) is a crucial tool in the field of network security. There are a lot of scopes for research in this pervasive field. Intrusion detection systems are designed to uncover both known and unknown attacks. There are many methods used in intrusion detection system to guard computers and networks from attacks. These attacks can be active or passive, network based or host based, or any combination of it. Current research uses machine learning techniques to make intrusion detection systems more effective against any kind of attack. This survey examines designing methodology of intrusion detection system and its classification types. It also reviews the trend of machine learning techniques used from past decade. Related studies comprise performance of various classifiers on KDDCUP99 and NSL-KDD dataset.

This chapter presents a design proposal for low-cost speed control and electrical fault mitigation of three-phase induction motors. The proposed system can control and monitor TIMs (three-phase induction motors) from far-flung areas. Here authors have proposed a relay-free system for fast fault clearance. IoT technology and low-cost microcontrollers have helped in achieving a system that is more reliable, economical, user friendly, and fast. It can be controlled by mobile application at the comfort of home. Data related to fault occurrence can be stored and analyzed for preventive maintenance. V/f scalar control method is used for speed control of TIM and able to control it in a wide range. Electrical faults such as over-current, over-temperature, over-voltage, and under-voltage are considered in this chapter. Simulation of the proposed design is done using Proteus 8 software. ESP32 is used to runs a web server that connects the mobile app with simulation.

Smart grid has changed power systems and their reliability concerns. Along with that, cyber security issues are also introduced due to the use of intelligent electronic devices (IEDs), wireless sensory network (WSN), and internet of things (IoT) for two-way communication. This chapter presents a review of different methods used from 2010 to 2020 focusing on citation as the main criteria for reliability assessment of smart grids and proposals to improve reliability when it comes to assessing a practical transmission system. It shows that evolutionary techniques are the latest trend for smart grid security.

In this chapter, the advantage of distributed generation can be seen in terms of system reliability and reliability of customer load. The solar photovoltaic (SPV) system is one of the distributed generations that may lead to the supply of electrical energy. The customer at the site of load demand mainly uses the SPV system. The installation of the SPV system is advantageous for the electrical load demand. Solar systems have greater efficiency for supplying both types of load (i.e., thermal and electrical) simultaneously. The modeling of two power system components (i.e., generation and distribution) can be performed using the Monte Carlo simulation (MCS) technique. The data used for generation modeling is taken from IEEE-RTS (reliability test system) and data for the distribution system is obtained from IEEE-RBTS (reliability busbar test system). The reliability parameters such as average energy not supplied (AENS) and loss of energy expectation (LOEE) are evaluated for the analysis of individual customer reliability and overall system reliability simultaneously.

This chapter discusses the current situation of renewable energy and the growing need for renewable energy. The present and past research revealed that the Ministry of New and Renewable Energy has done a great job and heading India towards renewable energy, but this is not done yet. India has not only sufficient climate condition, but also a large surface area which set a good chance for India to lead in the renewable sector in the world. An effort has been made to summarize the current scenario, benefits, and recent development of renewable energy.

Conventional generation is the most reliable option to meet the increased energy consumption in terms of operating performance. However, the increased greenhouse gas emission is a major threat from the conventional generating units due to fuel pollution. Although to meet the increased energy consumption, reliably conventional generating units are inevitable. So the government has taken the initiative to shut down the conventional generating units with higher pollution levels than the defined norms. This imposes the overall load burden to the other state generating units. As Delhi is sufficiently rich with solar radiation, the chapter proposes the solar PV installation to meet the generation gap of shutdown units.

Breast thermography is an emerging adjunct tool to mammography in early breast cancer detection due to its non-invasiveness and safety. Steady-state infrared imaging proves promising in this field as it is not affected by tissue density. The main aim of the present study is to develop a computational thermal model of breast cancer using real breast surface geometry and internal tumor specification. The model depicting the thermal profile of the subject's aggressive ductal carcinoma is calibrated by variation of blood perfusion and metabolic heat generation rate. The subject's IR image is used for validation of the simulated temperature profile. The thermal breast model presented here may prove useful in monitoring the response of tumor post-chemotherapy for female subjects with similar breast cancer characteristics.

The performance and efficiency of a solar PV cell are greatly dependent on the precise estimation of its current-voltage (I-V) characteristic. Usually, it is very difficult to estimate accurate I-V characteristics of solar PV due to the nonlinear relation between current and voltage. Metaheuristic optimization techniques, on the other hand, are very powerful tools to obtain solutions to complex non-linear problems. Hence, this chapter presents two metaheuristic algorithms, namely particle swarm optimization (PSO) and harmony search (HS), to estimate the single-diode model parameters. The feasibility of the metaheuristic algorithms is demonstrated for a solar cell and its extension to a photovoltaic solar module, and the results are compared with the numerical method, namely the Newton Raphson method (NRM), in terms of the solution accuracy, consistency, absolute maximum power error, and computation efficiency. The results show that the metaheuristic algorithms were indeed capable of obtaining higher quality solutions efficiently in the parameter estimation problem.

A compact rectangular slotted antenna fed through coplanar waveguide for rectenna system is proposed in the application of radio frequency (RF) energy harvesting at center frequency of 2.45 GHz in the wireless local area network (WLAN) band. Three unequal widths of rectangular slots with equal distance have been created step by step to maximize the peak gain to 3.6 dB of the antenna. Radiation plot of the proposed antenna has been depicted to be omnidirectional for RF energy harvesting with maximum radiation efficiency characteristics. The dimension of the antenna is reduced up to 28 × 17 mm2 with better reflection coefficient of -34.6dB.

Preface

This book is aimed at both engineering students and practicing professionals specializing in the domain of computational methodologies; Electronics and Electrical field to provides useful and concise information concerning the latest technology used. There are a lot of challenges in the real-world scenario say medical field, communication system or be it the drastic increase in the computing devices bringing IoT concept in picture, that could be answered through the revolutionary technological approaches. One of the key objectives of the research should be its usability and application. This book attempts to document and spark a debate on the research focused on technology in the context of emerging fields such as Computational technologies, Electrical and electronic engineering field. The application of technology is a key theme in every chapter of this book. This book intends to showcase technologies that could bring about a fundamental change in the field of engineering and technology.

The chapters included in this volume bring together recent research of more than twenty authors in the field of neural networks and pattern recognition, machine learning, vascular cardiology, breast cancer detection, smart grid, and RF communication system. This volume describes the application of pattern recognition schemes to the classification of various types of waveforms and images. Power electronics is the technology behind the conversion of electrical energy from a source to the requirements of the end-user. Therefore, it is of vital importance to both industry and the individual citizen. Energy conversion techniques are now a primary focus of the power electronics community with rapid advances being made in conversion technologies in recent years that are detailed in this book. The technology of DC/DC conversion is making rapid progress and, according to incomplete statistics, there are more than 600 topologies of DC/DC converters in existence with new ones being created every year.

The chapters span a large variety of problems in signal and image processing, using mainly machine learning for IoT and biomedical field, smart grid, and RF and microwave domain.

The book contains the latest technological research work from three different domains. The work has been done by using various simulation tools such as MATLAB, COMSOL, HFSS, CST, ADS, and NETSIM.

Chapter 1 describes the design of a basic power-efficient object dropping game named “Catch me if you can” which works on the ARDUINO PLATFORM. A dropping mechanism is developed using the servo motors interfaced to Arduino. The mechanism includes an object holder attached to the servomotor and a loading tube. The dropping of the objects is controlled by any wireless Bluetooth controlling device like mobile phone or joystick and by the keypad interface installed in the design. The game is about catching the objects dropping randomly as a challenge which is controlled by the operator. The overall simulated design can be done in the Easy EDA platform. The overall game can be controlled by an app which is designed by MIT App Inventor. This game can be implemented in Amusement parks, exhibitions, kid schools, and shopping malls. Besides this entertainment aspect, commercially it works as a small-scale product packaging system by involving DC motors, which are needed in moving packaging belt. This mechanism is efficient in packaging products in some specific count. This system is helpful for the small-scale candy, biscuit, ice-cream, beverages, and other packaging industries. The main advantage of this design is cheap and affordable to young industrialists.

Chapter 2 describes the reconstruction of the 3D image using CNN. It focuses on activities that use in-depth learning Strategies for measuring the 3D status of common things from one or more RGB images. It organizes based literature in the layout presentations, network structures, and training methods they use. As the survey was conducted for methods of reconstructing common objects, this paper also evaluates some of the latest efforts that emphasize particular categories of an object such as the shape of the human body and face. This provides an examination and correspondence of the execution of some important papers, summarizing some open-ended issues in the field, and exploring encouraging indications for subsequent research.

Chapter 3 describes the Machine learning-based Intrusion Detection System (IDS) which is a research field of network security depending on the effective and accurate training of models. The models of IDS must be trained with new attacks periodically, therefore it can detect any security violations in the network. One of the most frequent security violations that occur in the network is the Denial of Service (DoS) attack. Therefore, training of IDS models with the latest DoS attack instances is required. The training of IDS models can be more effective when it is performed with the help of machine learning algorithms because the processing capabilities of machine learning algorithms are very fast. Therefore, the work presented in this paper focuses on building a model of a machine learning-based Intrusion Detection System for Denial-of-Service attacks. Building a model of IDS requires a sample dataset and tools. The sample dataset which is used in this research is NSL-KDD while WEKA is used as a tool to perform all the experiments.

Chapter 4 focuses on reliable and fast communication networks for the effectiveness of smart grid automated substations. The communication network is a prerequisite to connecting different protection, control, and monitoring equipment are within the power substation. Ethernet fulfills all the requisites such as reliable, fast, secure, interoperable, LAN-based communication system for smart substations. Therefore, the main aspect is to improve the reliability of the network by prioritizing the critical components by using the knowledge of component importance measures (CIM). This critical ranking of the components is significant in influencing the performance of the networks. In this paper, analysis of IEC 61850 Ethernet-based substation communication network (SCN) architectures has been examined using various reliability importance measures.

Chapter 5 focuses on preventing attacks on Big Data systems ranges from DDoS to Information theft, ransomware to end user-level security. So, implementing security to Big Data system is multiple phase-based ongoing processes in which security is imposed from perimeter level to distributed file system security, cloud security to Data security, storage to data mining security, and so on. In this paper, the authors have identified some key vulnerable points related to Big Data and also proposed a Security Model.

Chapter 6 emphasizes that in medical science, proper medical analysis and examination of cardiovascular disease is a very crucial, convincing, and sophisticated task for saving a human life. Data analytics rises because of the absence of sufficient practical tools for exploring the trends and unknown relationships in e-health records. It predicts and achieves information that can ease the diagnosis. This survey examines the cardiovascular disease prediction system developed by different researchers. It also reviews the trend of machine learning approaches used in the past decade with results. Related studies comprise the performance of various classifiers on distinct datasets.

Chapter 7 examines the function of the electromyography signal implant and its performance in the health care sector from a technical control perspective. The purpose of the myoelectric prosthesis is to restore the basic functions of the lost organs in the joint using neural signals produced by the muscles. But unfortunately, the use of such myosignals is difficult and complicated. Besides, once detected, it usually requires a computational force strong enough to convert it into a user-controlled signal. Its modification of the actual function of the implantis limited by several factors especially those associated with the fact that each amputee has a different muscle movement. Modified Artificial Intelligence systems designed for pattern recognition have the potential to improve the size of implants but still fail to provide a system in which artificial arms can be controlled by brain signals. Although recent advances in pattern recognition techniques have been able to distinguish many degrees of freedom with very high accuracy, their level of efficiency was easily accessible and expressed in working applications.

Chapter 8 proposes a switched inductor configuration based non-isolated DC-DC converter with high voltage gain for EV applications. The proposed converter has two output capacitors instead of a single output capacitor for voltage boosting capabilities. Enhancement of the output voltage with the addition of a greater number of switched inductor cells is also possible in this configuration. The most advantageous factor of this proposed converter is the use of low-voltage semiconductor devices as they don't require large heat sinks. The converter operation in the steady-state is fully analysed. In addition to that, for stability analysis, the small-signal model for the proposed converter has also been developed. The frequency response using the small-scale transfer function of the converter has also been done by employing MATLAB. A suitable controller with suitable parameters has also been designed to improve the overall stability of the DC-DC converter in consideration. The results obtained after simulation verifies the feasibility of the converter.

Chapter 9 focuses on identifying the best machine learning algorithm by comparing their performances for predicting cardiovascular diseases in a reasonable time. The machine learning algorithms which have been used in the presented work are Naïve Bayes, Support Vector Machine, K-Nearest Neighbours and Random Forest. The dataset which has been utilized for this comparison is taken from the University of California, Irvine (UCI) machine learning repository named "Heart Disease Data Set".

Chapter 10 focuses on providing a simple, efficient, cost-effective, and reliable communication system for traffic management. Keeping in view the aim of smart city, after cleanliness, traffic is a major concern nowadays. A case study is presented through the proposed model in this work that will help in improving the traffic condition of the city. The available data is analyzed and processed through Raspberry-pi. This data is simultaneously being updated at the webserver through the cloud. Based on the data available in real-time, the system enables to control traffic system dynamically. This helps in reducing congestion and provides a fast going way for heavy vehicular traffic. The system can be clubbed with the existing centralized traffic control system in the Indore city to manage traffic conditions in a better way.

Chapter 11 explains that security is one of the fundamental issues now a day for both computer systems and computer networks. Intrusion Detection System (IDS) is a crucial tool in the field of network security. There are a lot of scopes for research in this pervasive field. Intrusion Detection Systems are designed to uncover both known and unknown attacks. There are many methods used in intrusion detection systems to guard computers and networks from attacks. These attacks can be active or passive, network-based or host-based, or any combination of them. Current research uses machine learning techniques to make intrusion detection systems more effective against any kind of attack. This survey examines the designing methodology of the intrusion detection system and its classification types. It also reviews the trend of machine learning techniques used in the past decade. Related studies comprise the performance of various classifiers on the KDDCUP99 and NSL-KDD dataset.

Chapter 12 presents a design proposal for low-cost speed control and electrical fault mitigation of three-phase induction motors. The proposed system can control and monitor TIMs (Three-phase induction motors) from far-flung areas. Here authors have proposed a relay-free system for fast fault clearance. IoT technology and low-cost microcontrollers have helped in achieving a system that is more reliable, economical, user friendly, and fast. It can be controlled by mobile application at the comfort of home. Data related to fault occurrence can be stored and analysed for preventive maintenance. V/f scalar control method is used for speed control of TIM and able to control it in a wide range. Electrical faults such as over-current, over-temperature, over-voltage, and under-voltage are considered in this chapter. Simulation of the proposed design is done using Proteus 8 software. ESP32 is used to runs a web server that connects the mobile app with simulation.

Chapter 13 explains that the Smart grid has changed power systems and its reliability concerns. Along with that, cybersecurity issues are also introduced due to the use of Intelligent Electronic Devices (IEDs), Wireless Sensory Network (WSN), and Internet of Things (IoT) for two-way communication. This paper presents a review of different methods used from 2010 to 2020 focusing on citation as the main criteria for reliability assessment of smart grids and proposals to improve reliability when it comes to assessing a practical transmission system.

Chapter 14 presents the advantage of distributed generation in terms of system reliability and reliability of customer load. The Solar Photovoltaic (SPV) system is one of the distributed generations, which may lead to the supply of electrical energy. The customer at the site of load demand mainly uses the SPV system. The installation of the SPV system is advantageous for the electrical load demand. Solar systems have greater efficiency for supplying both types of load (i.e. Thermal and electrical) simultaneously. The modeling of two power system components i.e., generation and distribution can be performed using the Monte Carlo simulation (MCS) technique. The data used for generation modeling is taken from IEEE- RTS (Reliability test system) and data for the distribution system is obtained from IEEE- RBTS (Reliability Busbar test system). The reliability parameters such as Average Energy Not Supplied (AENS) and Loss of Energy Expectation (LOEE) are evaluated for the analysis of individual customer reliability and overall system reliability simultaneously.

Chapter 15 discusses the current situation of renewable energy and the growing need for renewable energy. The present and past research revealed that the Ministry of New and Renewable Energy has done a great job and heading India towards renewable energy, but this is not done yet, India has not only sufficient climate condition but also a large surface area which set a good chance for India to lead in the renewable sector in the world. An effort has been made to summarize the current scenario, benefits, and recent development of renewable energy.

Chapter 16 explains that the Conventional generation is the most reliable option to meet the increased energy consumption in terms of operating performance. However, the increased greenhouse gas emission is a major threat from the conventional generating units due to fuel pollution. Although to meet the increased energy consumption, reliably conventional generating units are inevitable. So the government has taken the initiative to shut down the conventional generating units with higher pollution levels than the defined norms. This imposes the overall load burden on the other state generating units. As Delhi is sufficiently rich in solar radiation, the paper proposes the Solar PV installation to meet the generation gap of shutdown units.

Chapter 17 presents a study is to develop a computational thermal model of breast cancer using real breast surface geometry and internal tumor specification. The model depicting the thermal profile of the subject's aggressive ductal carcinoma is calibrated by variation of blood perfusion and metabolic heat generation rate. The subject's IR image is used for validation of the simulated temperature profile. The thermal breast model presented here may prove useful in monitoring the response of tumor post-chemotherapy for female subjects with similar breast cancer characteristics.

Chapter 18 presents the performance and efficiency of a solar PV cell are greatly dependent on the precise estimation of its current-voltage (I-V) characteristic. Usually, it is very difficult to estimate accurate I-V characteristics of solar PV due to the nonlinear relation between current and voltage. Metaheuristic optimization techniques, on the other hand, are a very powerful tool to obtain solutions to complex non-linear problems. Hence, this book chapter presents two metaheuristic algorithms namely, Particle swarm optimization (PSO), and Harmony Search (HS), to estimate the single-diode model parameters. The feasibility of the metaheuristic algorithms is demonstrated for a solar cell and its extension to a photovoltaic solar module and the results are compared with the numerical method namely, the Newton Raphson method (NRM) in terms of the solution accuracy, consistency, absolute maximum power error, and computation efficiency. The results show that the metaheuristic algorithms were indeed capable of obtaining higher quality solutions efficiently in the parameter estimation problem.

Chapter 19 presents a compact rectangular slotted antenna fed through a coplanar waveguide for a rectenna system which is proposed in the application of radiofrequency (RF) energy harvesting at the center frequency of 2.45 GHz in the Wireless local area network (WLAN) band. Three unequal widths of rectangular slots with equal distance have been created step by step to maximize the peak gain to 3.6 dB of the antenna. The radiation plot of the proposed antenna has been depicted to be omnidirectional for RF energy harvesting with maximum radiation efficiency characteristics.

Since this book covers relatively wide areas and various contents related to complex scientific issues, due to the limited expertise and competence of the authors, errors and omissions may be inevitable, so we deeply appreciate the readers' criticism and comments. We feel optimistic that the significant area and its practical applications will be highly valued for its documentation.

Rajiv Singh
G.B. Pant University of Agriculture and Technology, Pantnagar, India

Ashutosh Kumar Singh
Indian Institute of Information Technology, Allahabad, India

Ajay Kumar Dwivedi
SIET, India & Dr. A.P.J. Abdul Kalam Technical University, Lucknow, India

P. Nagabhushan
Indian Institute of Information Technology, Allahabad, India

Chapter 1
Catch Me If You Can Game as Well as Packaging System Efficient Design Using Arduino UNO

Trinadh Manikanta Gangadhar Dangeti
RGUKT Srikakulam, India

Manikanta Prem Kumar Bheemuni
RGUKT Srikakulam, India

Naga Subrahmanyam Boddeda
RGUKT Srikakulam, India

Pavan Kumar Kachala
RGUKT Srikakulam, India

Sai Ram Pavan Taneeru
RGUKT Srikakulam, India

Vara Durga Siva Sai A.
RGUKT Srikakulam, India

ABSTRACT

This chapter is about a basic power-efficient object dropping game named "Catch Me If You Can," which works on the Arduino platform. A dropping mechanism is developed using the servo motors interfaced to Arduino. The mechanism includes an object holder attached to the servomotor and a loading tube. The dropping of the objects is controlled by any wireless Bluetooth controlling device like a mobile phone or joystick and by the keypad interface installed in the design. The game is about catching the objects dropping randomly as a challenge which is controlled by the operator. The overall simulated design can be done in EasyEDA platform. The overall game can be controlled by an app which is designed by MIT App Inventor. This game can be implemented in Amusement parks, exhibitions, kid schools, and shopping malls. Besides this entertainment aspect, commercially it works as a small-scale product packaging system by involving DC motors, which are needed in moving packaging belts. This mechanism is efficient in packaging products in some specific count.

DOI: 10.4018/978-1-7998-3327-7.ch001

1. INTRODUCTION

Road fighter, Snake game, Tetris and Pacman are few of the 90's video games that were the kid's favorite. Pac-man and Tetris are played on the electronic screens and the player will have the controls of the game. These games are famous for creating fun in places like amusement parks, carnivals and game stations. Children are more likely to get attracted to these kinds of games. So the games were designed to show attractive to the children. These machines are powered by electricity and most of the system designed using microcontroller (Chen and Lin, 2009), so power handling is one of the aspects that are to be considered. Here we have created such a kind of game named "Catch me if you can". The design of the game includes boxes equipped with a dropping mechanism that holds a few objects in them and each box is given a number. An object will be released from a box at a time according to the instruction given by the game operator. The person playing the game has to catch the object falling to get the score in the game. We implemented our system using wireless connectivity using Bluetooth module (Debnath et al., 2019). The overall design implemented using Arduino (Badamasi, 2014).

2. OVERVIEW AND HISTORY

We are using Microprocessors and Microcontrollers for the implementation of the embedded systems and many other Automatic devices from the late 1970.But there are limitations to the devices such as Less RAM and ROM, Less power efficient and High Latency. That is why we decided to design our system using an Open Source platform called "Arduino". It was created by Massimo Banzi at LVREA in 2005, with the intention to produce a less expensive way for neophyte and maestro make devices which interacts with the environment using sensors and actuators (Kushner, 2011).

The Arduino UNO board have 1.**14 Digital I/O pins(pins 0-13)**,Zero and Five are the values for the input and output for these digital pins, and it also consists of **6 Analog inputs pins(pin 0-5)**,,these pins can measure voltage from 0 to 1024 different levels of voltage, The Analog output pins are digital pins(pins 3,5,6,9,10,11),these pins are also called as PWM pins.

The Board can be powered by 5volts to 7 volts, also with a USB port of a Computer. It has a KBs RAM, 32 Kb flash memory, 1Kb EEPROM, and 16 MHz clock speed. We can design a variety of systems such as embedded systems and Robots using the Arduino platform. (Banzi and Shiloh, 2014).

We can interface a lot of Transducers and Sensors with Arduino, like LEDs, LCDs, motors and Sensors etc. To establish a wireless communication with Arduino we can use Blue tooth, GSM and other wireless modules (McRoberts, 2011).

In our system we use servo motors to store and release the objects. Servo motor consists three pins 1.digital pin 2. VCC (supply) 3.Ground, these pins connected to Arduino digital pin of servo motor connected to PWM pin of Arduino (Bhargava and Kumar, 2017). Servo motors are widely used in many real-life applications.

Servo motor is a type of Transducer which works on feedback, which helps to control the position of the motor. Generally, these motors have the capability to have a rotation of **0 to 180 degrees**, but full rotating servos are also available. We can find Servo motors in Remote controlled Airplanes, cars and boats. Servo motors are used to build robots also. Servos control the joint moments in Robots.

Figure 1. Arduino (Banzi and Shiloh, 2014)

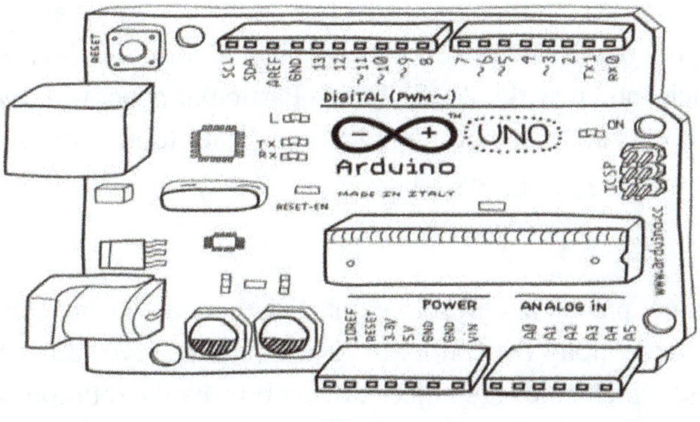

A Servo is constructed using a DC motor, Gears which are connected between the motor and shaft, Arduino controls the position of the servo motor by giving. **Pulse Width Modulation (PWM) signals**. The longer the pulse width is, the higher the angle servo moves. Servo.h library helps do this thing.

In our design we use Bluetooth module for communication purpose. Bluetooth Module design for transparent wireless serial connection setup, **HC05 module** (electronica studio) can be used to establish wireless communication using **Bluetooth Serial Port Protocol (SPP) module**. The transfer speed is of 3Mbps with complete radio transceiver and with the baseband of 2.4GHz.It have Default Baud rate: 9600 and Typical -80dBm Sensitivity (Singh et al., 2018).

Figure 2. Servo Motor (Components101)

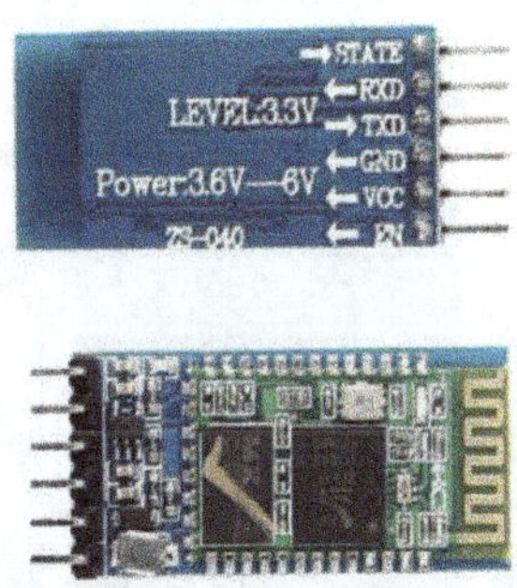

Figure 3. Bluetooth Module (HC-05) (electro some)

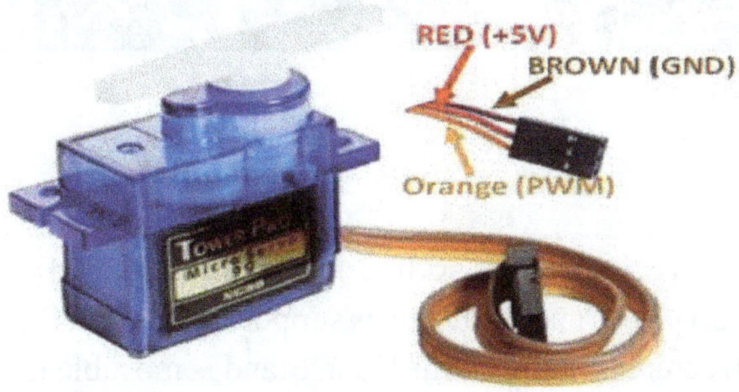

3. PROPOSED METHOD

Most of the Research papers based on Games are made using common devices such as joystick, LED, keyboards and LCDs (Duch and Jaworski, 2018). In this Particular paper we have designed a game which is equipped with Servo Motors and radio communication (Blue- tooth module) with some mechanism.

Game Design

Game consists a long frame, placed at certain height. Six rectangular boxes fixed at certain places in the frame and Wiring is made along the frame. In each box one servo motor is attached. Servo motor equipped with circular disc which holds the objects. In the box, by the rotation of servo motor the objects are loaded and released out from the box. The Box construction and Frame construction and overall Setup of our design is explained in below sections.

Box Construction

For a rectangular empty cardboard box, two holes are made one for loading objects and another for discharging at top of the box and bottom of the box respectively. For loading objects a cylindrical tube is attached at one hole at top of the box. At the bottom of the box a servo motor is attached, and a circular disk with a holder is fixed to the servo motor. The servo motor with circular disc with holder is present inside the rectangular box. Every time servo motor rotates the circular disk, which holds objects and release objects at bottom hole of the box as show in Fig.4 the Box construction is done.

Figure 4. Circular disk with servo motor holder, circular disk, circular disk with servo motor, cylindrical tube for loading, overall box with mechanism, loading objects, releasing objects

Frame Construction

The frame is the body of the game, it holds the six boxes, which is equipped with mechanism as explained above. Boxes placed at certain places and frame gives support to the game. The frame is tied to the three stands, which enables the structure to stand on the floor. Stand is movable it consists wheels, a long pole is attached to movable stand .The frame is attached to poles. The overall wiring and connection is made along with (attached to) the frame. Frame is very flexible to bend. We designed the frame with stands in such a way that the frame is moving up and down as our convenient. The three stands consists wheels, we can move the overall system easily as shown in Fig.5.

Overall Setup

The overall Setup consists movable stands with poles, plastic frame, rectangular boxes and servo motor with circular disk. At first the poles are fixed to movable stands, and then frame is attached as described above section and then boxes are attached to frame. The rectangular box with mechanism and frame with stands, all put together the overall system is designed as shown in Fig.7.Boxes are made with cardboard, the frame is made with plastic, and this works on the Arduino platform, which is open-source and cheap.

Figure 5. Frame construction with one box, frame with six boxes, overall frame

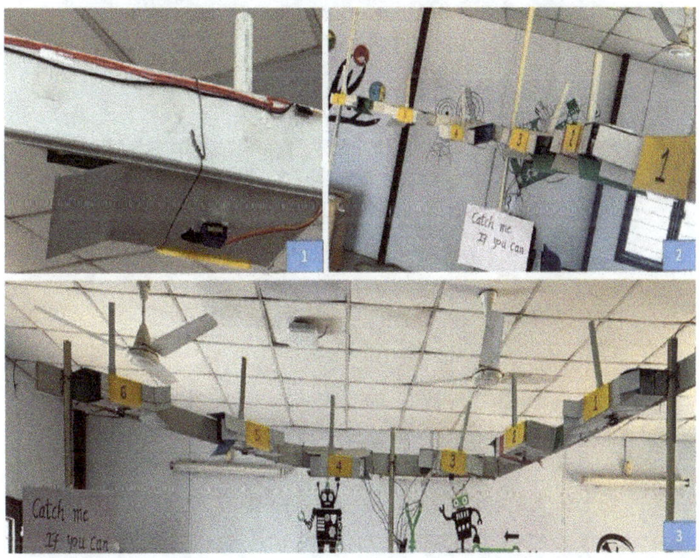

a. Hardware Connections

The Overall system designed using Arduino UNO Board. Arduino is interfaced with Bluetooth module (HC-05) and Servo motor. Operated with mobile by giving instruction to Arduino board using Bluetooth module. By giving instructions to Arduino, it operates servo motors to load and release objects. The Overall system controlled with mobile using an app. The overall setup with Arduino, Servo motor and Bluetooth module connections explained below sections.

Figure 6. Overall Setup

Figure 7. Arduino connections

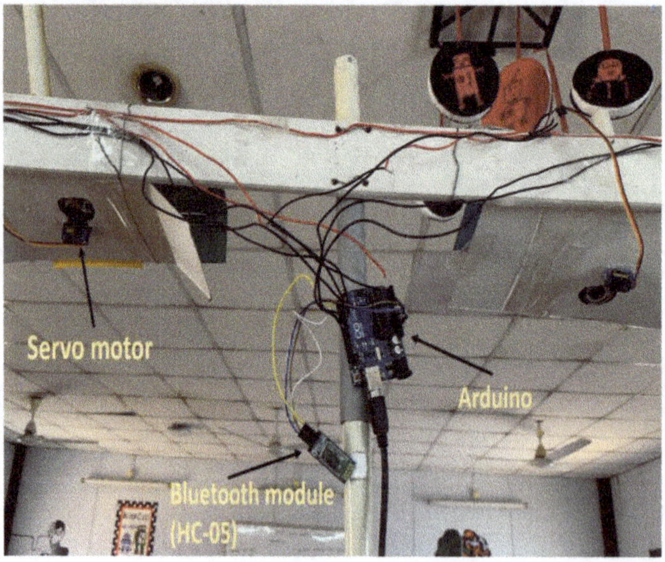

Servomotor Connections

As shown in Fig8.Servo motor (SG90) consists of three pins. 1. Control pin, it is connected to any PWM pin, 2.Power supply pin, it is connected to 5v terminal, 3.Ground pin is connected t to GND terminal in the Arduino. Arduino controls the rotation of servo motor by using servo library. Arduino controls angle of rotation and speed of servo motor by using PWM pins **Servo motor Specifications:** As shown in Table. I the main specifications in servo motor are torque and speed. The amount of time required by the motor to respond is always in proportion with the distance that it is going to travel. A pulse is produced every 20 ms. Width of pulses which determine the shaft position. As shown in Fig.8. The connections are done with Arduino and servo motor these are attached to the frame.

Figure 8. Servo motor connections with Arduino

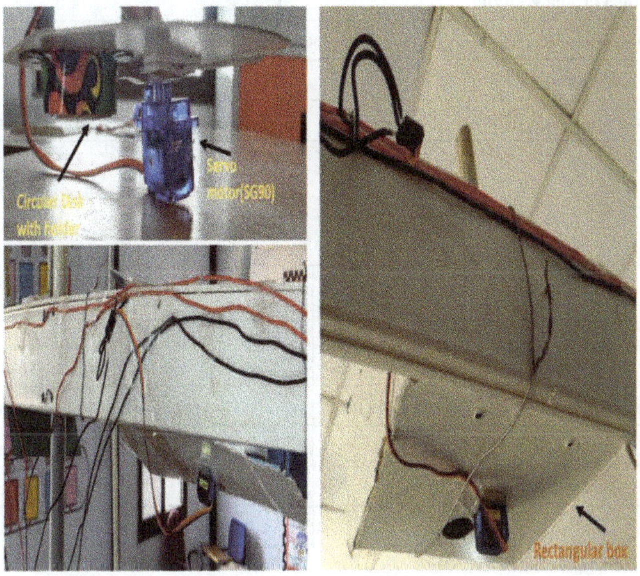

Bluetooth Module(HC-05) Connections

As shown in Table. II, Bluetooth module consists 6 pins and these pins configured as described in the below table. As shown in Fig.9 the Bluetooth module (HC-05) is interfaced with Arduino board (HC-05). Now connect Bluetooth network using mobile. After connection establish give instructions to Arduino using an app. App design discussed in below section. Mobile sends the instruction to Bluetooth. Bluetooth sends these instructions to Arduino finally Arduino controls the servo according to instructions.

Overall Connections:

The overall system designed using Arduino. As explained in above sections, the servo motor and Bluetooth module are interfaced together as shown Fig.12.From Arduino, 5V and GND pin is connected commonly to six servo motors VCC pin, Bluetooth module VCC pin and six servo motors GND pin, Bluetooth module Ground pin respectively. To controlling rotating angle of the six servo motors control pins are connected to six digital pins in Arduino i.e,D2,D3,D4,D5,D6,D7 .The transmission between Arduino and Bluetooth is a serial communication i.e., is as shown in Fig.10.

This overall simulated circuit can be designed in EasyEDA platform(EasyEDA).

Table 1. Specifications of Servo Motor (Components101)

Dimensions:	23.2 x 12.5 x 22 mm
Weight	**9 g**
Operation Speed:	**0.12sec/60degree(4.8v)**
Stall Torque:	**2.5kg.cm/20.87oz.in(4.8v)**
Operating Voltage& angle	4.8V&120degree
Control System Direction	Analog & CCW
Required Pulse:	900us-2100us
Type of Bearing, Gear, Motor:	None, Plastic, Metal

Table 2. Pins and configuration bluetooth module

Pin	Description	Function
VCC	+5V	Connected supply 5V in Arduino
GND	ground	Connected to ground in Arduino
TXD	Transmitter Pin(Bluetooth Serial Sending Pin)	Connect the pin to the Arduino Recover pin(RX)
RXD	Receiver pin(Bluetooth serial receiving pin)	Connect the pin to the Arduino Transmitter Pin(TX)
KEY	Mode Switch Input	AT Command mode
State	Indicator	It shows HC-05 On/Off

Figure 9. Bluetooth module connections with Arduino

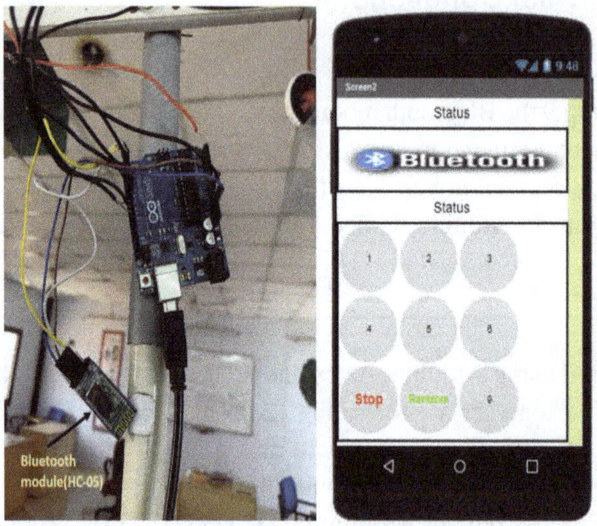

Figure 10. Transmitting & receiving data from Bluetooth to Arduino vice versa (fritzing)

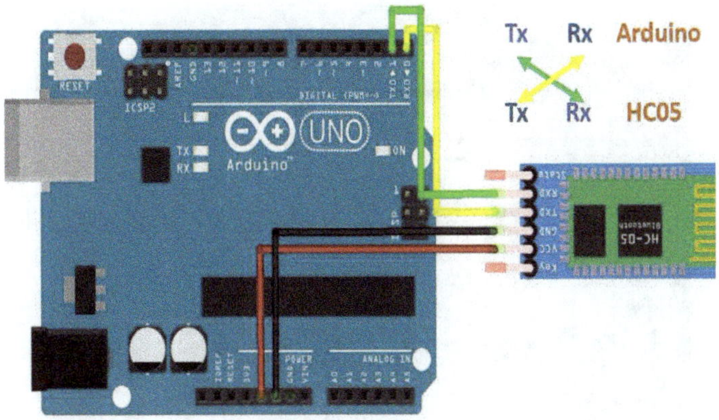

Figure 11. Overall Connections stimulated(EasyEDA

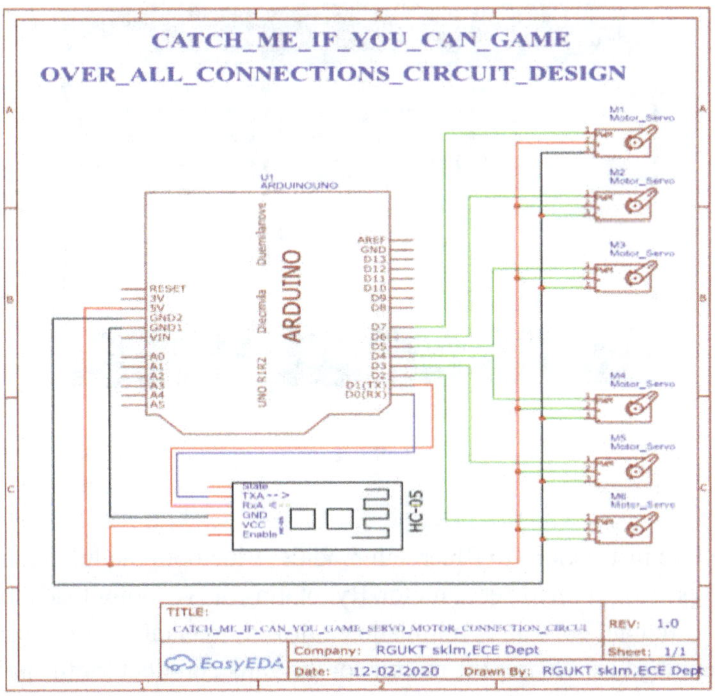

App Design

We are developed app using MIT App Inventor (Inventor) Instructions to Arduino finally Arduino controls the servo according to instructions. For controlling the system using Bluetooth wireless connection.Fig.13 shows app interface. In app inventor, first design interface by placing things what we need and arrange them in a way as shown in Fig.13.And next place blocks according to your design in background as shown in Fig.14

Figure 12. Overall Connections

Figure 13. Main screen, manual control screen, automatic control screen (inventor)

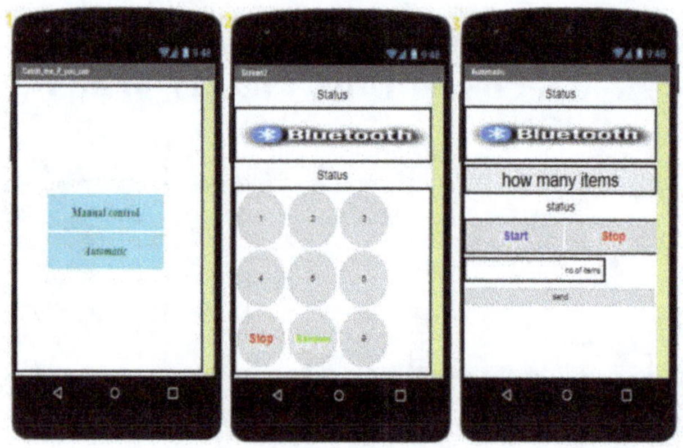

Operation

By using the app we give instructions to Bluetooth module, Bluetooth module sends the instructions to Arduino, Arduino gives instruction to Servo motor. By rotating Servo motor load the objects and release. As shown in Fig.13 1.Manual control 2.Automatic control in manual control user need to select one number that corresponding box releases the marble for the game, product for the packaging machine. In random control boxes are respond and release according to randomly generated number. In automatic control depends upon entered number, that corresponding box will respond.

These instructions can also be given from the keyboard installed in the design itself. The dropping mechanism is designed in a way that the holder will catch the object perfectly and release perfectly without any fail.

Figure 14. Blocks (Inventor)

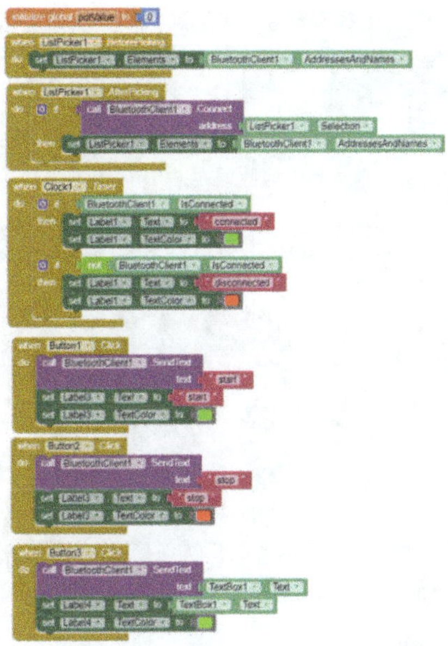

Figure 15. loading position, dropping position, return to loading position

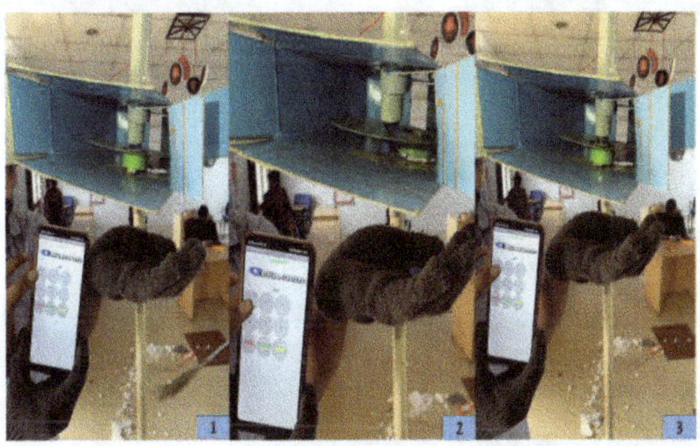

4. APPLICATIONS

The primary application of the design is "Catch me if you can" game. In practice, this can be installed at places like Kid schools, amusement parks, shopping malls and parks. Another application is that it is useful as a **packaging machine** in the industries. There will be a conveyor belt which consist of empty boxes need to fill items. Those empty boxes exactly located under Add this the packaging machine boxes. 1st we need to load machine boxes with items with certain count then load those items into empty boxes placed on the conveyor belt. Then rotate belt. As this is cheap and beneficial, more people can avail this product in future. These type of designs can increase the students programming ability.

Figure 16. Overall Operation

Figure 17. A person playing this game

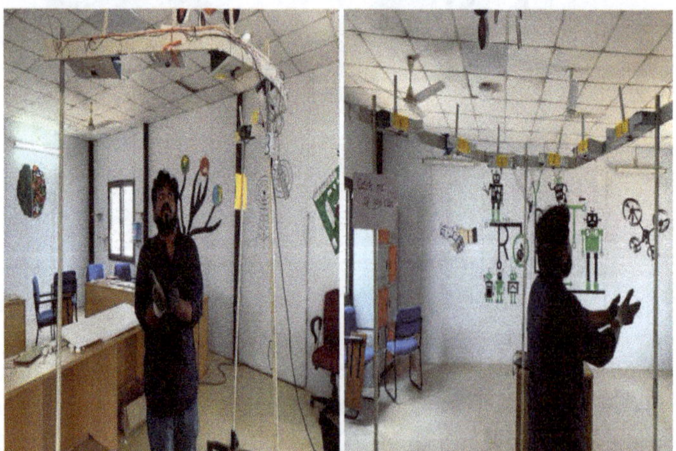

5. FUTURE SCOPE

This design can become the replacement of manual packaging techniques in small scale industries in future. By increasing the holder's count in the circular disk, the system can best act as a packaging machine. By interfacing size detecting sensors to the Arduino, the holder can separate fruits of different sizes into separate boxes; this feature is much helpful for the fruit cultivators. By using color sensors also, separating objects and fruits efficiency is increased .As every component is simple and easy to assemble manufacturing and bringing this to availability is easy and cheap. In the game point of view by installing a camera to the game and using artificial intelligence will increase the effectiveness of the play .Artificial intelligence observes the players attention and releases objects accordingly.

6. CONCLUSION

We constructed this machine using cardboards, water pipes, scissors, glue gun, threads, plasters, iron poles, chair stands, papers, connecting wires, Arduino Uno, Bluetooth module and Servo motors. We spend around 1500 rs on constructing this in- credible machine. Except for the electronic components, other stuff used for construction is mostly waste materials and easily available. So in mass manufacturing also product cost is considerably low to others. The assembling of the parts is very easy and has a simple structure. So it is effortless to replace any component which is damaged. So the lifetime of the machine is also long. Our main intention of building this machine is to provide both a low-cost entertainment game and a power-efficient packaging machine in a single device, which is affordable for the poor also.

ACKNOWLEDGMENT

The authors would like to convey their gratitude to Mr.N.Sri Sai,Ms.U.A.L.Satya Sri of Electronics and Communication Department for their absolute help.This work was supported by our institute RGUKT Srikakulam,AP.

REFERENCES

Badamasi, Y. A. (2014). The working principle of an arduino. *2014 11th International Conference on Electronics, Computer and Computation (ICECCO)*, 1–4. 10.1109/ICECCO.2014.6997578

Banzi, M., & Shiloh, M. (2014). *Getting started with Arduino: the open source electronics prototyping platform*. Maker Media, Inc.

Bhargava, A., & Kumar, A. (2017). Arduino controlled robotic arm. In *2017 International conference of Electronics, Communica- tion and Aerospace Technology (ICECA)*, *2*, 376–380. doi: 10.1109/ ICECA.2017.8212837

Chen, X., & Lin, C. (2009). Tetris game system design based on at89s52 single chip microcomputer. In *2009 Third Inter- national Symposium on Intelligent Information Technology Application, Shanghai*, (pp. 256–259). IEEE.

Components101. (n.d.). *Servo motor pin config*. Available: https://components101.com/servo-motor-basics-pinout-datasheet.

Debnath, B., Dey, R., & Roy, S. (2019). Smart switching system using bluetooth technology. 2019 *Amity International Conference on Artificial Intelligence (AICAI)*, 760–763. doi: 10.1109/AICAI.2019.8701298

Duch, P., & Jaworski, T. (2018). Enriching computer science programming classes with Arduino game development. *2018 11th International Conference on Human System Interaction (HSI)*, 148–154.

EasyEDA. (n.d.). *Easyeda circuit editor*. Available: https://easyeda.com/editor

electronicaestudio. (n.d.). *Bluetooth module hc-5 datasheet*. Available: http://www.electronicaestudio.com/docs/istd016a.pdf

electrosome. (n.d.). *hc-05-serial-bluetooth-module*. Available: https://electrosome.com/hc-05-serial-bluetooth- module/

fritzing. (n.d.). *fritzing circuit editor*. Available: https://fritzing.org/building-circuit/

Inventor, A. (n.d.). *MIT app inventor*. Available: https://appinventor.mit.edu/

Kushner, D. (2011). The making of arduino. *IEEE Spectrum*, 26.

McRoberts, M. (2011). *Beginning Arduino*. Apress.

Singh, V. K., Sahu, A., Beg, A., Khan, B., & Kumar, S. (2018). Speed direction control of dc motor through bluetooth hc-05 using Arduino. *2018 International Conference on Advanced Computation and Telecommunication (ICACAT)*, 1–3.

Chapter 2
Reconstruction of 2D Images Into 3D

Gaurav Agarwal
Raj Kumar Goel Institute of Technology, India

Viraat Saaran
Raj Kumar Goel Institute of Technology, India

Vaishali Kushwaha
Raj Kumar Goel Institute of Technology, India

Shraddha Singh
Raj Kumar Goel Institute of Technology, India

Paurush Mudgal
Raj Kumar Goel Institute of Technology, India

ABSTRACT

3D reconstruction is a long-standing complication when comes to testing happening from decades from machine learning, computer graphics, and computer perspective environments. Using CNN for the reconstruction of the 3D image has enchanted growing attentiveness and shown spectacular execution. Emerging in the new era of abrupt development, this chapter lays out an in-depth study of the latest developments in the field. Its focuses on activities that use in-depth learning strategies for measuring the 3D status of common things from one or more RGB images. It organizes based on literature in the layout presentations, network structures, and training methods they use. As the survey was conducted for methods of reconstructing common objects, this chapter also evaluates some of the latest efforts that emphasize particular categories of an object such as the shape of the human body and face. This provides an examination and correspondence of the execution of some important papers, summarizing some open-ended issues in the field, and exploring encouraging indications for subsequent research.

DOI: 10.4018/978-1-7998-3327-7.ch002

1. INTRODUCTION

This Reconstruction of 2D images into 3D is a particularly associated topic in Computer Vision. Computer Vision and the augmented reality industry emphasize reconstructing 3D view to extemporize the virtual outcomes. 3D imaging is also called stereoscopy, which is broadly used for the intensification of a 2D image by expanding the illusion of depth. The fundamental aim of 3D reconstruction which depends on the image is to deduce the arena from 2D images, the object's structure, and 3D geometry.

2D images consist of rich information about the face encoding which can be utilized in another plane. The data of the presentation of the lower dimension from the higher dimension departed when the image of a 3D object is projected on a 2D plane. Hence, for building 3D constitute there will never be sufficient data from the 2D image. The preceding information about the 3D shape is necessary to construct 3D interpretation from a single image.

The main problem is the plane of projection, in a 2D image if an image from one projection then the data cannot be retrieved (or sometimes hidden) for the other plane. But in the case of the face, if the image is taken well enough then it can have all the position vectors for the facial encoding of the image. The generated output thus can be placed or synchronized for the 3D plane with new positions. The main objective is to reconstruct the 3D shape of an object from multiple or even single 2D images. The study is arranged as: Section 2 describes the applications. Section 3 depicts the background and literature review. Section 4 discusses various taxonomy used for the 3D reconstruction of images. Section 5 concludes the paper.

2. APPLICATIONS

Numerous applications need 3D models. One such application is 3D face reconstruction and face alignment. When work on face alignment started, it focused on observing special fiducial points. Face alignment serves as a condition for various facial chore-like face identification, 3D face regeneration, and so on.

3D face reconstruction works by evaluating 3D vertices and equivalent colors from a particular image. During reconstruction, the non-visible area's texture got wrenched because of self-occlusion. Other applications that can be done or to be expected in the future are 3D pose estimation, depth image, texture editing, or face swapping. In texture editing, we can exchange the texture and modify a particular part or blend images.

Facial Puppetry or duplication is a type of video-based animation that is used in games and movie industries as well as teleconferencing to animate virtual characters. Here, the main purpose is the transfer of expressions from the user to an animated character.

Face Replacement is a process in which swapping the face is done. The main challenge is to generate realistic output after the face-swapping is completed.

Facial Re-enactment is also a process of transferring the facial expression from the source to target in a video, but here both examples are real and no animation is done.

Speech-driven Animation-Strategies in this method are typically partner units of sound, i. e., phonemes, to their visual partner, referred to as viseme, to integrate new mouth movements.

Video Dubbing is a process of mouth movement with the voice of some other person. A good video dumbing should consist of synced movement of the sound and the mouth.

Virtual Makeup, depending on the 3D reconstruction algorithm, we can change the generated face. We can add makeup, a tattoo, or any other accessories that are not only realistic but adaptable to the construction of the face.

3. BACKGROUND

CNN's are then started to be used for re-establishing the 3D details from one 2D facial image which ultimately gives the result for dense alignment of face and reconstruction of the 3D face. Although the 3D space has a certain limitation given by the face model which reduces their performance.

The tasks performed by two end-to-end works attained state-of-the-art execution by eliminating the constraints of model space. In first model, 2D coordinates and 68 facial landmarks are being used for directing a complicated network that requires a supplementary network to estimate the depth value (Bulat et al., 2017). Although this technique does not allow dense alignment. While in another model, 2D image is regressed by using a network and also produces the 3D image's volumetric presentation (Jackson et al., 2017). But this network has the limitation; it destroys the semantic information of the points, due to which it has to re-impose the shape of the face consisting of the full volume in order. Therefore, it requires a complicated network for regression because of the inability to represent the resolution of the recovered shape. To summarize, the works based on models have semantic information of the point but are lacking in representation space. On the contrary, techniques without the model have modernization execution; but do not provide the connotation meaning

3.1 Related Work

M. Aharchi and M. Ait Kbi (Aharchi & Kbi, 2019), presented the construction of a 3D image using 2D images which are pragmatic because feasibility is also a fundamental concern of computer vision and image-based models. Presently, there are various approaches to the reconstruction of 3D images from single or multiple 2D images, every algorithm has its acknowledgment state of implementation. 3D reconstruction has various applications like medicine, robotic mapping, city planning, robot navigation, augmented reality. 3D Reconstruction from Image Requirements was conducted by getting the points of the desired coordinates of the picture, it is important to resolve a various kind of issues which are: Calibration issue - How the points are projected on the image; Matching issue - Potential to acknowledge and match the points that emerge on the pictures and Reconstruction issue - Examination of regulating 3D feature points taken from the matching done and the variables of normalization. There are two reconstruction Methods as Active methods: Active methods hinder the item to be regenerated uninterrupted by mechanical and radiometric techniques (structured light, laser rangefinder, and other active recognition techniques). And, submissive methods of reconstruction don't inhibit with items to be reconstructed. To deduct the 3D structure via image transformation, only sensors are used to calculate the fluorescence released by the side of the object. To execute an effectual estimation for the presentation of the distinct approaches from 2D into 3D transformation needs an attentive outline of sets and standard data. It is uneven to suppose that an approach shows less execution time and a decreased error rate than other methods. One more drawback is that for each independent algorithm, each possessing various attributes and presentation. It is concluded that a big amount of 2D to 3D construction methods was committed to the retrieval of the 3D image using a 2D one. It is important to use and integrate numerous depth cues for the aim to attain a rigorous transformation algorithm.

In 2019, Krishna Sai Joga and K. Kavya Sreen (Joga et al., 2019) used deep learning techniques. This paper layout a wide look over the previous five-year evolution in the area of reconstruction of the 3D object which was based on an image. This categorized method into point-based surface-based, and volumetric techniques. Reconstruction of the 3D human body by volumetric approaches straight conclude the 3D tenancy framework. Instead, of using analytical replica this approach speaks to or allowed to retrieve complete geometry even with cloth on human bodies still it is bounded to the easy framework. Point-based representation is uncomplicated but systematic in terms of memory stipulation. It constitutes by an illogical set S = {xi, yi, zi} from range i=1 to N points. This portion talks about the unique point situated delineation and their proportionate network architecture. Point clouds are not in well-ordered form and they do not comfortably set into the convolutional planning. So that makes use of contiguous consistency. According to capture, network architecture, and the tutoring appliances, it then talks about methods in each category. The paper also talks about and contrasts the accomplishment of some key techniques. This look-over centers on approaches that delineate 3D recreation from at least one RGB pictures as the issue of recovering the 3D calculation of items. The proposed method gave very promising results. There is a lot of scope for extension and better accuracy.

Ronald (Yu et al., 2017), proposed the network works on per-pixel motion and a potential comparison between 3D face model textured projection and 2D input photographs of a person. The training of a network is done by generating a huge dataset with dense labels of synthetic faces by using a morphable face model with slight changes in occlusions, expressions, and pose. It is established that for upgrading the capability to operate real images, refinement in training should be done using real photographs. The author does not bound to the regression-based approaches of the model space and to the visibility of sparse facial landmarks, which are predicted by dense correspondences. Relying on speed and accuracy, the proposed approach outshine the modernization face alignment techniques, when the 3D face fitting step and facial detection are mixed. The approach is also tested on the video frames which outperforms per-frame processing even in uttermost lighting conditions, occlusions, and pose-variations.

Amin Jourabloo (Jourabloo et al., 2016), gives a combination of 3DMM and CNN regressor method is used for the alignment of the face which generally works for large-pose face images. The preparation of the technique relies on the parameters of 3D shape and matrix of camera projection through which the 3DMM fitting problem is regulated and predicted by a cascade of CNN-based regressors. For productive CNN training, designs of a pose-invariant illusion of features are permitted by dense 3D shape. In contrast to the modernization, immense evaluation over the exacting databases is conducted to produce better results.

Wenbin (Hu et al., 2020), presents the novel framework called "Reinforced Differentiable Attributes" (REDA), for increased accuracy for the reconstruction of the face along with the accuracy of face density. The authors claim it to be more effective and general than any other differentiable rendering. The author also proposes the unoccupied deformation layer and can upgrade the precision of the model by pertaining this layer on the head of the 3DMM (3D Morphable Model). This end-to-end proposed technique maintains both RGB-D and RGB and training based on deep learning. Various evaluation proves that this approach is both efficient and effective.

These days 3D television is a major landmark of visual media. In recent years, researchers are focusing on developing imaging and transformation skills in the 3D model using depth analysis (Priya et al., 2017). The third dimension can often be seen only from a human perspective. The eyes visualize the depth and brain of the third-largest reconstruction with the help of various visual representations. Researchers have used this tactic to rebuild the 3D model from different views with the help of certain contrast parameters to measure. In recent days there are special cameras that help take a 3D viewing model directly. There are a few examples of the dual stereoscopic camera, deep-range camera range, etc. Here the cameras usually take an RGB portion of the image and its corresponding depth map. A depth map is defined as a function that helps to assess the depth of an object in a particular area (i.e. instead of a pixel). Often the pressure is considered to be deep.

CNN is working with a single 2D face image (Jackson et al., 2017), it does not need precision to align or establish a strong connection between photographs, working for facial expressions without thinking and expressions, and can be used to recreate all 3D face geometry (including invisible facial parts) by-passing construction (during training) and installation (during testing) of the 3D Morphable Model. This can be achieved with a simple CNN format that enables direct file format 3D to face geometric volume representation from one 2D image. It also shows how related work is related locally made visible space can be installed the proposed framework and assistance in the development of reconstruction quality, especially in cases of great looks and facial expressions speeches.

4. TAXONOMY OF 3D RECONSTRUCTION

There has been many approach, methods, algorithms, and techniques used for the 3D reconstruction of images from 2D images. As illustrated and proposed by many authors, some of them are described below.

4.1 Watershed Algorithm

M. Arulselvi and B. Gnana Priya (Gnana et al., 2019), proposed that there was an ascent in the ubiquity of 3D equipment, for example, gaming devices, 3D printing, Smartphone's, TV's, clinical hardware, and others. The conversion of 2D to 3D is utilized at different stages to obtain 3D statistics. The people have a stereoscopic view that can be utilized to notice the depth of an item. The depth discrimination is because of the binocular incongruities of our eyes. Additionally, the 3D impact can likewise be made by parallax impact by considering dual pictures and mounding them. An image taken from a 2D or 3D camera can be transformed into a 3D image. A watershed algorithm based on the marker is used in which automatic and semiautomatic conversion modes can be used to convert 2D to 3D.

Figure 1. The architecture of Watershed Algorithm

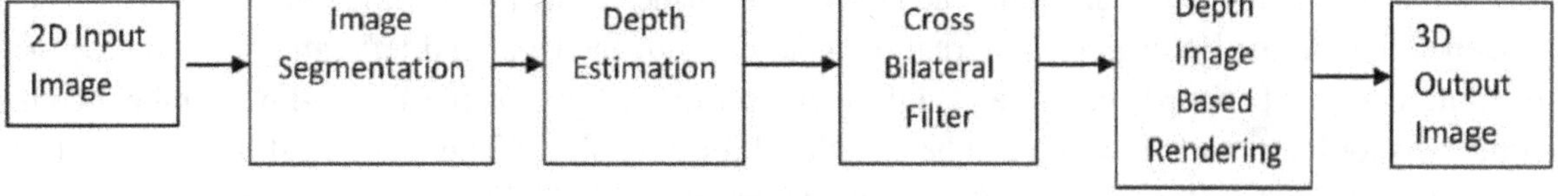

The depths of the various fragments of an image are allocated in a semiautomatic way. The determination of depth is very crucial, it can be improved by using a good quality 3D image. The depth map should be opaque and precise to deliver a standard picture. Depth is communicated as a grayscale picture made out of dim pixels in the extent 0-255. The value 0 speaks to the 3D pixels are situated at the distance in the 3D scene and 255 speaks to that the 3D pixels are found closer at the scene. The 3D reconstruction of images is held by utilizing the depth-based rendering methods to change over 2D image alongside the created depth map into a 3D image. The subsequent stage is picture distorting which guides the right or left view pixel by pixel dependent on depth esteems. An overall answer to transform any 2D picture to 3D isn't available.

4.2 SIFT Algorithm

SIFT is a Scale-Invariant Feature Transform (Hu et al., 2019). Because of the negative effects of natural illumination along with distortion of cameras, while capturing images, it becomes hard to equivalent pixel characteristic points in dual compatible digital pictures. The SIFT characteristic shows a parallel method, algorithm of scale-invariant feature transform, which was given by David G. Lowe in 2004. The SIFT algorithm matching feature can match solid texture points precisely. It can proceed when both compatible digital images transform, rotate, and parallel transform. The procedure followed by the SIFT algorithm is defined as -

- **Detection of scale-space extrema:** It is the first step of the computation. In this, all the images and scales' locations are identified.
- **Localizing key points:** 3D quadratic function is used for determining the scale and location for every candidate key point. Simultaneously, the bordering key points which are not stable gets deleted.
- **Assigning the orientation:** Consistent orientation is assigned to a key point prior to their local image properties. It helps in the presentation of the key point descriptor.
- **Generating Descriptor for local key point:** The creation of key point descriptor also known as the feature vector of the SIFT algorithm. It is done by the orientation and gradient magnitude at each image sample.

After the creation of SIFT feature values for key points, Euclidean separation between quality vectors make use of to calculate the resemblance of key points in various computerized images. An element spot from a single image is selected, and then the other side two-element spots are obtained by going through entirely element spots in the other side alternative image with the minimum and next minimum separation in the Euclidean. Between these two places, if the divider between the smallest and smallest distance in the Euclidean is below the limit value, then they are considered the points of the homologous element.

Depending on the internal and external boundary rate of the right and left cameras, and the corresponding two-pixel spot p1, p2 left and right images arranged in 3D spot P, two-line numbers constituting 3D spot P are settled, which can be calculated for 3D point P. The diagram shows a suitable 3D reconstruction model. But in real statistics, due to errors created by camera measurement along with similar characteristic points, right and left cameras blocked two lines which may not break at the same time; because of which it is required to embrace alike models to measure global links for 3D spot P.

Figure 2. Ideal Model for 3D reconstruction

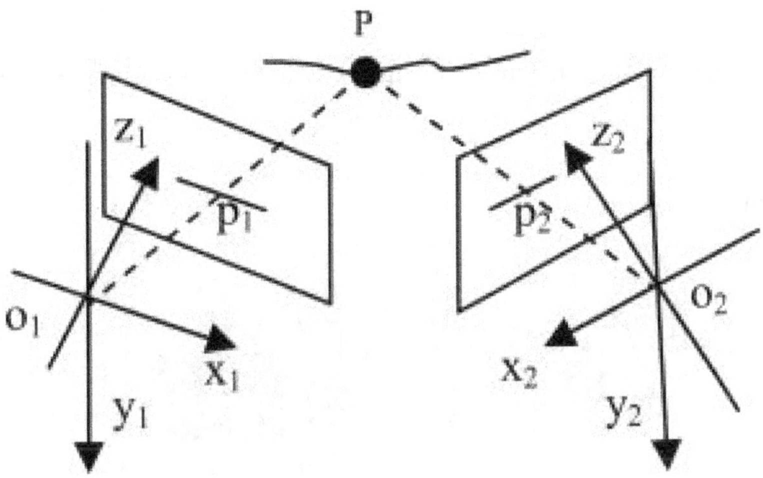

4.3 SURF Algorithm

For object recognition SIFT algorithm is used, but it was pertinently steady and people require to get into the high gear version. To overcome this problem SURF (Speeded-up Robust Features) algorithm is used (Sykora et al., 2014). SURF depends on the addition of products of the elements of the Hessian matrix for both lamina as well as for position also. This algorithm uses wavelet responses in both horizontal and vertical directions for feature explanation. For dimensional analysis, there are two techniques for computing measurements of an object from an image: Focal length and the Pixel density method.

The focal length method works on a formula.

$x/f = X/D$

Where, x =Object's length (in pixels).

X =Object's real-world length.

f = Len's Focal length.

D = Separation between camera lens and object.

The pixel density method is based on a reference object. The reference object is chosen according to these two conditions.

1. The reference object is lying in a similar plane to the object in the image.
2. The reference object is simply recognizable in the image.

Surf consists of two steps- the first is feature extraction and the second is feature description. The fast-hessian detector is used for feature extraction. Hessian matrix is used for the detector as it performs very well in terms of accuracy and time. The scale and the location are selected by the means of the determinant of hessian. The Hessian matrix $\mathfrak{H}(x, \sigma)$: which is at scale σ in an image *I* at point x can be written as:

$$\mathfrak{H}(x, \sigma) = \left[L_{xx}(x,\sigma) L_{xy}(x,\sigma) L_{xy}(x,\sigma) L_{yy}(x,\sigma) \right]$$

Where, the above parameters show the image I complexity of the Gaussian second-order by-product at point x.

Figure 3. SURF box filter

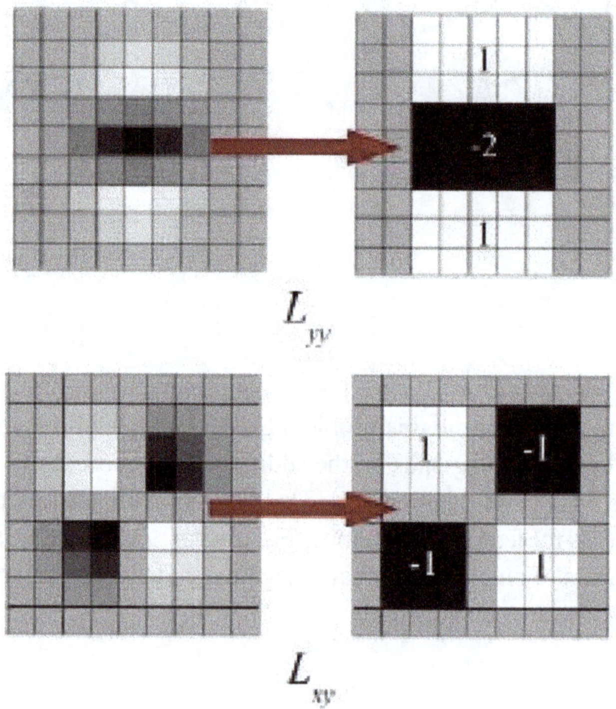

In feature description, for generating descriptors for SURF, two steps are required. The first step takes input as a piece of information from having reproducible orientation from the circular region. This circular region is converted back into the square region which is used for extracting the SURF descriptor.

4.4 PCA (Principle Component Analysis)

The acquaintance of the thought to utilize Principal Component Analysis (PCA) to regulate face pictures and the ensuing utilization of the following technique for actualizing a face acknowledgment framework, incited the advancement of PCA-build face handling frameworks (Blanz et al., 1999). PCA-build models are created by pertaining PCA on the deviations of training of the covariance matrix example of the mean example, to separate the eigenvectors that characterize the fundamental methods of changeability inside a training set. Because of the inspection, training models can be regulated towards a less dimensional space speaking to the weights related to every eigenvector. To construct shape and surface PCA-based 3D models, PCA decay is pertained doubly to demonstrate shape fluctuation and surface inconstancy utilizing two particular models. PCA is regulated on the covariance matrix of the shape and surface divergence. Because of the outcome of the examination, the generation of new 3D shapes or textures can be done using:

$$Y = \bar{Y} + \sum_{i=1}^{m} x_i \alpha_i$$

where m is the number of significant eigenvectors, $\bar{Y}$ is the mean 3D shape or texture among the training set, Y is a new 3D texture or shape vector, α_i are weights for eigenvectors, x_i is the ith eigenvector and inside a training set. Therefore, the presentation of a 3D face via a group of texture models and parameters of the shape models relying on this respective framework.

For 3D face reconstruction, many researchers use 3D face models based on PCA, and for executing reconstruction of 3D face algorithm apply as a basis a PCA-based 3D model (Atick et al., 1996). But they use cylindrical coordinates rather than using Cartesian coordinates.

Figure 4. Block Diagram of 3D PCA-Based models

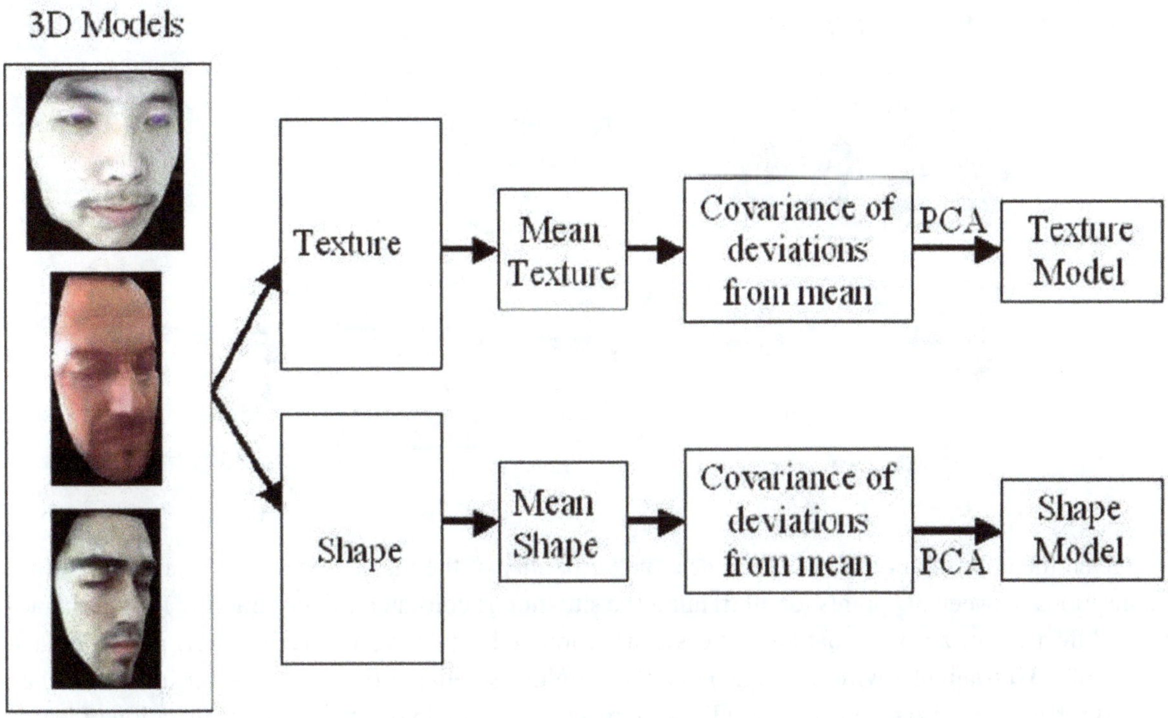

4.5 3DMM (3D Morphable Model)

Rather than applying a PCA model, Blanz and Vetter (Blanz et al., 1999) used 3DMM a prototypes-based model, of three-dimensional (3D) reconstruction from one image. The 3DMM is a productive model of facial features along with appearance based on two key concepts: First, all faces are in dense points, which are usually set on the model face set in the registration process and stored in any further steps. As a result of this correspondence, facial lines can be interpreted logically, producing real-life morphs. The second idea is to distinguish facial features from colour and to distinguish these from external factors such as lighting and camera parameters. The Morphable Model could include a mathematical model of facial contouring, which was the main analysis of objects in the first task and included other learning strategies in the next task.

3D faces can be generated using two-approach- first automatically from single or many mages and later is by the intuitive user interface. The two major problems faced by users in computer-assisted face modeling. The first is to follow the approach in which improbable features are avoided and more focussed on the naturalness of faces. The second problem is that new 3D face models get spontaneously recorded by one-to-one compatibility of an inner face model.

3D faces can be generated using two-approach- first automatically from single or many mages and later is by the intuitive user interface. The two major problems faced by users in computer-assisted face modeling. The first is to follow the approach in which improbable features are avoided and more focussed on the naturalness of faces. The second problem is that new 3D face models get spontaneously recorded by one-to-one compatibility of an inner face model.

Figure 5. Visual Abstract of 3DMM

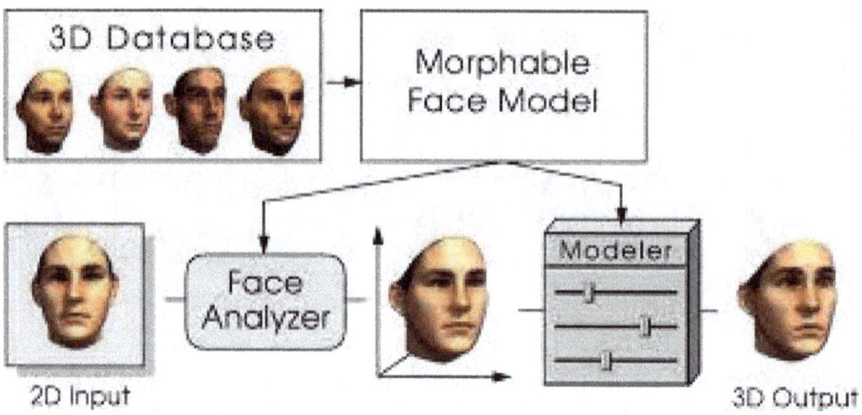

Given the collection of 3D facial modes, the variability of the visual approaches is applied to find connections between 3D points for all training the situation is consistent. Every training face is characterized by a vector accommodating vertices connection and texture vector having RGB power to each vertex of a 3D triangular wire. By organizing the weights used in the file texture and texture elements, new 3D shapes (S_{new}) and new features (T_{new}) are made according to the weighted organization of learning illustrations:

$$S_{new} = \sum_{i=1}^{M} S_i c_i \text{, } T_{new} = \sum_{i=1}^{M} T_i b_i$$

Where, M signifies total available samples in the training set, and S_i is a vector containing the directions of the state of the ith training test and T_i is the feature vector c_i and b_i are the weights registered to each learning specimen in direction to produce this model occurrence. Throughout the procedure of 3D face reconstruction, the task to evaluate the weights needed for integrating.

4.6 Deep3DFace Reconstruction

This reconstruction approach was given by Microsoft researchers (Yu Deng et al., 2010). In this, they proposed a CNN-root for one-image countenance reconstruction technique that makes use of blend-level image information for weakly-supervised learning. In this, the hybrid loss function is used which comprises perception-level loss and image-level loss. By using this Hybrid loss function this method outperforms multiple datasets when compared to models trained in a fully supervised manner.

Figure 6. Instruction channel for single image 3D face reconstruction

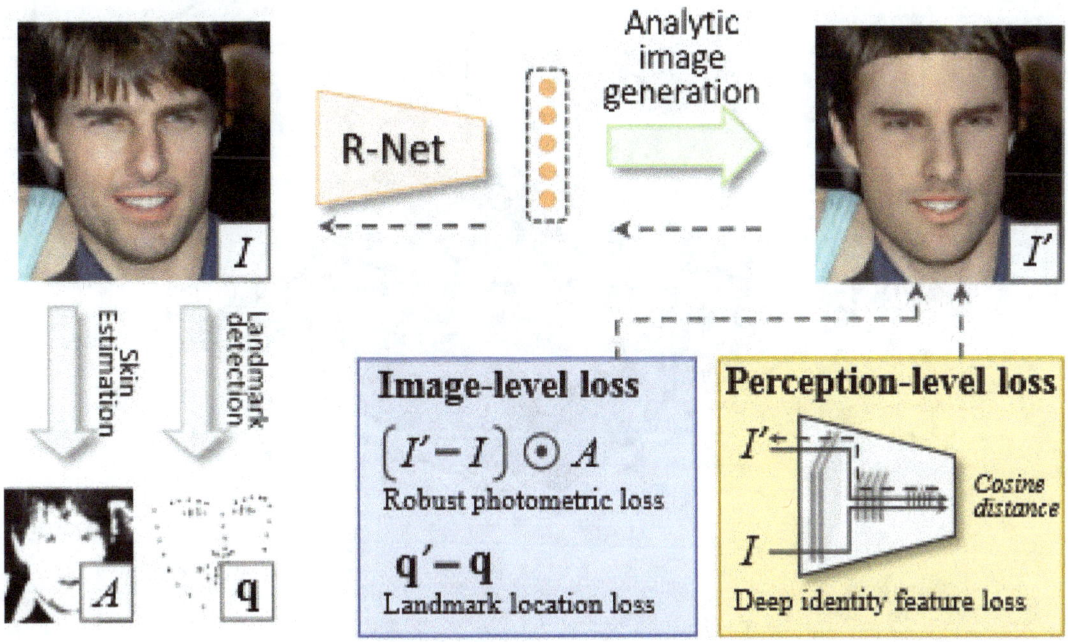

They put forward a novel structure dependence learning program for various-image face reconstruction gatherings. Their faith forecast subnet is also instructed in a weakly-supervised vogue minus a terrestrial-observation mark. This is considered to be the only attempt in the generation of a 3D face by using CNN from an unconstrained image set.

4.7 PRNet

This reconstruction approach was given by Yao Feng (Feng et al., 2018). This approach answers the problem of reconstruction of 3D face and alignment of the face together without any barrier of low-dimensional object space. To render the dense alignment and 3d facial structure UV position map, a novel representation, helps in keeping the track of all the information of the 3D face along with its semantic explanation.

For training weight mask has been allotted to different weights to every point on the UV map and then calculated the total weight loss. They also provide the framework that is capable of running at more than 100fps to directly obtain the reconstruction of a 3D face from a 2D image.

Figure 7. Illustration of UV position map

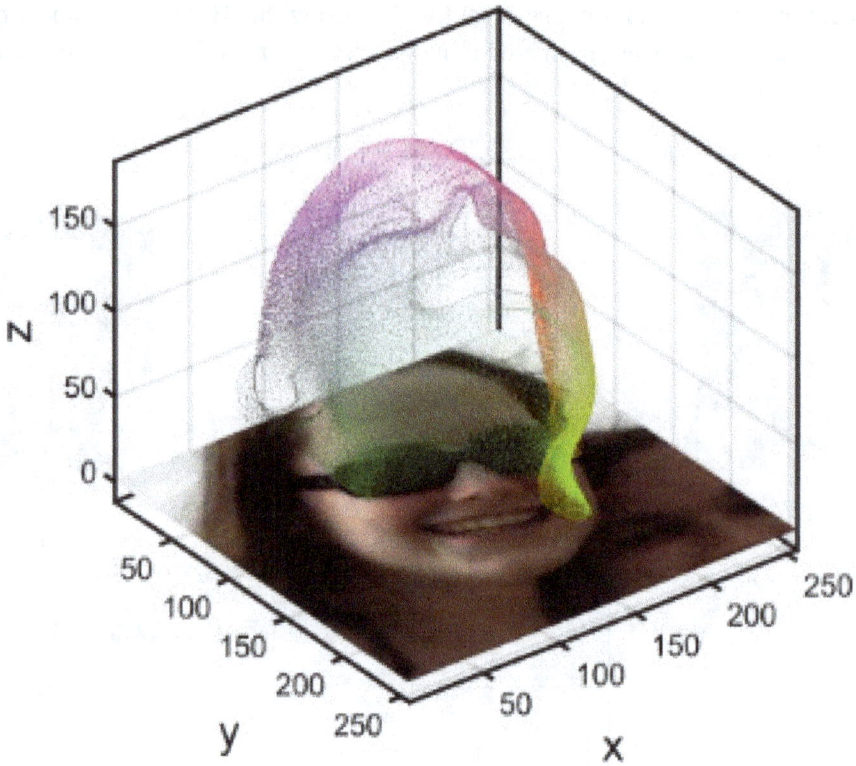

5. DISCUSSION

The evolution of 3D data operating algorithms is presently a very strong tendency instead of processing a face image. Investigators employed in the field intended to take over initialize the advanced capacity of 3D data representation to improve systems accomplish higher execution contrast to standard 2D faces image processing methods. Nevertheless, an important query that needs to be acknowledged is therefore the evolution of 3D face image processing methodology is accurately supported, vigorous, low cost, real-time, and real-time data possession technology Cursor rent there are two major ways to capture 3D face data. The initial way is to utilize committed 3D scanners and secondly to utilize 2D imaging devices now integrated with 3D face-building methodologies. Another way to start does not meet all the specifications as 3D scanners are often costly, large, they do not work in real-time, and they generally need managed imagery circumstances to efficient and co-operation of the topic are needed for a long time to install the scanning process. Looking at the other side, image capture utilizing 2D imaging devices can be made in real-time, which requires no user interference applying affordable appliances. But the central problem is if the modernization of 3D graphic reconstruction techniques is sufficient to maintain the requirement of 3D real-life 3D data accession for face processing.

6. CONCLUSION

3D face dependent on the image remaking procedures gives a financially efficient method of getting 3D facial information. This sort of innovation has been applied in various fields consisting of the processing of 3D face images. Notwithstanding, barely any main points of interest are as yet should have been explained, with the goal that is relevant to the picture relied on 3D reconstruction can be reached out to a more extensive scope of the utilization. In this regard, the fundamental problems that should be settled include the advancement of exact and completely mechanized 3D reconstruction frameworks that can work in uninhibited conditions.

REFERENCES

Aharchi, M. (2019). A Review on 3D Reconstruction Techniques from 2D Images. *Springer Third International Conference on Smart City Applications (SCA).*

Atick, J. J., Griffin, P. A., & Redlich, N. A. (1996). Statistical approach to shape from shading: Reconstruction of 3d face surfaces from single 2d images. *Neural Computation*, *8*, 1321–1340. doi:10.1162/neco.1996.8.6.1321 PMID:8768397

Blanz, V., & Vetter, T. (1999). A morphable model for the synthesis of 3d faces. *International Conference on Computer Graphics and Interactive Techniques*, 187–194. 10.1145/311535.311556

Bulat, A., & Tzimiropoulos, G. (2017). How far are we from solving the 2d and 3d face alignment problem? (and a dataset of 230,000 3d facial landmarks). *2017 IEEE International Conference on Computer Vision (ICCV).* arXiv:1703.07332

Feng, Y., Wu, F., Shao, X., Wang, Y., & Zhou, X. (2018). *Joint 3D Face Reconstruction and Dense Alignment with Position Map Regression Network.* arXiv:1803.07835,2018.

Gnana Priya, B., & Arulselvi, M. (2019). 3d Image Generation from Single 2d Image using Monocular Depth Cues. *International Journal of Engineering and Advanced Technology.*

Hu. (2019). Research on a three-dimensional reconstruction method based on the feature matching algorithm of a scale-invariant feature transform. In *Elsevier Mathematical and Computer Modelling Book.* Academic Press.

Jackson, A., Bulat, A., & Tzimiropoulos, Y. (2017). Large Pose 3D Face Reconstruction from a Single Image via Direct Volumetric CNN Regression. *2017 IEEE International Conference on Computer Vision (ICCV).* arxiv:1703.07834

Jackson, A. S., Bulat, A., Argyiou, V., & Tzimiropoulos, G. (2017). Large pose 3d face reconstruction from a single image via direct volumetric CNN regression. *2017 IEEE International Conference on Computer Vision (ICCV)*, 1031-1039. 10.1109/ICCV.2017.117

Joga, Sree, Spandana, & Pushpa. (2019). 3D reconstruction of regular objects from multiple 2D images using a reference object. International Journal of Advance Research, Ideas and Innovations in Technology.

Jourabloo, A., & Liu, X. (2016). Large-pose face alignment via CNN-based dense 3d model fitting. Computer Vision and Pattern Recognition.

Swarna Priya, R. M., Aarthy, S. L., Gunavathi, C., & Venkatesh, P. (2017). 3D Reconstruction of a Scene from Multiple 2D Images. *International Journal of Civil Engineering and Technology*, *8*(12), 324–331.

Sykora, P., Kamencay, P., & Hudec, R. (2014). Comparison of SIFT and SURF Methods for Use on Hand Gesture Recognition based on Depth Map. *AASRI Procedia*, *9*, 19–24. doi:10.1016/j.aasri.2014.09.005

Yu, Saito, Li, & Ceylan. (2017). *Learning dense facial correspondences in unconstrained images.* arXiv:1709.00536.

Yu, D., Wang, S., & Deng, L. (2010). Sequential labeling using deep-structured conditional random fields. *Journal of Selected Topics in Signal Processing*, *4*(6), 965–973. doi:10.1109/JSTSP.2010.2075990

Zhu, W., Wu, H. T., & Chen, Z. (2020). ReDA: Reinforced Differentiable Attribute for 3D Face Reconstruction. *Proceedings of the IEEE/CVF Conference on Computer Vision and Pattern Recognition (CVPR)*, 4958-4967.

Chapter 3
Machine Learning Based Intrusion Detection System for Denial of Service Attack

Ashish Pandey
Kamla Nehru Institute of Technology, Sultanpur, India

Neelendra Badal
Kamla Nehru Institute of Technology, Sultanpur, India

ABSTRACT

Machine learning-based intrusion detection system (IDS) is a research field of network security which depends on the effective and accurate training of models. The models of IDS must be trained with new attacks periodically; therefore, it can detect any security violations in the network. One of most frequent security violations that occurs in the network is denial of service (DoS) attack. Therefore, training of IDS models with latest DoS attack instances is required. The training of IDS models can be more effective when it is performed with the help of machine learning algorithms because the processing capabilities of machine learning algorithms are very fast. Therefore, the work presented in this chapter focuses on building a model of machine learning-based intrusion detection system for denial of service attack. Building a model of IDS requires sample dataset and tools. The sample dataset which is used in this research is NSL-KDD, while WEKA is used as a tool to perform all the experiments.

INTRODUCTION

Network intrusion or cyber-attack (Kumar et al., 2006) is an activity which prevents normal functioning of a system or computer network. To redress network intrusions, a mechanism called intrusion detection is used which mitigates or reports these intrusions. Basically, Intrusion Detection System (IDS) is an application that monitors, detects and prevents the network or the system against any suspicious activity (Sabhnani & Serpen, 2004) (Shyu et al., 2003). An IDS can be more effective if the training and testing

DOI: 10.4018/978-1-7998-3327-7.ch003

of its models are performed with machine learning classifiers because machine learning helps to gain insight from a bulky amount of data which is very inconvenient to humans and sometimes also useless.

With the traditional intrusion detection methods, we have to keep on watching the networks efficiently. To defeat these issues, in the current years, there have been various attempts to build an efficient IDS based on machine learning for large scale attacks like DoS (Su, 2011) (Aneetha & Bose, 2012) (Koc et al., 2012) (Chandrashekhar & Raghuveer, 2013) (Lin et al., 2015) (Srimuang & Intarasothonchun, 2015) (Hadri et al., 2016) (Syarif & Gata, 2017). But, still there are some limitations in the existing works. The work presented in this paper has also proposed a model of IDS for DoS attacks. The proposal of model has utilized a sample dataset named NSL-KDD (Effendy et al., 2017) (Shrivas & Dewangan, 2014) (Subba et al., 2016).

Next section presents the proposed work including methodology of proposed work, experimental setup, classifier used for training and testing of models and performance metrics. After this, the first part of proposed work which is class-wise model comparison is presented. In the same sequence, the second part of proposed work is also presented in which a model of machine learning based IDS for DoS attack is provided. Last section presents the conclusion of this research work.

PROPOSED WORK

The work presented in this paper focuses to propose a model of IDS for Denial of Service attack which is based on machine learning. To achieve the goal, the proposed work is divided into two parts. These two parts are:

- Class-wise model comparison of machine learning based IDS to identify the 'best class model' against Denial of Service attack using sample dataset; and
- Proposing a model for machine learning based IDS for Denial of Service attack with the optimal attributes of 'best class model'. These optimal attributes are selected from 'best class model' after performing the 'Attribute Selection'.

The proposed model can detect Denial of Service attack better than or equal to the 'best class model'. Overall, the proposed work is able to produce "Machine Learning based Intrusion Detection System for Denial of Service Attack".The methodology of proposed work, experimental setup including tool and sample dataset, machine learning classifier used for training and testing of models and performance metrics which have been utilized to evaluate the performance of models are described in following sub-sections.

Methodology of Proposed Work

The steps pursued in achieving the goal of proposed work are as follows:

Step 1: NSL-KDD is selected as sample dataset.
Step 2: Weka tool is selected for experiments.
Step 3: Preprocessing of dataset files (training and test) to remove irrelevant attack classes from dataset files i.e. Probe, User to Root (U2R) and Remote to Local (R2L).
Step 4: Generation of 14 new dataset files (training and test) for each combination of attributes group.

Step 5: Training and testing of models with dataset files using Random tree (Aldous, 1991) (Breiman et al., 1984) which is used as a binary classifier. It classifies the instances as normal or attack.
Step 6: Comparing the experimental results of 15 models on the basis of performance metrics (Bramer, 2013).
Step 7: Choosing the 'best class model' for attribute selection which performs better than the others.
Step 8: Applying Correlation based Feature Selection (CFS), Information Gain (IG) and Gain Ratio (GR) algorithms in dataset files of 'best class model' to perform attribute selection.
Step 9: Training and testing of proposed model with the optimal attributes which have been selected after performing attribute selection.
Step 10: Comparing the performance result of proposed modelwith the'best class model' which has been obtained after class-wise model comparison of initial 15 models.

Experimental Setup

The experimental setup includes the hardware and software resources which have been utilized in the proposed work.

Weka

Waikato Environment for Knowledge Analysis (Weka) is a suite of machine learning software written in Java and runs on almost any platform. To solve the real-world data mining problem on a given dataset, they can be enforced straight or called from your own Java code. The version used in this research work is 3.6.9.

Sample Dataset

The sample dataset which has been utilized in this paper is NSL-KDD. The dataset is available in a number of versions. In this paper, the version KDDTrain+ is used for training of models while the testing of models is performed with version KDDTest+. Each version contains different number of instances but number of attributes in each version is forty-two. The 'attribute number 42' in the sample dataset describes the 'class' of an instance (normal or attack). The remaining forty-one attributes are grouped into four classes which are:

- Basic (B) group: The attributes related to individual TCP connections are included in this group.
- Content(C) group: The attributes related to domain knowledge in a connection are included in this group.
- Traffic (T) group: The attributes which have been calculated using a two-second time window are included in this group.
- Host (H) group: The attributes related to assessment of attacks which last for more than two seconds are included in this group.

The name of individual attributes is mentioned in table 1.

Table 1. Attributes of NSL-KDD dataset

Basic Group		Content Group		Traffic Group		Host Group	
Sr. No.	**Attribute Name**	**Sr. No.**	**Attribute Name**	**Sr. No.**	**Attribute Name**	**Sr. No.**	**Attribute Name**
1	Duration	10	Hot	23	Count	32	Dst_host_count
2	Protocol_type	11	Num_failed_logins	24	Srv_count	33	Dst_host_srv_count
3	Service	12	Logged_in	25	Serror_rate	34	Dst_host_same_srv_rate
4	Flag	13	Num_compromised	26	Srv_serror_rate	35	Dst_host_diff_ srv_rate
5	Src_bytes	14	Root_shell	27	Rerror_rate	36	Dst_host_same _src_ port_rate
6	Dst_bytes	15	Su_attempted	28	Srv_rerror_rate	37	Dst_host_srv_ diff_ host_rate
7	Land	16	Num_root	29	Same_srv_rate	38	Dst_host_serror_rate
8	Wrong_fragment	17	Num_file_creations	30	Diff_srv_rate	39	Dst_host_srv_serror_ rate
9	Urgent	18	Num_shells	31	Srv_dif_host_rate	40	Dst_host_rerror_rate
		19	Num_access_files			41	Dst_host_srv_ rerror_rate
		20	Num_outbound_cmds				
		21	Is_hot_login				
		22	Is_guest_login				

Used Classifier

Machine learning (Wu et al., 2007) is a powerful craft based on artificial intelligence and comprises a lot of algorithms to develop a model. These models are trained with input data known as training dataset and examined by unknown input termed as testing dataset (Lei et al., 2010). Training of these models also depends on classifier used to develop the model. The proposed work has used random tree classifier algorithm because it does not require much tuning of parameters and produces better accuracy in results.

The random tree algorithm is given by Leo Breiman and Adele Cutler (Aldous, 1991) (Breiman et al., 1984). The algorithm is suitable for classification as well as regression. Basically, it performs ensemble for tree predictors which are referred as forest. The steps used by random tree in classification are:

- Select the input feature vector;
- Perform classification based on every tree of forest;
- Calculate the number of votes;
- Label the class in output which has gained majority of "votes".

Performance Metrics

The comparison of models is evaluated by performance metrics (Bramer, 2013) i.e. Accuracy, Detection Rate (DR) and False Alarm Rate (FAR). All these performance metrics are fundamentally calculated from the four key elements of the confusion metrics. These key elements depict the actual and predicted classes. The key elements are True Positive (TP), False Positive (FP), False Negative (FN) and True Negative (TN).

Accuracy (Bramer, 2013) can be calculated as:

$$Accuracy = \frac{\mathrm{TP} + \mathrm{TN}}{\mathrm{TP} + \mathrm{FP} + \mathrm{FN} + \mathrm{TN}} \tag{1}$$

Basically, it is a fraction of all the instances which are correctly predicted or classified.

Detection Rate (DR) (Bramer, 2013) can be calculated as:

$$Detection\ Rate = \frac{\mathrm{TP}}{\mathrm{TP} + \mathrm{FN}} \tag{2}$$

Basically, it is a fraction of instances which are predicted as true positive from all the instances which are exactly positive instances.

False Alarm Rate (FAR) (Bramer, 2013) can be calculated as:

$$False\ Alarm\ Rate = \frac{\mathrm{FP}}{\mathrm{FP} + \mathrm{TN}} \tag{3}$$

A graphical plot known as the Receiver Operating Characteristic (ROC) curve is also used for plotting the DR vs. FAR.

The next section describes the first part of proposed work which is class-wise model comparison.

CLASS-WISE MODEL COMPARISON

The purpose of class-wise model comparison is to identify the group of attributes or combination of groups from sample datasetwhich can detect DoS attack more accurately than others. To perform class-wise model comparison, fifteen pairs of dataset files are prepared with the help of four groups Basic, Content, Traffic and Host. The experiments which have been performed for class-wise model comparison are explained with its results and analysisin following sub-sections.

Experiments

Since the class-wise model comparison is performed for machine learning based IDS for Denial of Service attack, so the first task is to separate all records of DoS attack class and normal class from the KDDTrain+

and KDDTest+ files. This is the preprocessing part of experiments which have been performed in this research. The experiments which have been performed for class-wise model comparison are:

Preparation of Dataset files

The first experiment is to prepare15 pairs of dataset files for training and testing of models separately. Since the 42nd attribute is used to predict the status of attack, therefore it must be included in all the prepared files whether it is training file or test file. These fifteen dataset files are mentioned in following table 2.

Table 2. Prepared fifteen files

Sr. No.	Combination of Attribute Classes	Total Number of attributes	Attributes taken from Individual class			
			B(9)	C(13)	T(9)	H(10)
1	B	9	9	0	0	0
2	C	13	0	13	0	0
3	T	9	0	0	9	0
4	H	10	0	0	0	10
5	BC	22	9	13	0	0
6	BT	18	9	0	9	0
7	BH	19	9	0	0	10
8	CT	22	0	13	9	0
9	CH	23	0	13	0	10
10	TH	19	0	0	9	10
11	BCT	31	9	13	9	0
12	BCH	32	9	13	0	10
13	BTH	28	9	0	9	10
14	CTH	32	0	13	9	10
15	BCTH	41	9	13	9	10

Training and Testing of Models

Training and testing of models are performed with random tree classifier in Weka tool. These models are built with 15 training filesof table 2 using 10-cross validation. Further, these models are tested against the respective test dataset filescontaining the same class of attributes.

Experimental Results

The experimental results of class-wise model comparison of proposed work are expressed in the form of confusion matrix which has been simulated on Weka tool. The confusion matrix gives the result which shows the relation of actual classification to the predicted classification. The experimental results produced for fifteen models are as follows:

Table 3. Experimental results of 15 models

Sr. No.	Number of Attribute Classes	Class Models	TN	FN	FP	TP
1	1	B(9)	8895	1214	816	5953
2	1	C(13)	7354	537	2357	6630
3	1	T(9)	9163	1772	548	5395
4	1	H(10)	9173	2020	538	5147
5	2	BC(22)	9583	842	128	6325
6	2	BT(18)	9622	1184	89	5983
7	2	BH(19)	9624	1060	87	6107
8	2	CT(22)	9106	1346	605	5821
9	2	CH(23)	9093	1522	618	5645
10	2	TH(19)	9193	1566	518	5601
11	3	BCT(31)	9008	1110	703	6057
12	3	BCH(32)	9628	1114	83	6053
13	3	BTH(28)	9618	1223	93	5944
14	3	CTH(32)	9156	1787	555	5380
15	4	BCTH(41)	9489	1409	222	5758

The calculated performance metrics for all the models of table 3 are tabulated in table 4.

Table 4. Calculated performance metrics of 15 models

Sr. No.	Number of Attribute Classes	Class Models	Accuracy (%)	False Alarm Rate (%)	Detection Rate (%)
1	1	B(9)	87.9725	8.402	83.061
2	1	C(13)	82.8534	24.271	92.507
3	1	T(9)	86.2542	5.643	75.275
4	1	H(10)	84.8441	5.540	71.815
5	2	BC(22)	94.2528	1.318	88.251
6	2	BT(18)	92.4576	0.916	83.479
7	2	BH(19)	93.2041	0.895	85.209
8	2	CT(22)	88.4405	6.230	81.219
9	2	CH(23)	87.3207	6.363	78.763
10	2	TH(19)	87.6525	5.334	78.149
11	3	BCT(31)	89.2582	7.239	84.512
12	3	BCH(32)	92.9079	0.854	84.456
13	3	BTH(28)	92.2028	0.957	82.935
14	3	CTH(32)	86.1239	5.715	75.066
15	4	BCTH(41)	90.3365	2.286	80.340

After calculating the performance metrics of all 15 models, the comparison of these models is performed. The comparison of models is based on accuracy, detection rate and false alarm rate.

Comparison of Results

The first comparison is performed on the basis of DR and FAR of each model. The values of DR and FAR are taken from table 4. The comparison of models is presented by bar graph in figure 1. The red bar represents the detection rate while blue bar represents the false alarm rate.

Figure 1. DR and FAR of 15 models

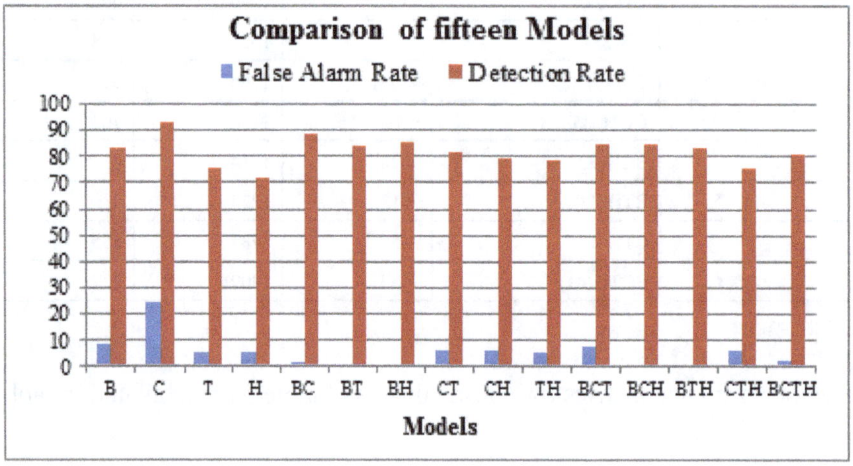

Key observations from bar graph of figure 1 are:

- Content class model has highest detection rate which is significantly higher than model BCTH;
- False alarm rate of content class model is also highest which is also significantly higher than model BCTH;
- Detection rate of BC class model is on second position and also significantly higher than model BCTH;
- False alarm rate of BC class model is also lower than model BCTH;
- BCH class model has lowest false alarm rate.

The second comparison is performed on the basis of number of attribute classes in a model. This comparison is presented by ROC curve which identifies the best combination of DR and FAR. Figure 2 shows all the models which contains single group of attributes. Similarly, figure 3 and figure 4 depict all the modelscontaining two groups and three groups of attributes respectively. Since these three comparisons are performed against model BCTH (containing all groups of attributes), so model BCTH is also included in these figures. The aim of these comparisons is to identify the best single, double and triple class models which produce detection rate better than or equal to model BCTH and minimum false alarm rate better than or equal to model BCTH at the same time. The red arrow which is pointing a class model on curves manifests the best combination of DR and FAR of each comparison.

Figure 2. ROC curve for single class models

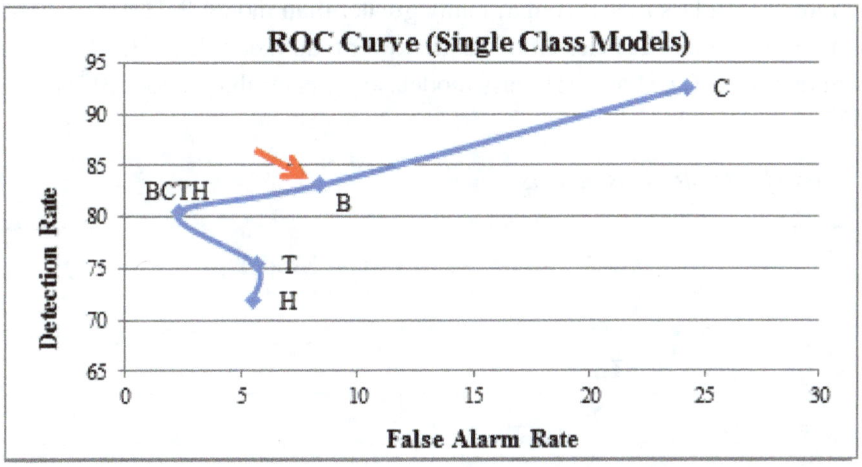

The key informations which are obtained from ROC curve of figure 2 are:

- DR of Content class model is the highest but FAR is also very high at the same time;
- Both DR and FAR of Basic class model is little greater than model BCTH;
- Detection rate of Traffic class model is little lower than model BCTH but false alarm rate is little grater at the same time;
- Detection rate of Host class model is the lowest and false alarm rate is almost equal to Traffic class model.

Figure 3. ROC curve for double class models

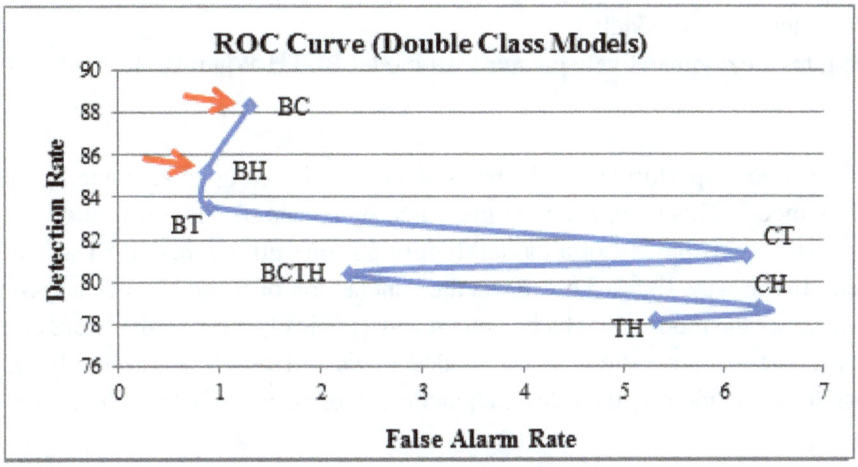

Similarly, key informations which are obtained from ROC curve of figure 3 are:

- Detection rates of BC, BH and BT class models are greater than model BCTH but false alarm rates are lower than model BCTH at the same time. It shows that the presence of Basic class improves both DR and FAR simultaneously;
- Detection rate of CT class model is marginally greater than model BCTH;
- Detection rates of CH and TH class models are lower than model BCTH;
- False alarm rates of CT, CH and TH class models are greater than model BCTH.

Figure 4. ROC curve for triple class models

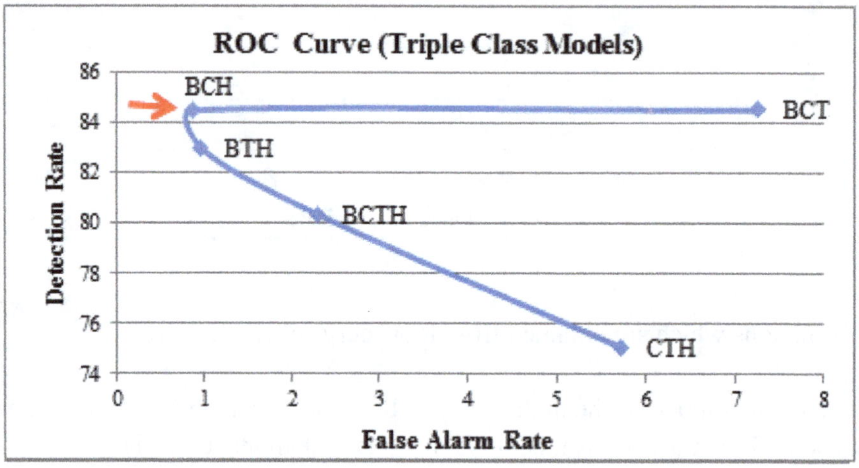

Following the figure 2 and figure 3, the key informations which are obtained from ROC curve of figure 4 are:

- False alarm rates are comparatively greater than model BCTH When both Content and Traffic class are present together in class models
- False alarm rates are comparatively lower than model BCTH When both Basic and Host class are present together in class models
- Detection rates are comparatively greater than model BCTH When Basic class is present in class models.

The third comparison is performed on the basis of a particular group of attributes which is definitely present in a class model. This comparison is presented by bar graph in which blue bar represents the FAR and red bar represents the DR of a model. Figure 5 shows all the models in which Basic class is certainly present. In this way, figure 5 manifests the benefaction of Basic class clearly.Similarly, figure 6 and figure 7shows all the models in which Content and Traffic class respectively are certainly present. Following the sequence, figure 8 shows all the models in which Host class is certainly present. The aim of these comparisons is to identify the role and benefaction of each attribute group in detection rate and false alarm rate.

Figure 5. DR and FAR of basic class with other classes

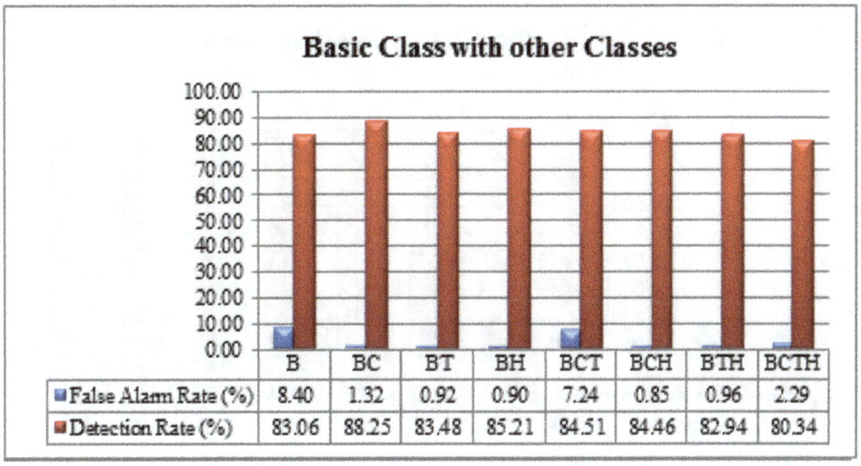

	B	BC	BT	BH	BCT	BCH	BTH	BCTH
False Alarm Rate (%)	8.40	1.32	0.92	0.90	7.24	0.85	0.96	2.29
Detection Rate (%)	83.06	88.25	83.48	85.21	84.51	84.46	82.94	80.34

The key informations which are obtained from bar graph of figure 5 are:

- Detection rate is greater than 80% for all the models;
- False alarm rate improves where Host class is present;
- Detection rate improves where Content class is present.

Figure 6. DR and FAR of content class with other classes

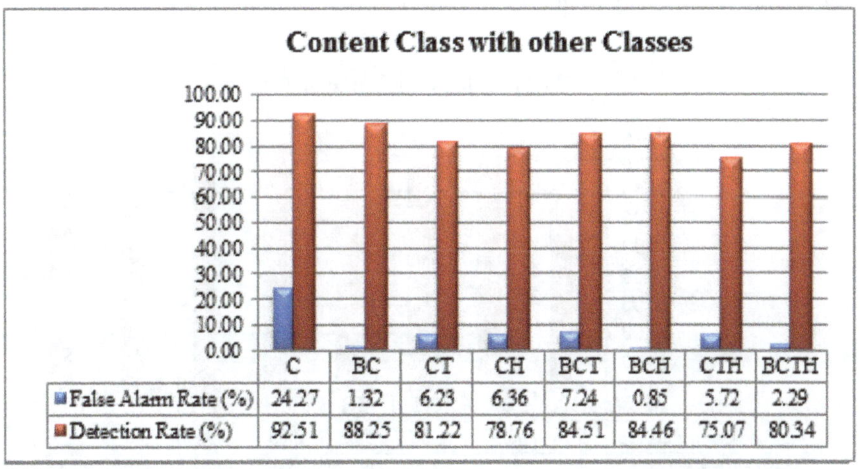

	C	BC	CT	CH	BCT	BCH	CTH	BCTH
False Alarm Rate (%)	24.27	1.32	6.23	6.36	7.24	0.85	5.72	2.29
Detection Rate (%)	92.51	88.25	81.22	78.76	84.51	84.46	75.07	80.34

The key informations which are obtained from bar graph of figure 6 are:

- Both DR and FAR are highest for individual Content class model;
- Detection rate revamps if Basic class is the part of models;
- False alarm rate revamps if the combination of Basic class and Host class is present in the model.

Figure 7. DR and FAR of traffic class with other classes

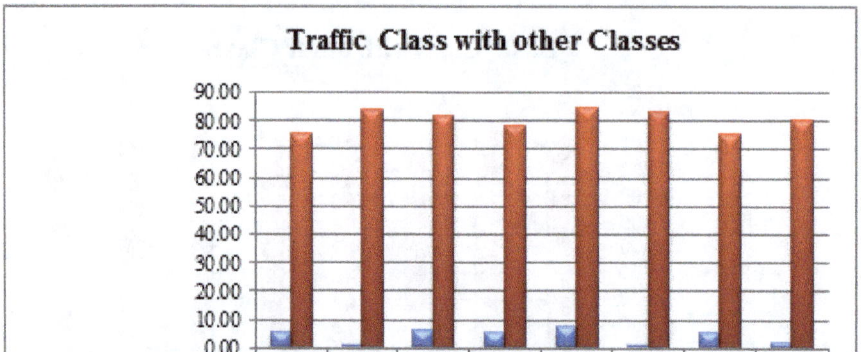

	T	BT	CT	TH	BCT	BTH	CTH	BCTH
False Alarm Rate (%)	5.64	0.92	6.23	5.33	7.24	0.96	5.72	2.29
Detection Rate (%)	75.28	83.48	81.22	78.15	84.51	82.94	75.07	80.34

The key informations which are obtained from bar graph of figure 7 are:

- Detection rate is low when Traffic class has no combination with other classes;
- Detection rate is also low when combination of Content class and Host class is present in the model;
- False alarm rate revamps if Basic class is present but Content class is absent in the model.

Figure 8. DR and FAR of host class with other classes

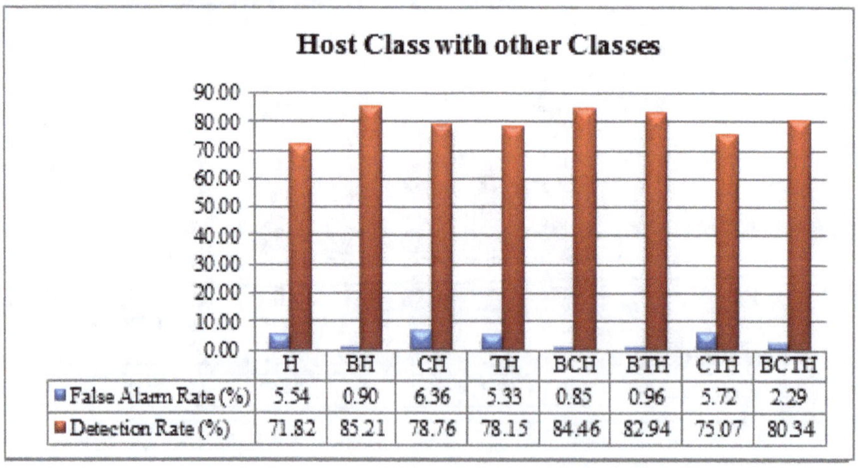

	H	BH	CH	TH	BCH	BTH	CTH	BCTH
False Alarm Rate (%)	5.54	0.90	6.36	5.33	0.85	0.96	5.72	2.29
Detection Rate (%)	71.82	85.21	78.76	78.15	84.46	82.94	75.07	80.34

The key informations which are obtained from bar graph of figure 8 are:

- Detection rate is lowest when Host class has no combination with other classes;
- Detection rate revamps if Basic class is the part of models;

- False alarm rate also revamps if Basic class is the part of models.

Analysis of Results Comparison

The facts which have been gathered after the comparison of results are tabulated in table 5. Table 5 is created with the help of figure 5 to 8. Table 5 provides the values of best DR (maximum) and FAR (minimum) with the values of worst detection rate (minimum) and false alarm rate (maximum) for the class models. In table 5, second row contains the values which have been obtained from figure 5. Figure 5 shows all the models in which Basic class is certainly present. Similarly, the best and worst values of figure 6 to 8 are recorded in row number 3 to 5 respectively. Figure 6, figure 7 and figure 8 shows all the models in which Content, Traffic and Host class respectively are certainly present. The values mentioned in the cells of table, also provides the name of class model in parenthesis to which it belongs.

Table 5. Analysis of comparison for best class model

Attribute Class (certainly present)	Best Detection Rate (%)	Best False Alarm Rate (%)	Worst Detection Rate (%)	Worst False Alarm Rate (%)
Basic Class	88.251 (BC)	**0.854 (BCH)**	80.340 (BCTH)	8.403 (B)
Content Class	**92.507 (C)**	**0.854 (BCH)**	75.066 (CTH)	**24.271 (C)**
Traffic Class	84.512 (BCT)	0.916 (BT)	75.066 (CTH)	7.239 (BCT)
Host Class	85.209 (BH)	**0.854 (BCH)**	**71.815 (H)**	6.363 (CH)

The gist of table 5 are:

- Content class provides the best detection rate but it cannot be used alone because it also provides the worst FAR at the same time;
- The rest three best detection rates are from BC, BH and BCT class models in which Basic class is certainly present. It means that best detection rate can only be produced if Basic class is the part of model but BCT class model also produces worst false alarm rate. Therefore, only BC and BH class models are considered for best detection rate;
- The best false alarm rate is obtained from BCH class model which is present in row-2, row-3 and row-5. Row-5 also contains the worst false alarm rate which is produced by CH class model. It means that best false alarm rate can only be produced if Basic class is combined with either Content class or Host class or the combination of both. Therefore, BC, BH and BCH class models are considered for best false alarm rate;
- Since, the best class model is one which produces best DR and best FAR at the same time. Therefore, only BC and BH class models are considered for 'best class models' in class-wise model comparison.

Finally, class-wise model comparison shows that best class model is one in which 'Basic' class must be present. Further Basic class combination with Content class maximizes the DR and minimizes the

FAR. Similarly, Basic class combination with Host class presents improved DR but FAR is reduced greatly. Therefore, class-wise model comparison provides two best models BC and BH.

PROPOSED MODEL

In previous section, class-wise model comparison has been performed. As a result, two 'best class models' have been identified for machine learning based IDS for DoS attack in terms of accuracy, DR and FAR. This section presents the second part of proposed work which focuses on proposing a model for machine learning based IDS for Denial of Service attack with the optimal attributes of 'best class model'. These optimal attributes are selected from 'best class model' after performing the 'Attribute Selection'. The best class models which have been identified after 'class-wise model comparison' are BC and BH. The number of attributes in BC and BH class models is 22 and 19 respectively; therefore, time taken to build these class models is also long.

Since the proposed model is built with the optimal attributes of 'best class models'. Therefore, the goal of attribute selection is to find those attributes in BC and BH class models which maximizes the detection rate and minimizes the FAR for proposed model i.e. machine learning based IDS for Denial of Service attack. The time taken to build the proposed model must also be smaller than the 'best class models'. The experiments which have been performed for building the proposed model is explained with its results and analysis in following sub-sections.

Experiments

The proposed model is built with selected attributes of 'best class models' after performing attribute selection. Attribute selection focuses on selecting those attributes in 'best class models' which can enhance the computational performance like detection rate, FAR and time taken to build the model. To perform attribute selection, the combination of three different attribute selector algorithms CFS, IG and GR are chosen with two search methods namely ranker search and best first search. All experiments are explained in below sub-sections.

Attribute Selection in Best Class Models

The first step is to apply CFS, IG and GR algorithms on training files of BC and BH class models. Attributes selected by CFS algorithm are evaluated on the basis of information gain and gain ratio which have been calculated by IG and GR algorithms. Those attributes which have good information gain and gain ratio are included to prepare new training and test files for proposed model. Though, there are many algorithms available to perform attribute evaluation but CFS, IG and GR produces optimal results.

Correlation based Feature Selection (CFS) utilizes a famous method known as correlation. In statistics, it is known as Pearson's correlation coefficient.CFS selects the most compatible and mandatory attributes in a given dataset.

The Information Gain describes the information provided by an attribute during the classification. Basically, it measures the difference of information where the value of an attribute is known from where the value of attribute is unknown. The common method which is used in calculation of information gain is Shannon entropy.

Gain ratio is the result of modification which has been performed in information gain which reduces high-branch attributes bias. It utilizes the size and number of branches during the selection of an attributes. It also performs split operation to correct information gain and calculate split ratio of an attribute.

Model Building and Testing

To build the model, random tree classifier is used in Weka tool. These models are built withnew training set files using 10-cross validation. Further, these models are tested against the respective new test dataset filescontaining the same class of attributes and using same classifier algorithm random tree.

Experimental Results

After performing attribute selection, the proposed model which is built with selected attributes must maintain following conditions:

- Detection rate should be greater than or almost equal to BC and/or BH class models;
- False alarm rate should be less than or almost equal to BC and/or BH class models;
- Time taken to build the model should be smaller than BC and BH class models.

CFS, IG and GR algorithms have selected five attributes in BC class model after performing attribute selection. The selected attributes are:

- Flag;
- Src_bytes;
- Dst_bytes;
- Num_failed_logins;
- Logged_in.

In the same way CFS, IG and GR algorithms have also selected five attributes in BH class model after performing attribute selection. The selected attributes are:

- Flag;
- Src_bytes;
- Dst_bytes;
- Dst_host_srv_diff_host_rate;
- Dst_host_srv_serror_rate.

After selecting the attributes in best class models BC and BH, two new models are built. These new models are BC (5) and BH (5). The performance results of new model BC (5)and BH (5) are expressed in the form of confusion matrix which has been simulated on Weka tool. The experimental results produced for new models are tabulated in table 6. These results are in terms of performance metrics.

After the performance results of new models, the comparison of these models is performed with best class models BC and BH. The comparison of models is based on accuracy, DR and FAR.

Table 6. Performance results of new models BC(5) and BH(5)

Sr. No.	Number of Attributes	Model Name	Accuracy (%)	False Alarm Rate (%)	Detection Rate (%)	Time Taken (seconds)
1	5	BC(5)	93.429	0.895	85.740	22.41
2	5	BH(5)	92.167	0.823	82.670	30.99

Comparison of Results

The comparison of results is presented by ROC curve which identifies the best combination of DR and FAR. Figure 9 shows the proposed models and best class models in terms of their DR and FAR. The aim of this comparison is to check whether the proposed models have fulfilled all the requirements of the best IDS or not. The red arrow which is pointing a class model on curve manifests the best combination of DR and FAR for this comparison.

Figure 9. ROC curve for proposed model

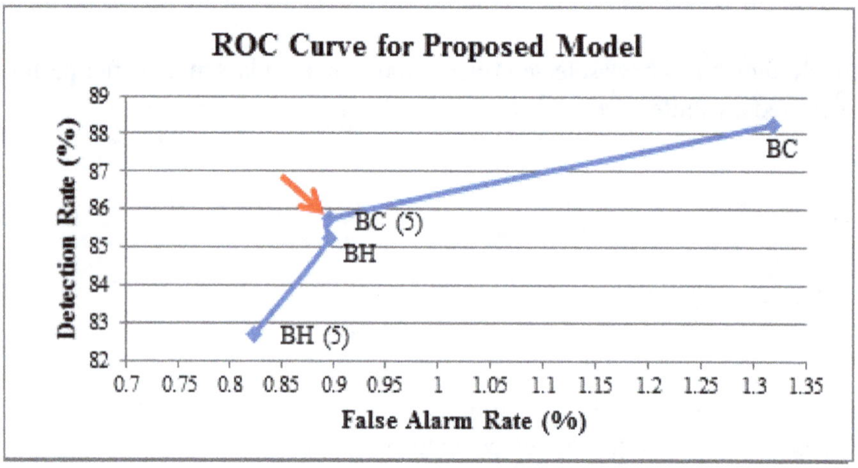

The key informations which are obtained from ROC curve of figure 9 are:

- BC (5) model has detection rate little lower than BC class model but false alarm rate is less than the same.
- BC (5) model has detection rate little greater than BH class model but FAR is equal to the same.
- BH (5) model has lowest DR and lowest FAR.
- Time taken to build the models BC(5) and BH(5) are 22.41 seconds and 30.99 seconds respectively.

Analysis of Results Comparison

Although, class-wise model comparison has identified BC and BH as best class model for machine learning based IDS for DoS attack. But proposed model has presentedthe better solution with less attributes i.e. BC (5) model. The goal of attribute selection was to identify minimum attributes which can enhance computational performance. The results comparison of models BC (5) and BH (5) against 'best class models' has revealed three major points:

- The accuracy and DR of proposed model BC (5)is almost parallel to the best class model BC;
- The false alarm rate of proposed model BC (5) is equal to the best class model BH;
- The time taken to build the modelBC (5) is very less in comparison to time taken by the best class models BC and BH. It is also less than model BH (5).

Overall, BC (5) is the optimal class model with only five attributes which has qualities of both 'best class models' BC and BH. Therefore, proposed model for machine learning based IDS for Denial of Service attack is BC (5).

CONCLUSION

The work presented in this paper has focused to propose a model of machine learning based IDS for DoS attack. To achieve the goal, it has performed two tasks i.e. class-wise model comparison using sample dataset and proposing a model after attribute selection. As a result of class-wise model comparison, class models BC and BH are selected as best class models in terms of accuracy, DR and FAR against DoS attack. After class-wise model comparison, attribute selection has been performed in best class models. The selected attributes of best class models have been used to build the new models BC (5) and BH (5). The performance of these new models is compared against best class models. As a result, new model BC (5) has provided the combined performance of best class models BC and BH. Therefore, BC (5) has been selected as proposed model of machine learning based IDS for DoS attack.

REFERENCES

Ahmad, E. D., Kusrini, K., & Sudarmawan, S. (2017). Classification of intrusion detection system (IDS) based on computer network. In *Proceedings of 2017 2nd International conferences on Information Technology, Information Systems and Electrical Engineering (ICITISEE)*. IEEE.

Aldous, D. (1991, January). The continuum random tree. I. *Annals of Probability*, *19*(1), 1–28.

Amal, H., Khalid, C., & Rajae, T. (2016). Intrusion detection system using PCA and Fuzzy PCA techniques. In *Proceedings of International Conference on Advanced Communication Systems and Information Security (ACOSIS)*. IEEE.

Aneetha, A. S., & Bose, S. (2012, August). The combined approach for anomaly detection using neural networks and clustering techniques. *Computer Science & Engineering: An International Journal*, *2*(4), 37–46. doi:10.5121/cseij.2012.2404

Basant, S., Santosh, B., & Sushanta, K. (2016). Enhancing performance of anomaly-based intrusion detection systems through dimensionality reduction using principal component analysis. In *Proceedings of IEEE International Conference on Advanced Networks and Telecommunications Systems (ANTS)*. IEEE.

Chandrashekhar, A. M., & Raghuveer, K. (2013, January). Fortification of hybrid intrusion detection system using variants of neural networks and support vector machines. *International Journal of Network Security and Its Applications*, *5*(1), 71–90. doi:10.5121/ijnsa.2013.5106

Correlation based Feature Selection. (n.d.). Available at: https://machinelearningmastery.com/perform-feature-selection-machine-learning-data-weka/

Gain Ratio. (n.d.). *Hong Kong University of Science and Technology*. Available at: www.cse.ust.hk/~qyang/537/PPT/dtrees1.ppt

Information Gain. (n.d.). Available at: https://www.saedsayad.com/decision_tree.htm

Kumar, S. A., & Kumar, D. A. (2014, August). An ensemble model for classification of attacks with feature selection based on KDD99 and NSL-KDD data set. *International Journal of Computers and Applications*, *99*(15), 8–13. doi:10.5120/15560-4109

Kumar, S. M., & Gursel, S. (2004). *Why Machine Learning Algorithms Fail in Misuse Detection on KDD Intrusion Detection Data Set. ACM Transactions on Intelligent Data Analysis.*

Leo, B., Friedman, J.H., Olshen, R.A., & Stone, C.J. (1984). Classification and regression trees. Monterey, CA: Wadsworth & Brooks/ColeAdvanced Books & Software.

Levent, K., Mazzuchi, T. A., & Shahram, S. (2012). A network intrusion detection system based on a Hidden Naïve Bayes multiclass classifier. *Expert Systems with Applications*, *39*, 13492–13500.

Li, L., Yang, D.-Z., & Shen, F.-C. (2010). A Novel Rule-based Intrusion detection System Using Data Mining. *Proc. of 3rd IEEE International Conference on Computer Science and Information Technology*, 169–172.

Lin, W.-C., Shih-Wen, K., & Chih-Fong, T. (2015). CANN: An intrusion detection system based on combining cluster centers and nearest neighbors. *Knowledge-Based Systems,* 13–21.

Max, B. (2013). *Principles of Data Mining* (2nd ed.). Springer.

Nsl-kdd data set for network-based intrusion detection systems. (2014). Available at: http://nsl.cs.unb.ca/NSL-KDD/

Rama, S. A., & Windu, G. (2017). Intrusion detection system using hybrid binary PSO and K-nearest neighborhood algorithm. In *Proceedings of 11th International Conference on Information & Communication Technology and System (ICTS)*. IEEE.

Shyu, M., Chen, S., Sarinnapakorn, K., & Chang, L. (2003). A novel anomaly detection scheme based on principal component classifier. *Proceedings of the IEEE Foundations and New Directions of Data Mining Workshop, in conjunction with the Third IEEE International Conference on Data Mining (ICDM03)*, 172–179.

Su, M.-Y. (2011). Real-time anomaly detection systems for Denial-of-Service attacks by weighted k-nearest-neighbor classifiers. *Expert Systems with Applications*, *38*, 3492–3498.

Vipin, K., Jaideep, S., & Aleksandar, L. (2006). *Managing cyber threats: Issues, approaches, and challenges* (Vol. 5). Springer.

Waikato environment for knowledge analysis (weka) version 3.6.9. (2013). Available at: https://download.cnet.com/Weka-32-bit/3000-10254_4-75852610.html

Worachai, S., & Silada, I. (2015). Classification model of network intrusion using Weighted Extreme Learning Machine. In *Proceedings of 12th International Joint Conference on Computer Science and Software Engineering (JCSSE)*. IEEE.

Wu, & Kumar, Quinlan, Joydeep, Yang, Hiroshi, McLachlan, Ng, Liu, Yu, Zhou, Steinbach, Hand, & Steinberg. (2007). Top Ten Data Mining Algorithms. Knowledge and Information Systems Journal. *Springer-Verlag London*, *14*(1), 1–37.

Chapter 4
Reliability Importance Measures-Based Analysis of Substation Communication Network

Rakhi Kamra
National Institute of Technology, Kurukshetra, India

G. L. Pahuja
National Institute of Technology, Kurukshetra, India

ABSTRACT

Smart grid can work effectively only when a reliable, fast communication network is available. The communication network is a prerequisite to connect different protection, control, and monitoring equipment within the substation. Ethernet fulfills all the requisites such as reliable, fast, secure, interoperable, LAN-based communication system for smart substations. Therefore, the main aspect is to improve the reliability of the network by prioritizing the critical components by using the knowledge of component importance measures (CIM). In this chapter, analysis of IEC 61850 ethernet-based substation communication network (SCN) architectures has been examined using various reliability importance measures (RIM). The importance measures namely Birnbaum, improvement potential, criticality importance, and reliability achievement worth have given their justified rankings of the various components of SCN architectures. The practice of these CIMs works towards the identification of the components that can be allocation of resources for the improvement of system reliability.

INTRODUCTION

In today's era, human beings are highly dependent on complex networks like Computer networks, telecommunication, wireless networks, Smart Grid(SG), transportation, gas and oil networks, which may be for financial transactions, social connections, mission - critical systems, or in emergency situations. These

DOI: 10.4018/978-1-7998-3327-7.ch004

networks need to be ultra- reliable for the smooth and proper functioning of the society. The reliability of these networks not only has to be very high but trustworthy also. Power Sector is among the critical infrastructures, and has always witnessed advancements since its evolution. Over the past decades, the use and demand of electricity has increased rapidly, therefore engineers are trying best to fulfill the needs of world. The present-day engineering revolution and increasing electricity rate, shrinking fossil fuels and intensifying fears about global warming have aggravated energy utilities and governments to decide upon solutions for improving efficiency and reliability of their existing power grids for continuous supply of power (Faheem et al., 2018; Thomas & Ali, 2010). Nowadays, the electric power grid is transforming into an intelligent grid, called the 'Smart Grid' (SG). The intelligence added to the traditional power grid may be deterrent to maintain/enhance reliability due to complexity of the system. Reliability of such systems depends upon network structures and communication technologies that provide end to end data communication (Faheem et al., 2018).Network Reliability can be enhanced through optimization techniques and Importance Measures (IM) analysis. There is always the need of increasing the reliability of the systems and its components, this can be achieved through Reliability Optimization.

Optimum performance is the primary goal for any network design engineer. To achieve this goal, reliability optimization is the crucial tool to accomplish the demands of the customers. Reliability optimization is basically based on allocation of redundancy in the system. The redundancy allocation problem implicates optimal choice of components via redundancy subjected to various constraints. Further, as the resources are limited, it is a challenging task to allocate resources to the most critical component of the system/network. The most efficient approach of analyzing criticality/ranking of components of a given system is through IM analysis (Fang, Pedroni, & Zio, 2016; Kuo & Zhu, 2012b). Still, the system goes down and fail, and they require maintenance/repair which involves cost. Practically, the most important component should be maintained first, but system has large number of components. Hence, all the components should be prioritized/ranked on the basis of Reliability/Cost/Risk. So that, the system is restored fast to normal.

The key objective is to attain optimal balance between maintenance and reliability in an effective manner. This objective should be achieved at minimum cost. Hence, there is a need for CIMs analysis of the complex networks like SCN in SG. In general, systems consist of large number of components. Failure of each component may affect system reliability differently depending upon their structure in a system and on its reliability etc. i.e all the components have relative importance with regards to their effect on system reliability. This chapter aims to do reliability analysis of SCN using IMs that is Birnbaum Importance Measure (BIM), Reliability Achievement Worth (RAW) and Criticality Importance (CI).

The key points of the chapter presented here are as:

1. Defining the various reliability importance measures for analysis of any system.
2. Evaluating the defined importance measures for reliability analysis of SCN.

Background

Since the mid of the 20th century, various CIMs are being presented by the researchers. Literature exists on CIM analysis in which system's components are prioritized on the basis of their criticality to obtain a highly reliable, risk free, maintainable and performable system. A number of IMs have been described in studies, like BIM, CI, IP, RAW, and Risk Reduction Worth (RRW)(Almoghathawi & Barker, 2019; Amrutkar & Kamalja, 2017; Espiritu, Coit, & Prakash, 2007; Kuo & Zhu, 2012a; Wu, Chen, Wu, &

Wang, 2016). IMs have also been used to isolate system components that may be one of the reason for degradation in performance of the system/network (Almoghathawi & Barker, 2019). The core feature of this study is to recognize the critical network components which are responsible for the performance of the infrastructure networks during restoration and recovery phase. The authors have worked with resilience based IM for the problem (Almoghathawi & Barker, 2019). Espiritu et al. (Espiritu et al., 2007) have introduced new criticality based importance measures best suited for Electricity Transmission Systems (ETS) in power system. The new measures relate to the disruption rate of the system and its element rather than taking failure probability in the specified duration. A broad description on demonstrating the IMs to describe the problems of redundancy allocation, upgradation of system, and fault diagnosis and maintenance is presented in (Ramirez-Marquez, Rocco, Gebre, Coit, & Tortorella, 2006).

The reliability of a network is dependent on its components and on the structure of the network. These extremely intricate networks comprises of several interrelated working elements(Dharmaraja, Vinayak, & Trivedi, 2016). The failures of any network are directly related to the increase in maintenance/operation budget of any system. Therefore, lot of importance is given to system's reliability and its improvement [7]. In 2012, Kuo and Zhu(Kuo & Zhu, 2012b) have presented a detailed review on RIMs . Further, they have also discussed comparison between various IMs in complex systems. BIM be the most widespread IM applied on various applications. The authors have provided expressions of IMs like IP, CI, RAW, Operational Criticality Index derived in terms of BIM or Fussell Vesley IM for streamlined studies (Awadallah, Milanovic, Wang, & Jarman, 2015; Dui, Si, & Yam, 2017; Si, Liu, Jiang, Jin, & Cai, 2019; Zheng, Okamura, & Dohi, 2018).

Importance Measures have been studied for many complex systems such as Substation Automation Systems(SAS), all digital protection systems, wind turbine systems, computer network(Chauhan, Pahuja, Singh, & Rani, 2016; Hajian-Hoseinabadi, 2013; Salazar, Nejjari, Sarrate, Weber, & Theilliol, 2016; Zhang, Portillo, & Kezunovic, 2006). All these IMs express about the criticality of each component in the aforementioned systems and provide solution for the ranking of these components. This helps in minimizing maintenance, risk, safety and improvement in reliability optimally. To the best of our knowledge CIM analysis has not been implemented on networks like SG and SCN.

The survey benefits to implement the knowledge of various IMs on complex systems/networks which are still unexplored like SG and SCN. This provide valuable insights into the concerns of the modern day complex network to achieve the highest possible reliability. The following sections presents the definition of the various RIMs, and specifically then computed for IEC 61850 based SCN. In the last, results are discussed and conclusion is drawn.

MAIN FOCUS OF THE CHAPTER

The reliability of the whole system is related to the possible states (failure or operative) of its components. The organization of the components in order of their rank is established with the understanding of IMs. One of the best computational methodology for engineers to obtain reliability enhancement of the system is by RIM exploration. The detailed explanation with mathematical formulations of various RIMs is as follows:

1. Reliability Importance Measures (RIM)

Basically, all RIMs are evaluated when the mission time of the system is known and fixed and further the component reliability are calculated at a time within that mission time. These indices are significant in prioritizing maintenance efforts and suggest ways for effective reliability improvement of the structure of the system. Overall performance of the system can be enhanced in terms of reliability, risk, availability with the study of RIMs. Some of the Importance Measures (IM) s and their expressions are given below:

a. Birnbaum Importance Measure

The concept of importance of components in a system was first introduced by Birnbaum in 1969(Birnbaum, 1969). Birnbaum Importance Measure of component *k* at time *t* and is denoted by *BIM(k;t)*. It can also be defined as the rate at which system reliability $\mathcal{R}_s(t)$ is improved with the reliability of the component $\mathcal{R}_k(t)$. It signifies the prominence of the component *k* to the system reliability $\mathcal{R}_s(t)$ for a given interval of time, from 0 to *t*. Analytically, for a network system with *m* components, it is defined as in equation (1):

$$\left.\begin{aligned} BIM(k;\mathrm{t}) &= \frac{\partial \mathcal{R}_s(t)}{\partial \mathcal{R}_k(t)} \\ Also, BIM(k;\mathrm{t}) &= \mathcal{R}_s(t;\mathcal{R}_k(t)=1) - \mathcal{R}_s(t;\mathcal{R}_k(t)=0) \end{aligned}\right\} \quad (1)$$

for all $k=1,2,\ldots\ldots,m$

where $0<BIM(k;t)<1$

$\mathcal{R}_s(t)$ is the system reliability at time *t*
$\mathcal{R}_k(t)$ is the reliability of the component at time *t*

Also, $\mathcal{R}_s(t;\mathcal{R}_k(t)=1)$ is the reliability of the system at time *t* when the given component *k* is functioning perfectly, and $\mathcal{R}_s(t;\mathcal{R}_k(t)=0)$ is the reliability of the system at time *t* when the given component *k* is in the failed state.

Here, from equation (1), it is clear that *BIM(k;t)* do not depend upon the actual reliability of the components $\mathcal{R}_k(t)$. This comes as a drawback for *BIM*.

b. Criticality Importance Measure

Here, CIM of component *k* at time *t* and is denoted by *CIM(k;t)*. This measure considers the unreliability of each component, though BIM does not considers the same. Thus, less reliable component is given more importance and is found to be more critical. The mathematical expression of *CIM(k;t)* is depicted in equation (2):

$$\left.\begin{aligned} CIM(k;t) &= \left[\mathcal{R}_s\left(t;\mathcal{R}_k(t)=1\right)-\mathcal{R}_s\left(t;\mathcal{R}_k(t)=0\right)\right]\left\{\frac{1-\mathcal{R}_k(t)}{1-\mathcal{R}_s(t)}\right\} \\ CIM(k;t) &= \left[BIM(k;t)\right]\left\{\frac{1-\mathcal{R}_k(t)}{1-\mathcal{R}_s(t)}\right\} \\ CIM(k;t) &= \left[BIM(k;t)\right]\left\{\frac{\mathcal{F}_k(t)}{\mathcal{F}_s(t)}\right\} \end{aligned}\right\} \quad (2)$$

for all *k*=1,2,………*m*

where $\mathcal{F}_s(t)$ is the system unreliability at time *t*

$\mathcal{F}_k(t)$ is the unreliability of the component *k* at time *t*
m is the total number of components in the system

c. Reliability Achievement Worth

Reliability Achievement Worth (RAW) of component *k* at time *t* and is denoted by *RAW(k;t)* importance measures the effect of enhancement of reliability of a system. It quantifies the maximum possible percentage increase in system reliability made by a certain component. It is expressed as the ratio of the actual system reliability attained when component *k* is always in perfect working state $(\mathcal{R}_k(t)=1)$ to the actual value of the system reliability. Equation (3) gives the expression of RAW importance measure as:

$$\left.\begin{aligned} RAW(k;t) &= \frac{\mathcal{R}_s\left(t;\mathcal{R}_k(t)=1\right)}{\mathcal{R}_s(t)} \\ RAW(k;t) &= \frac{1+\mathcal{F}_k(t)\left(BIM(k;t)\right)}{\mathcal{R}_s(t)} \end{aligned}\right\} \quad (3)$$

for all *k*=1,2,……,*m*

Where $\mathcal{F}_k(t)$ is the unreliability of the component *k* at time *t*

m is the total number of components in the system

$\mathcal{R}_s(t)$ is the system unreliability at time t

The main focus of the chapter is to apply RIMs methodologies on architectures of communication network of the digital substations of SG. The next section give the detailed overview of the SCN.

2. Introduction to Substation Communication Network

The protection and control system of modern day grid with automated substation can work effectively only when a reliable, fast communication system is available. The communication system is required to connect different protection, control, and monitoring equipment's within the power substation. These various equipment's work on different protocols by different manufacturers, thus create the problem of interoperability. This issue is resolved by publishing a uniform standard for communication, protocol and configuration language in power substation that is ' Communication Networks and Systems in Substation' or called as IEC 61850 given by IEC working group TC57in 2003(Kanabar & Sidhu, 2009). Also, the standard lay down Ethernet communication network systems are based on OSI seven layer model. Ethernet fulfils all the requisites such as reliable, fast, secure, interoperable, LAN based communication system for smart substations(Thomas & Ali, 2010). The operations of substation like protection and control depend upon the communication at process level as shown in figure 1. The communication issue at the process level is concerned with topology of the Ethernet network (LANs). The different ways of work stations connected with each other refers to the different topologies mentioned in the following section. Reliability analysis of the Ethernet based communication had become a challenging task for the power system engineers. Various researchers have provided many solutions to reliability evaluation problem of substation and its communication systems(Kounev, Lévesque, Tipper, & Gomes, 2016; Thomas & Ali, 2010).Sidhu et al. (Kanabar & Sidhu, 2009) have provided reliability and availability analysis using RBDs for a practical substation automation system architectures layout with Ethernet. Apart from analysis of reliability, very less work has been reported on the importance measures (IM) of components of the smart substations. As discussed in the earlier part of this chapter, IMs are able to categorize weak components and which need to be first improved so as to enhance the overall system reliability. Also, IMs during operational phase of the system, guides to maintain the most critical component before the weaker one.

The performance of the network is dependent on these topologies of SCN(Ethernet LANs).Therefore RIM analysis of SCN is far more important to determine the performance of SAS and SG as a whole. In the following of the chapter, various SCN architectures and their Reliability Block Diagram (RBD) are discussed.

a. Substation Ethernet Communication Architectures

The hierarchical model of IEC 61850 based SAS constitutes of three levels as illustrated in figure 1.

Process level: It includes Current Transformers (CTs)/Potential Transformers (PTs), **Merging** Units (Mus), **etc**. The analog signals attained by CTs/PTs **are** converted to digital signals by MUs and sent over the Ethernet network to the bay level.

Figure 1. Block diagram representation of SAS based on IEC 61850

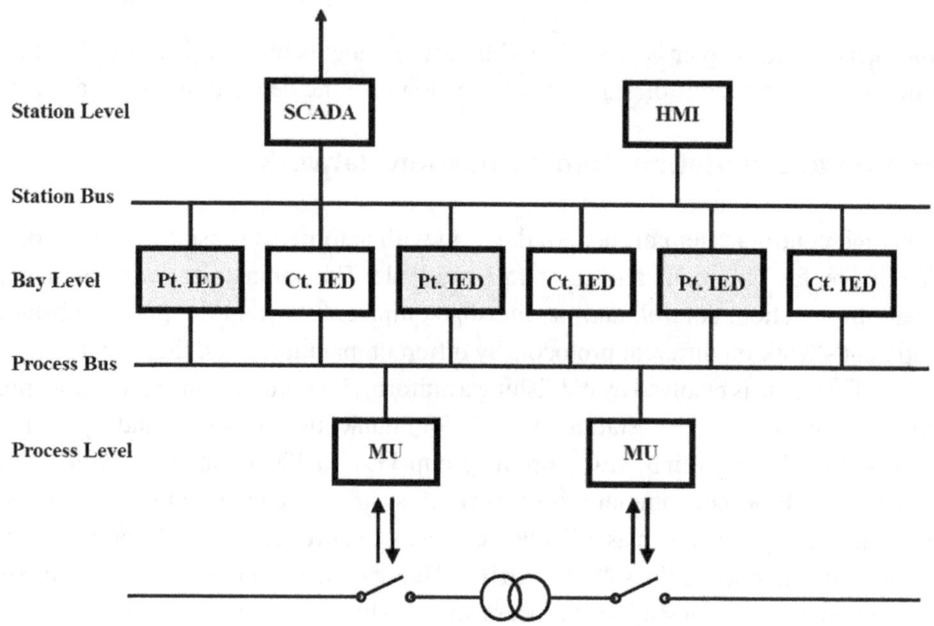

Bay level: This includes microprocessor based relays called Protection IEDs **(Pt.IED)** and bay controllers called control IEDs **(Ct.IED)**. Information are sent from the process level and received by Pt.IEDs and computations and decision signals are sent over the Ethernet network.

Station level: This comprises of the Human Machine Interface (HMI) and **SCADA system**. Here, the monitoring operations and are conducted from the status messages of different components of the substation.

Process bus: The two way communication among the process level and bay level is accomplished via process bus.

Station bus: This bus permits exchange of meaningful data concerning the bay level and station level, and allows data of all components of the substation accessible at the control hub.

b. Typical Substation Considered for the Reliability Analysis

The layout of the IEC 61850, 220/132 KV, single bus, T1-1 small substation is depicted in figure 2 (Kanabar & Sidhu, 2009). Table 1 shows the IEDs configuration for the foresaid substation. In total there are 5 bays considered, (line bays: 03, transformer bay: 01, feeder bus bay: 01.

The reliability analysis of various components of the selected substation has been performed using RBD technique. The Mean Time to Failure (MTTF) numerical values are presented in the Table 2(Kanabar & Sidhu, 2009). Communication links have very high MTTF, thus they are not considered for the calculations of network reliability.

The architecture of the line bay is of Star topology as presented in figure 3, here IEDs are interconnected to ESW by a single link. Reliability of the line bay is calculated using RBD as depicted in figure 4.

Figure 2. Layout of T1-1, single bus, small transmission substation

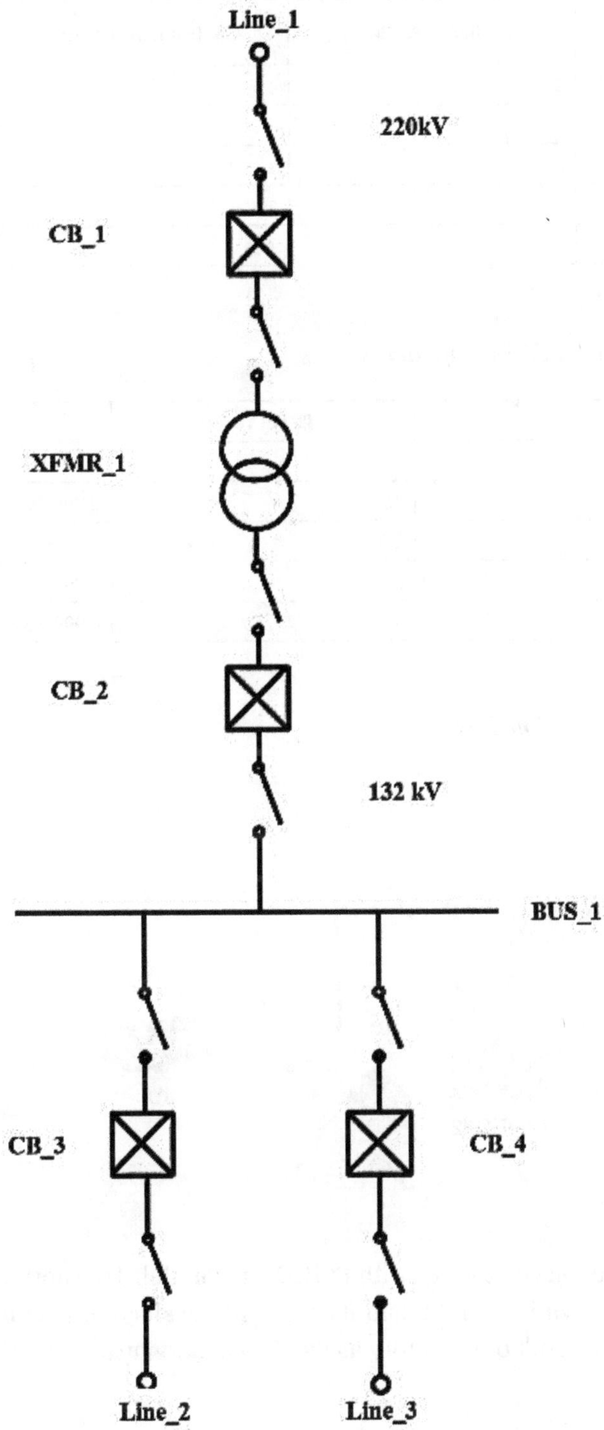

Table 1. IED Configuration of various bays of the substation

Name of the Bay	Protection IED (Pt.IED)	Control IED (Ct.IED)	Merging Unit (MU)
Line_1	2	1	1
Line_2	2	1	1
Line_3	2	1	1
TFM_1	2	1	2
Bus_1	2	1	2

Table 2. MTTF Values for all SAS components

Component	MTTF(Years)	Reliability
Pt.IED	150	0.999993912
Ct.IED	150	0.999993912
Merging Unit(MU)	150	0.999993912
Ethernet Switch (ESW)	50	0.999981735
Time Synchronization (TS)	150	0.999993912

Figure 3. Star topology of the line bays

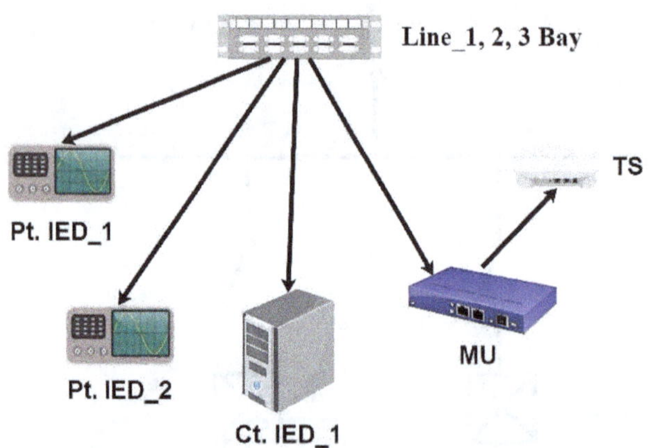

All the components are joined in series with Pt.IEDs in parallel. Transformer (TFM_1) and Bus Bay (Bus_1) architecture is shown in figure 5 and its reliability evaluation is done with RBD depicted in figure 6. Table 3 shows the result of reliability analysis evaluation using series - parallel configurations calculations using RBDs.

Figure 4. RBD for the line bays

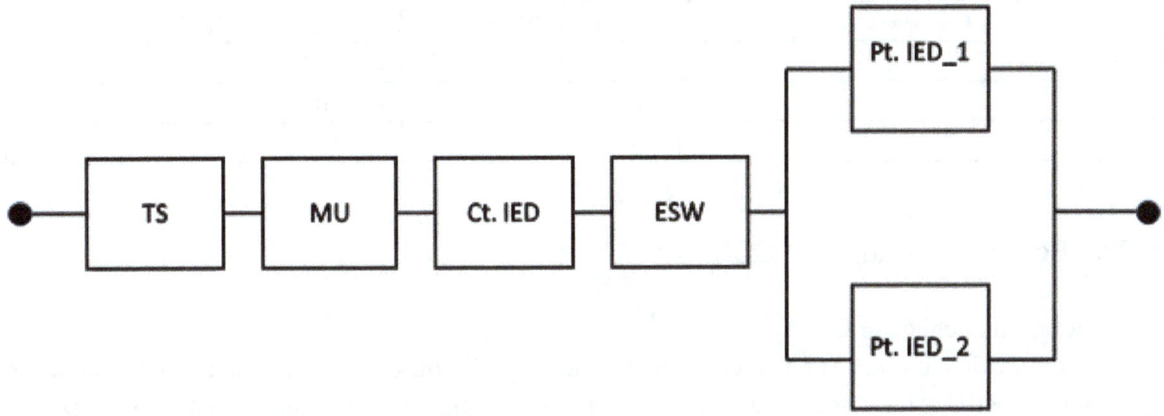

Figure 5. Transformer bay architecture (MU:2, TS:2)(TFM_1)

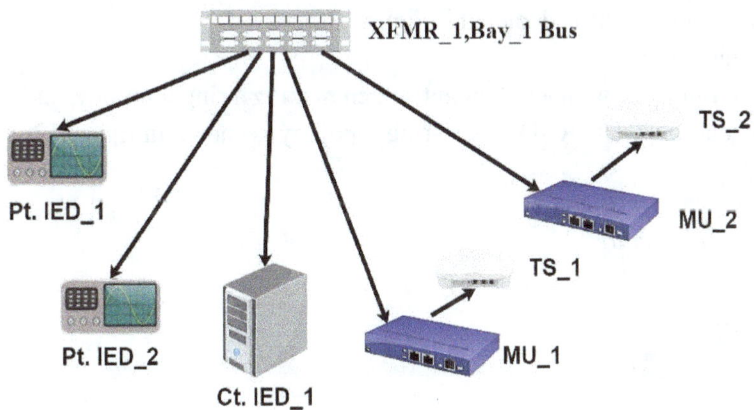

Figure 6. RBD for the Transformer bay architecture (MU:2, TS:2)(TFM_1)

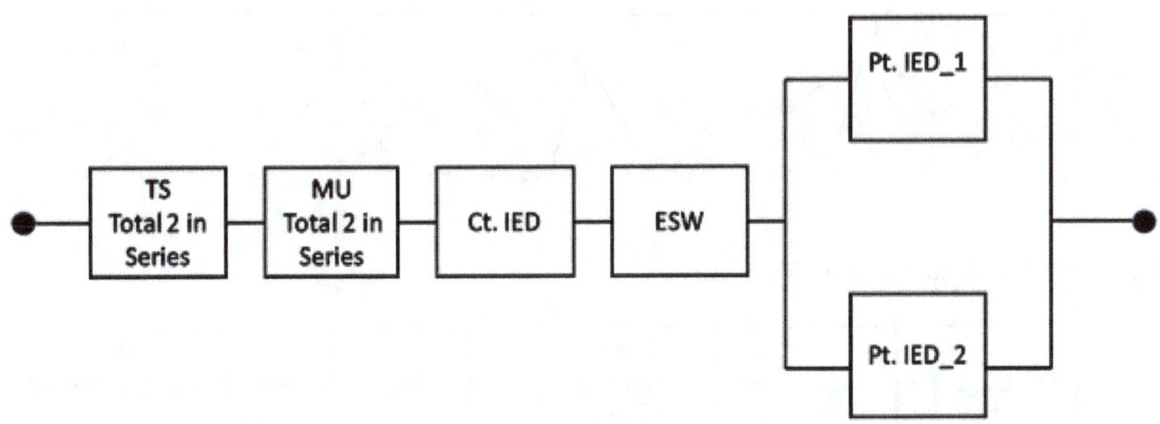

Table 3. MTTF and Reliability of Intra- Bay Communication

Component	MTTF(Years)	Reliability
Line_1,2,3	22.5	0.9999634
TFM_1	17.307	0.9999513
BUS_1	17.307	0.9999513

c. The Four Architectures of SCN

1. Cascaded Architecture:
   In this architecture, each ESW is connected to the next ESW making an open chain setup as shown in figure 7. RBD of the cascaded architecture is presented in figure 8, and depicts that for the successful working of the network, all ESWs should be working, that is, all switches are connected in series.
2. Ring Architecture:
   Here, the six ESWs are connected to one another to form a closed loop as demonstrated in figure 9 and its RBD arrangement is depicted in figure 10.
3. Star-Ring Architecture:
   It is depicted in figure 11 with a redundant switch making a ring topology, and rest six ESWs connected in star topology. The RBD of star-ring topology is shown in figure 12.

Figure 7. Cascaded architecture

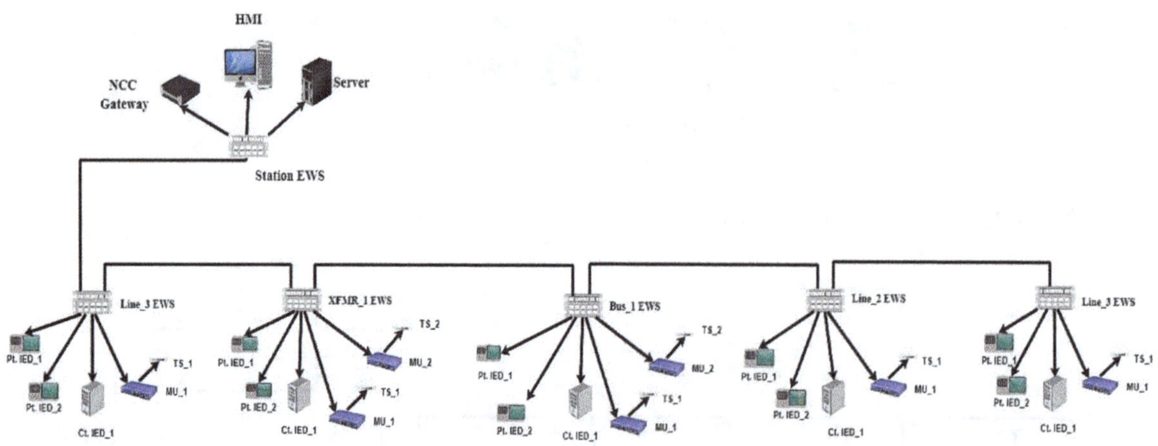

Figure 8. RBD for cascaded architecture

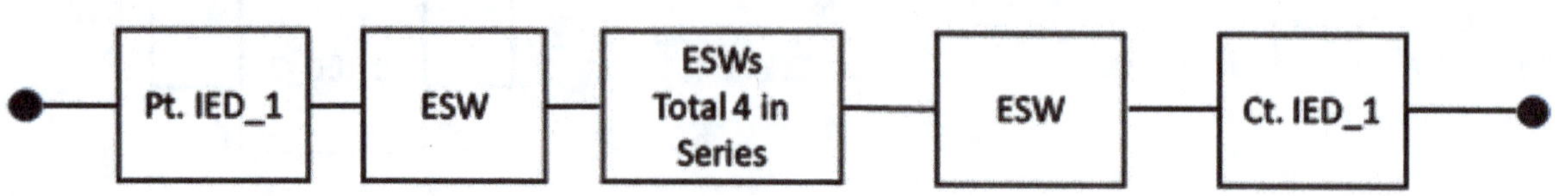

Figure 9. Ring architecture

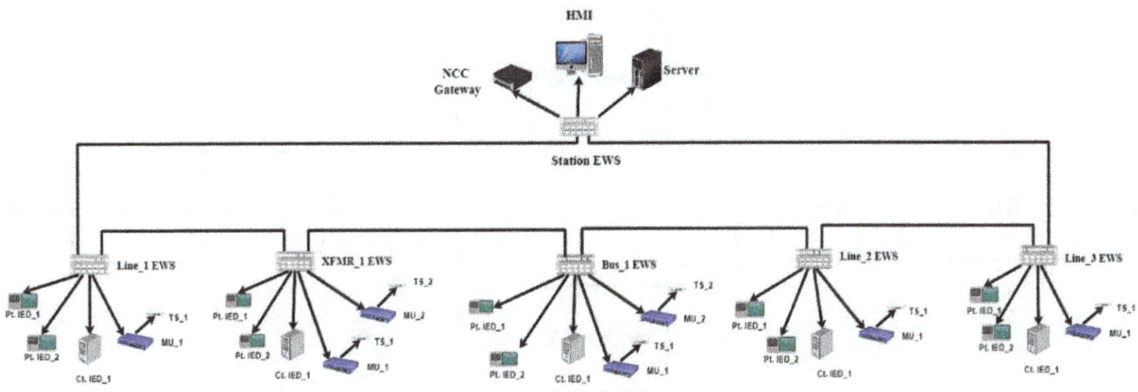

Figure 10. RBD of ring architecture

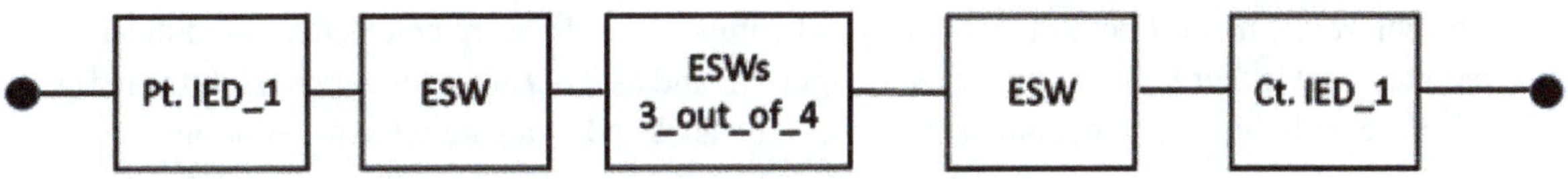

Figure 11. Star-ring architecture

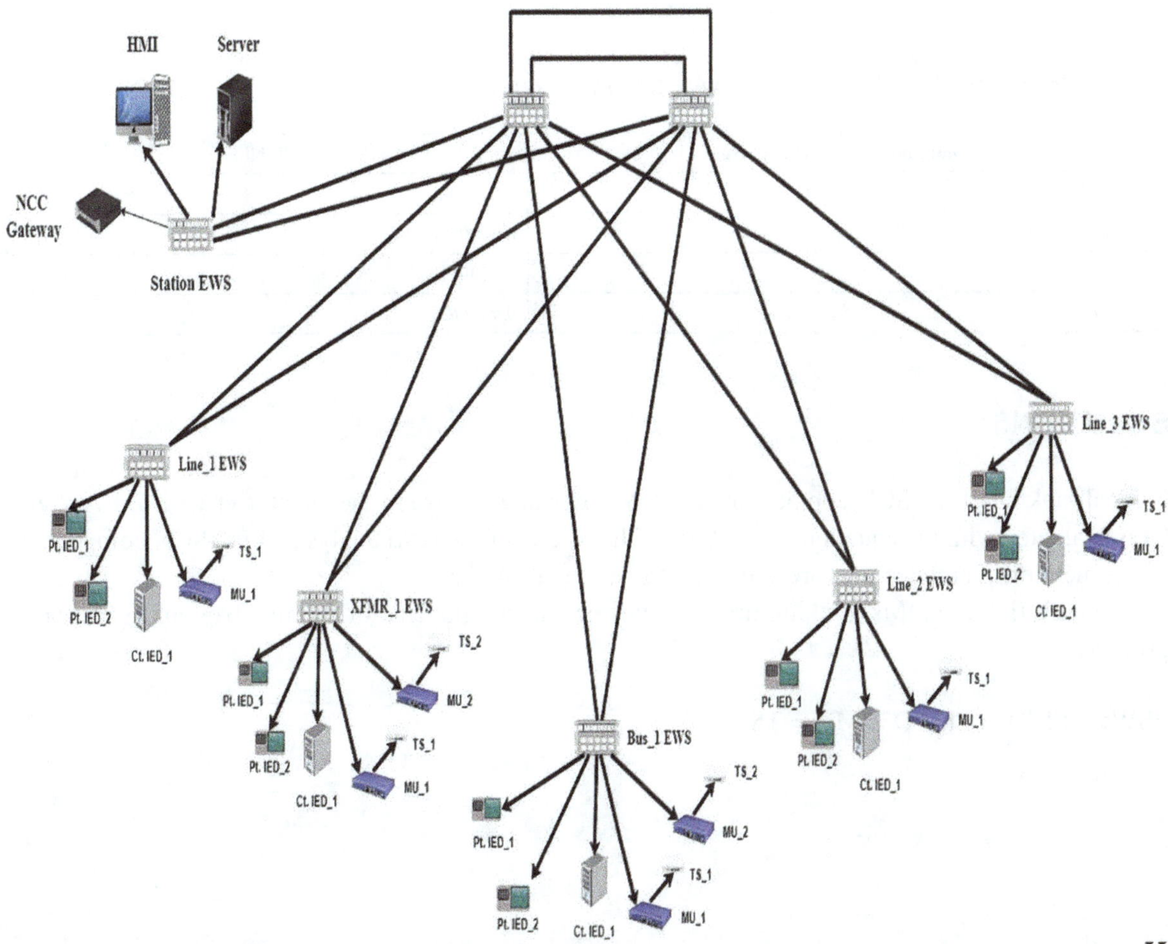

Figure 12. RBD for star-ring architecture

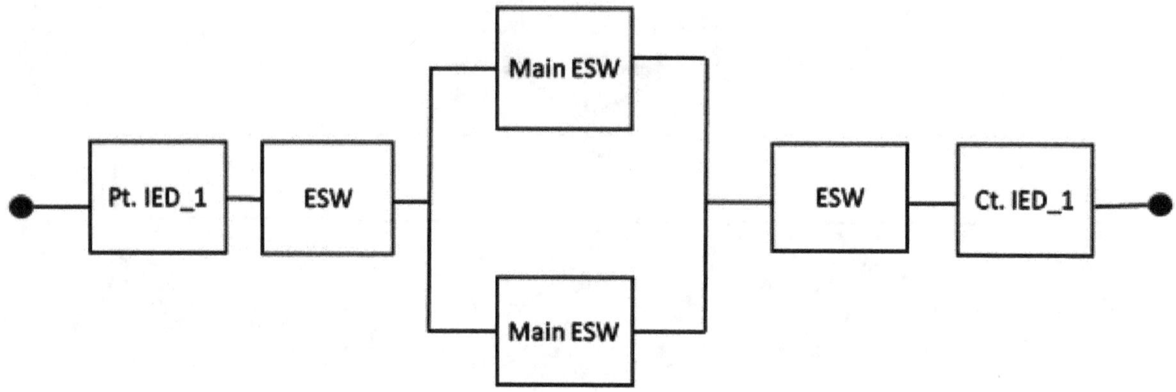

4. Redundant-ring architecture:
   To achieve the highest possible reliability, all components IEDs are connected in redundant ring architecture via four ESWs as presented in figure 13 and its RBD arrangement is exhibited in figure 14. Here both the rings are redundant, hence to be analysed as a parallel configuration.

Reliability evaluation for the four SCN architectures is tabulated in Table 4. The reliability evaluation of the SCN components has been obtained by RBD technique, where the communication is established between Pt.IED of one bay with Ct.IED of the other bay.

Table 4. MTTF and Reliability of inter-bay SCN architectures

Communication Architecture	Reliability
Cascaded	0.9999330
Redundant- Ring	0.9999878
Ring	0.9999513
Star- Ring	***0.9999621***

SOLUTIONS

Critical ranking of the SCN components by RIMs explained above in the chapter viz. BIM, RAW and CI is evaluated with the equations from 1 to 3. The results of the RIM analysis of the SCN components of the mentioned architectures are shown in Table 5 to Table 10.

From all the three IMs, ranking order of the various components of the line bay configuration is given as:

ESW> Pt.IED = Ct.IED = MU = TS

Figure 13. Redundant-ring architecture

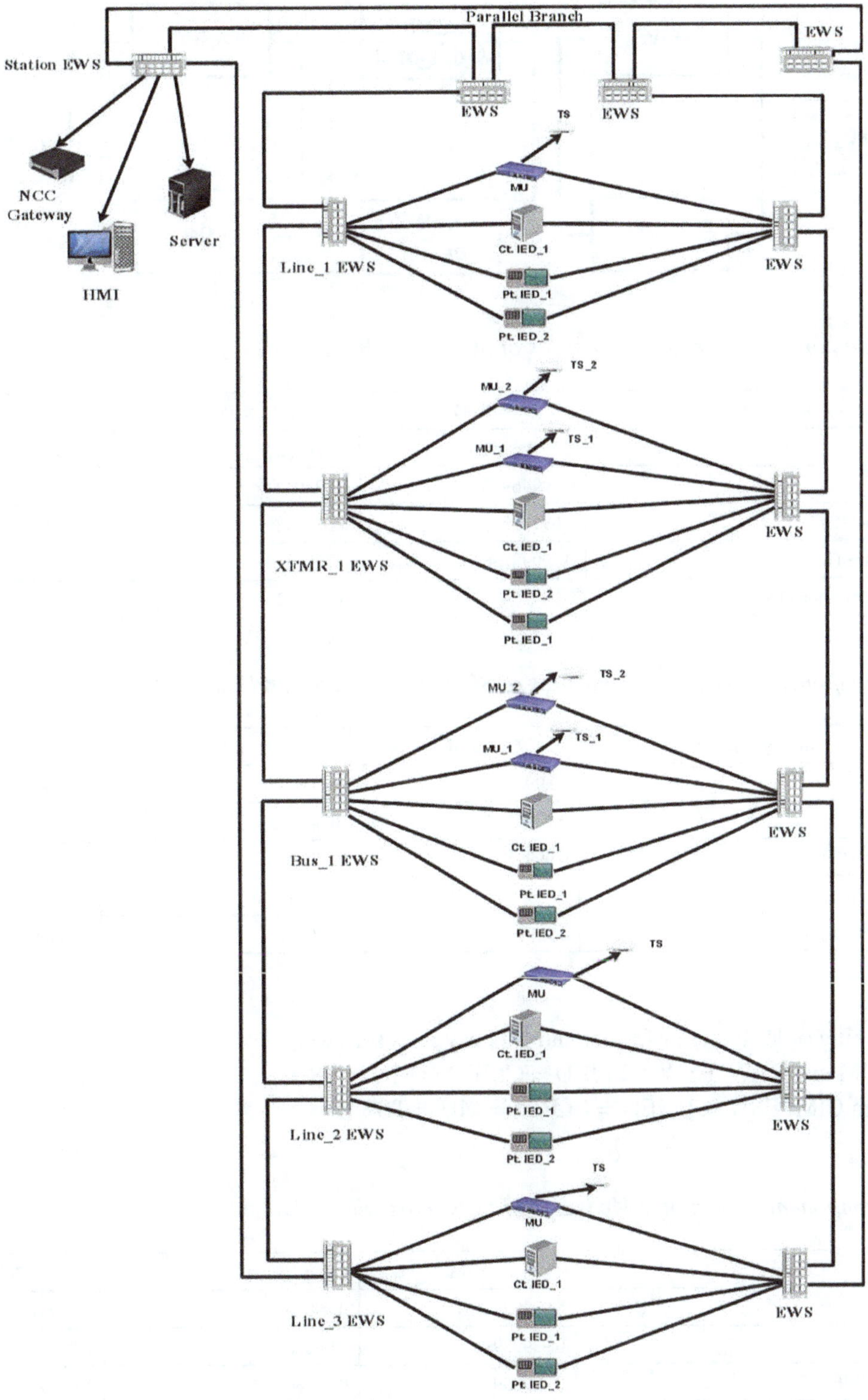

Figure 14. Redundant-ring architecture RBD

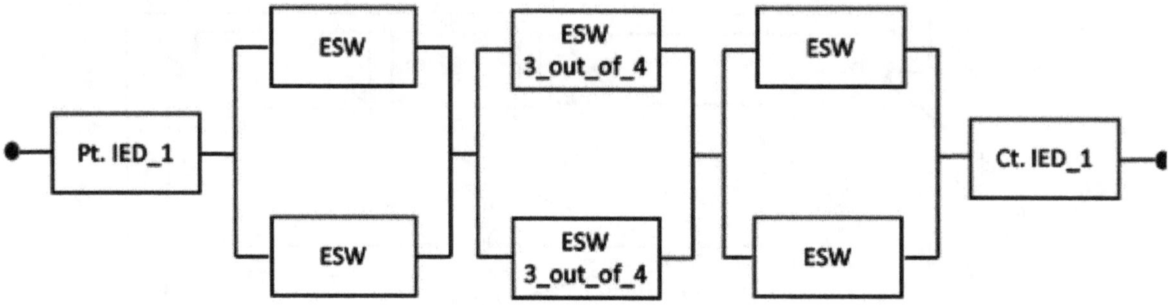

Table 5. Component Importance Measures of SCN (Line Bay)

Component	BIM	CIM	RAW
Pt.IED	0.999969488	0.024042	1.000042
Ct.IED	0.999969488	0.024042	1.000042
Merging Unit(MU)	0.999969488	0.024042	1.000042
Ethernet Switch (ESW)	0.999981665	0.49903	1.000055
Time Synchronization (TS)	0.999969488	0.024042	1.000042

Table 6. Component Importance Measures of SCN (Transformer and Bus Bay)

Component	BIM	CIM	RAW
Pt.IED	0.9999574	0.60877	1.000054
Ct.IED	0.9999574	0.60877	1.000054
Merging Unit(MU)	0.9999574	0.60877	1.000054
Ethernet Switch (ESW)	0.9999695	0.25975	1.000067
Time Synchronization (TS)	0.9999574	0.60877	1.000054

The ranking order for Transformer and Bus bay is as follows:
With BIM and RAW: ESW > Pt.IED = Ct.IED = MU = TS
But from CIM: ESW < Pt.IED = Ct.IED = MU = TS

Table 7. Component Importance Measures of SCN (cascade Architecture)

Component	BIM	CIM	RAW
Pt.IED	0.9999391	0.09086	1.000073
Ct.IED	0.9999391	0.09086	1.000073
Merging Unit(MU)	0.9999391	0.09086	1.000073
Ethernet Switch (ESW)	0.9999512	0.27259	1.000085
Time Synchronization (TS)	0.9999391	0.09086	1.000073

From all the three IMs, ranking order of the various components in the cascade architecture of the substation is as: ESW> Pt.IED = Ct.IED = MU = TS

Table 8. Component Importance Measures of SCN (Ring Architecture)

Component	BIM	CIM	RAW
Pt.IED	0.9999574	0.60877	1.000054
Ct.IED	0.9999574	0.60877	1.000054
Merging Unit(MU)	0.9999574	0.60877	1.000054
Ethernet Switch (ESW)	0.9999695	0.25975	1.000067
Time Synchronization (TS)	0.9999574	0.60877	1.000054

The ranking order for the components in the ring architecture of SCN is as follows:
With BIM and RAW: ESW > Pt.IED = Ct.IED = MU = TS
But from CIM: ESW < Pt.IED = Ct.IED = MU = TS

Table 9. Component Importance Measures of SCN (Star- Ring Architecture)

Component	BIM	CIM	RAW
Pt.IED	0.9999682	0.67642	1.000044
Ct.IED	0.9999682	0.67642	1.000044
Merging Unit(MU)	0.9999682	0.67642	1.000044
Ethernet Switch (ESW)	0.9999803	0.48191	1.000056
Time Synchronization (TS)	0.9999682	0.67642	1.000044

The ranking order for the components in the ring architecture of SCN is as follows:
With BIM and RAW: ESW > Pt.IED = Ct.IED = MU = TS
But from CIM: ESW < Pt.IED = Ct.IED = MU = TS

Table 10. Component Importance Measures of SCN (Redundant–Ring Architecture)

Component	BIM	CIM	RAW
Pt.IED	0.99999388	0. 49902	1.000018
Ct.IED	0.99999388	0. 49902	1.000018
Merging Unit(MU)	0.99999388	0. 49902	1.000018
Ethernet Switch (ESW)	1.00000615	0.91325	1.000030
Time Synchronization (TS)	0.99999388	0. 49902	1.000018

The ranking order for the components in the redundant ring architecture of SCN is as follows: With BIM and RAW: ESW > Pt.IED = Ct.IED = MU = TS

But from CIM: ESW < Pt.IED = Ct.IED = MU = TS

The above results demonstrates that the rank/order of all the components are at an identical position because the failure rates and reliability of all the components are taken as equal. Practically, all these components have very high reliability, amongst these Ethernet Switch (ESW) is the one which has lower failure rate. Also, it's an important inference that in Ring and its further architectures like Star-Ring, and Redundant Ring architecture, in comparison to Birnbaum Measure, the Criticality Importance Measure presents a more convincing metrics result, that ESW is less critical than other components. This is possible due to removal of single point of failure in ring topology, and by adding redundancies in the later hybrid architectures. RAW gives the same ranking of various components as by BIM. It also shows, that ESW is more critical component than others. It is also practically known, failure of ESW along with the communication media at any point in the network, will lead to catastrophic failures or may be a complete shutdown of the system, this is so because, all the information exchange within the protection and control components is through Ethernet itself. Further, ESW transfer the information to the servers, thus are the only channel for HMI also to operate. Even a delay of seconds in message transfer can lead to cascading failures in such mission critical systems as SG network, whose SAS is a vital part. Therefore, redundancy of ESWs are justifiable even though at an increased cost.

CONCLUSION

Reliability Importance Measures (RIMs) are valued for providing ways and prioritizing of methods for the upgradation in the working of the system, also are suggestive for the utmost proficient approach to work and sustain system status. RIMs like Birnbaum Importance Measure, Criticality Importance Measure, and Reliability Achievement Worth has been successfully been carried on SCN. Reliability analysis is a critical study in any system's design and then its maintenance. Improvement in system's reliability can very well reduce the maintenance budget of any system. Specifically, the foremost assignment is to rank the components on the basis of their importance. The methodologies presented here has been successful to find the solution to the problem of enumerating the prominence of components in reliability studies of SCN. Here, RIMs for architecture of SCN are analysed and provide an understanding to its most critical components. The analysis is helpful in enhancing the system's performance optimally. The RIMs analysis with RAW, BIM and CIM have concluded with rational ordering of components of the SCN structure. The practice of these IMs work towards the identification of the components that can be allocated of resources. This further reduces their vulnerability to attacks or to accelerate their recovery from faults. The analysis helps in minimizing maintenance and risk, and improvement in reliability optimally. The critical ranking of the components enumerated by using the measures deliver insightful meaning for where the investments ought to be done to achieve maximum possible reliability of the system.

REFERENCES

Almoghathawi, Y., & Barker, K. (2019). Component importance measures for interdependent infrastructure network resilience. *Computers and Industrial Engineering, 133*(September), 153–164.

Amrutkar, K. P., & Kamalja, K. K. (2017). An overview of various importance measures of reliability system. *International Journal of Mathematical. Engineering and Management Sciences, 2*(3), 150–171.

Awadallah, S. K. E., Milanovic, J. V., Wang, Z., & Jarman, P. N. (2015). Assessment of suitability of different reliability importance measures for prioritising replacement of transmission system components. *2015 IEEE Eindhoven PowerTech. PowerTech, 2015*, 1–6.

Birnbaum, Z. W. (1969). On the importance of different components in a multicomponent system. *Multivariate Analysis, 2*, 581–592.

Chauhan, U., Pahuja, G. L., Singh, V., & Rani, A. (2016). Reliability analysis of wind turbine system using importance measures. *12th IEEE International Conference Electronics, Energy, Environment, Communication, Computer, Control: (E3-C3), INDICON 2015.*

Dharmaraja, S., Vinayak, R., & Trivedi, K. S. (2016). Reliability and survivability of vehicular ad hoc networks: An analytical approach. *Reliability Engineering & System Safety, 153*, 28–38. doi:10.1016/j.ress.2016.04.004

Dui, H., Si, S., & Yam, R. C. M. (2017). A cost-based integrated importance measure of system components for preventive maintenance. *Reliability Engineering & System Safety, 168*(May), 98–104. doi:10.1016/j.ress.2017.05.025

Espiritu, J. F., Coit, D. W., & Prakash, U. (2007). Component criticality importance measures for the power industry. *Electric Power Systems Research, 77*(5–6), 407–420. doi:10.1016/j.epsr.2006.04.003

Faheem, M., Shah, S. B. H., Butt, R. A., Raza, B., Anwar, M., Ashraf, M. W., Ngadi, M. A., & Gungor, V. C. (2018). Smart grid communication and information technologies in the perspective of Industry 4.0: Opportunities and challenges. *Computer Science Review, 30*, 1–30. doi:10.1016/j.cosrev.2018.08.001

Fang, Y.-P., Pedroni, N., & Zio, E. (2016). Resilience-Based Component Importance Measures for Critical Infrastructure Network Systems. *IEEE Transactions on Reliability, 65*(2), 502–512. doi:10.1109/TR.2016.2521761

Hajian-Hoseinabadi, H. (2013). Reliability and component importance analysis of substation automation systems. *International Journal of Electrical Power & Energy Systems, 49*, 455–463. doi:10.1016/j.ijepes.2010.06.012

Kanabar, M. G., & Sidhu, T. S. (2009). Reliability and availability analysis of IEC 61850 based substation communication architectures. *2009 IEEE Power & Energy Society General Meeting*, 1–8.

Kounev, V., Lévesque, M., Tipper, D., & Gomes, T. (2016). Reliable Communication Networks for Smart Grid Transmission Systems. *Journal of Network and Systems Management, 24*(3), 629–652. doi:10.100710922-016-9375-y

Kuo, W., & Zhu, X. (2012a). *Importance Measures in Reliability, Risk, and Optimization*. John Wiley & Sons. doi:10.1002/9781118314593

Kuo, W., & Zhu, X. (2012b). Some recent advances on importance measures in reliability. *IEEE Transactions on Reliability*, *61*(2), 344–360. doi:10.1109/TR.2012.2194196

Ramirez-Marquez, J. E., Rocco, C. M., Gebre, B. A., Coit, D. W., & Tortorella, M. (2006). New insights on multi-state component criticality and importance. *Reliability Engineering & System Safety*, *91*(8), 894–904. doi:10.1016/j.ress.2005.08.009

Salazar, J. C., Nejjari, F., Sarrate, R., Weber, P., & Theilliol, D. (2016). Reliability importance measures for a health-aware control of drinking water networks. *Conference on Control and Fault-Tolerant Systems, SysTol*, 572–578. 10.1109/SYSTOL.2016.7739810

Si, S., Liu, M., Jiang, Z., Jin, T., & Cai, Z. (2019). System Reliability Allocation and Optimization Based on Generalized Birnbaum Importance Measure. *IEEE Transactions on Reliability*, *68*(3), 831–843. doi:10.1109/TR.2019.2897026

Thomas, M. S., & Ali, I. (2010). Reliable, Fast, and Deterministic Substation Communication Network Architecture and its Performance Simulation. *IEEE Transactions on Power Delivery*, *25*(4), 2364–2370. doi:10.1109/TPWRD.2010.2042824

Wu, S., Chen, Y., Wu, Q., & Wang, Z. (2016). Linking component importance to optimisation of preventive maintenance policy. *Reliability Engineering & System Safety*, *146*, 26–32. doi:10.1016/j.ress.2015.10.008

Zhang, P., Portillo, L., & Kezunovic, M. (2006). Reliability and Component Importance Analysis of All-Digital Protection Systems. *2006 IEEE PES Power Systems Conference and Exposition*, 1380–1387. 10.1109/PSCE.2006.296504

Zheng, J., Okamura, H., & Dohi, T. (2018). Component Importance Analysis of Mobile Cloud Computing System in the Presence of Common-Cause Failures. *IEEE Access: Practical Innovations, Open Solutions*, *6*, 18630–18642. doi:10.1109/ACCESS.2018.2822338

Chapter 5
Solution to Big Data Security Issues

Prashant Srivastava
Shambhunath Institute of Engineering and Technology, India

Niraj Kumar Tiwari
Shambhunath Institute of Engineering and Technology, India

Ali Abbas
University of Allahabad, India

ABSTRACT

Organizations now have knowledge of big data significance, but new challenges stand up with new inventions. These challenges are not only limited to the three Vs of big data, but also to privacy and security. Attacks on big data system ranges from DDoS to information theft, ransomware to end user level security. So implementing security to big data system is a multiple phase-based ongoing process in which security is imposed from perimeter level to distributed file system security, cloud security to data security, storage to data mining security, and so on. In this chapter, the authors have identified some key vulnerable point related to big data and also proposed a security model.

INTRODUCTION

The authors Yang et al. (2016) discuss in their research Big data can be described through number of ways but the 'V' that define big data according to ISO 2015 are as follows: Volume which describes the size of the data is too big, Variability which define it has the property of varying too fast, Velocity which defines that the speed of data transmission is very fast, Veracity define have too much noise and Variety says data is too dissimilar in nature. These different V's are the basic characteristics of "BigData". Organizations are doing analytics using big data to find business prospects, improve performance, and motivate in decision-making. Apart from this many tools of big data are open source and it is not developed with proper security enforcement and attracts the cybercriminals. For organizations that drive through cloud

DOI: 10.4018/978-1-7998-3327-7.ch005

and big data, security problems are more complex. These security issues are even problematic in case of the web application which runs various weak content management systems which includes the theft of vital information kept online, XSS Attacks, ransomware, and DDoS attacks that leads towards the server crash. The attackers often targets these Websites and performs various kinds of attacks. The problem are still poor when the organizations store sensitive and confidential information, such as customer contact details information, credit card numbers, and so on. Furthermore, attacks on an enterprise's big data could generate serious financial effects. Implementing security to big data platforms could be a blend of tradition security practices and the standard which are newly designed.

Although, without a proper security policies, this technology will not be accepted and will be gain as much of popularity as it can. So, it is important to have proper methodologies, implementation techniques and guidance not only to the Big Data system, but also its security. In this paper authors have identified some major security loop holes and also proposed a model as a solution.

SECURITY ISSUE TO BIG DATA

As discuss by Ardagna et al. (2014) and Moreno et al. (2018), now a day's big data is a first choice of attackers. Organizations are accepting big data technologies, by using its great analytics tools helps in decision-making, identifying prospects, and enhance the performance. But with the enormous consumption and increased usage of data, rises security concern of the big data. At last, the adoption of the big data create a question marks for many enterprises: how can you control big data's potential while effectively mitigating big data security risks?

The authors identified some security issue of big data they are as follows.

Distributed Frameworks

For early result of the data, analysis should be done in a very faster way so all big data implementations actually distribute the huge processing jobs among different processing nodes which leads the security concern like all the nodes should have proper security policy. Hadoop is an example of open source software for distributed system for big data analysis so Hadoop security procedures must be installed and updated. The attackers can use the mapper for Map Reduce to show improper values or key pairs, for destruction. Distributed system may decrease the outstanding burden on a framework, yet in the long run more systems mean more security issues.

Non-Relational Databases

Till now the databases available for data management and storage follows relational database management system concepts and uses tablular structures which is nearly not applicable in case of big data. Big data is highly diversified and scalable in nature. Non-Relational databases or NoSQL databases are designed in such a way that they do not follow the tabular structure, and follows storage models which is basically data type based which results Non-relational databases as more scalable and flexible. NoSQL has its own common security protocols so the organization who are implementing the Big Data system must set up their database in their own secure environment with additional security measures.

Data Representation Compression and Redundancy Reduction

Nearly all data sets in big data contains heterogeneity in data arrangement, meaning of data type, and accessibility. The data representation data more semantic for information processing system and the human interpretation. However the lack of proper representation of data can trim down the value of master data and stymie the efficient data analysis.

Normally, Datasets have high altitude of redundancy. Compression of data and redundancy diminution is effectual to reduce the cost of the entire system on the assumption that the expected value of the data will not be increased. Big Data contains huge data volume and variety with variable speed means data generated from various sources with different formats and video data require more space than others so that compression techniques are required. For video data we can use lossy compression as well as lossless compression techniques according to need

BIG DATA MANAGEMENT LIFE CYCLE

According to Chaudhary et al. (2016) in general big data management consists of four phases:

1. Data Collection Phase
2. Storage Phase
3. Knowledge discovery phase
4. Data Decision Phase

Figure 1. Big Data Management Life Cycle

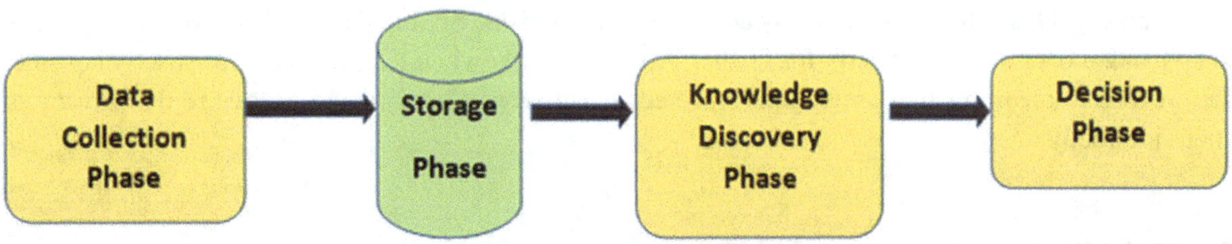

All these phase are designed to keep privacy as major concern. Security in each phase is major issue of the adoption of big data.

STORAGE

In big data, data is organized in multiple level of tiers, depending on business structure. For example top priority data usually be stored in top tier and if organization wants to secure this data then they need to lock down this storage tierwhich may leads to un-availability of data.

ENDPOINTS VULNERABILITY

End points are always vulnerable, in case of any big data the attackers always tries to get the vital information from these vulnerable end points. Security providers must analyze the logs from these points through handshake process to validate the authenticity of the end points.

REAL-TIME SECURITY/COMPLIANCE TOOLS

The security solutions and compliance tools, generate a great amount of knowledge; main problem is the way organization generate this dataand ignore the false positives.

DATA MINING

Data mining and knowledge discovery is the main component of big data; they discovers the patterns that help in decision making and also give suggestion for business strategies. So the main issue is that how to secure it from external environment to the insider threat who are inside the network and using it for destruction purpose.

DATA CONFIDENTIALITY

Now a days the big data service provider are not able to analyze and maintain this huge datasets proficiently as their size is limited. Organizations are bound to trust on the tools and the professionals for data analyses. This actively or passively adds the security risks. So, big data analysis May have chance to be handed over to another party for further processing only when the professionals strictly consider the preventive measures that are bound to protect sensitive raw information, to assure the privacy of client indirectly.

PROPOSED MODEL AS A SOLUTION

In case of Big Data the security cannot be implemented only at one stage. So the proposed security model has identified several domain and these domains can be further classified into various sub domains. All identified domains should have compliance with several standards and government policies like COBIT, SOX, ISO/IEC 27001/2, HIPPA, FISMA etc.

These tools are not new for big data. The only thing which is new, the method of securing various type of data and information in various stages.

Figure 2. Proposed Open Security Framework for Big Data Environment

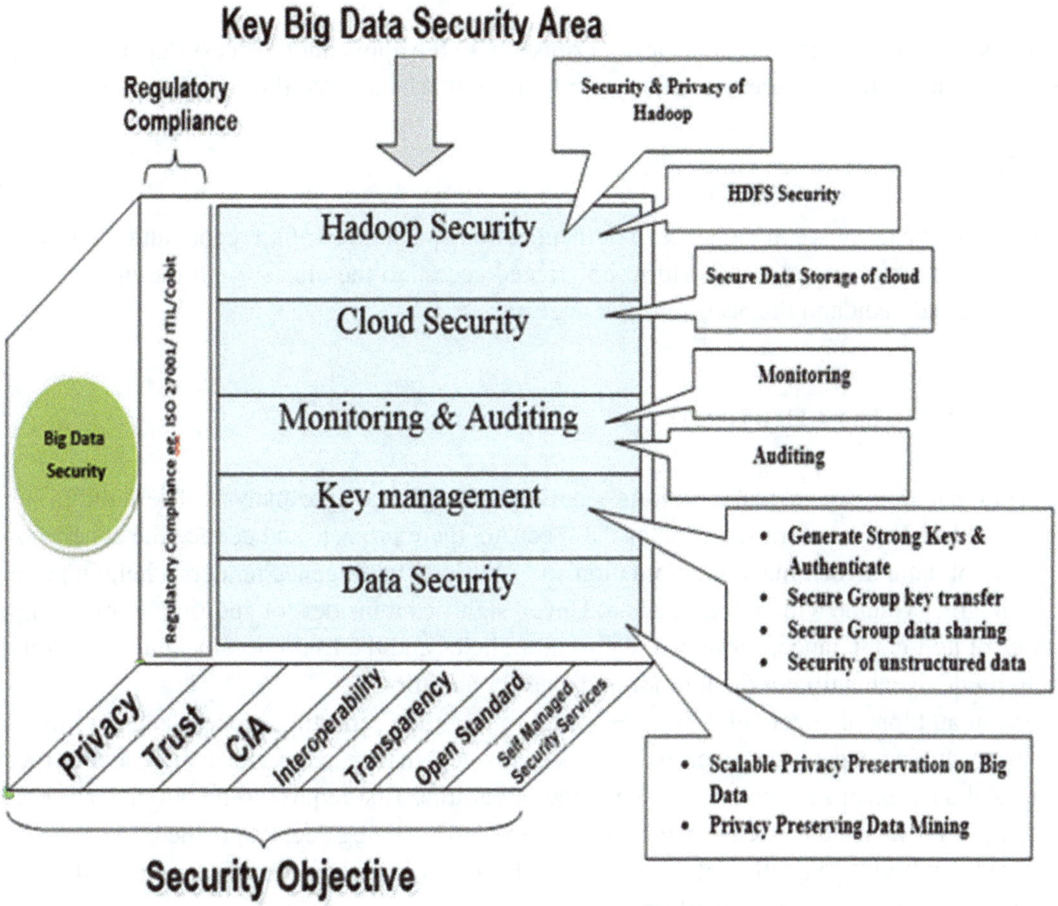

HADOOP SECURITY

Hadoop data storage unit can be secured by a procedure in which a security wall can be implemented virtually against a substantially strong cyber threat. Hadoop procures the security walls of such a high caliber with the help of protocols pertaining to security.

Authentication

Authentication in which verification of credentials of users is administered. The credentials contain dedicated user name and its password for security. The inputted credentials are verified with the stored details in the database if they arefound to be valid authentication will be delivered

To determine accurately the identity of a rendered user striving to access a Hadoop cluster or application depends on one of the factors given as

1. Confirming the participant identification
2. Typically carried out by checking credentials such as user name and password.

Authorization

In this the system sets a decision in favor of permission to the applicant to access data or not.Action of user depends on whether the user has been permitted and it is conducted through the access control list.

Auditing

Auditing being the last stage in sequence of authentication and authorization record and reports whether authenticated and authorized user perform its granted access to the cluster with the inclusion of data accessed / altered / edit and the occurrence of analysis.

MONITORING AND AUDITING

There occur many ways out to make auditors benefitted through real time analysis. The auditors becomes capable to utilize Big data for expansion of the spectrum there projects and deduce the conclusion over larger stock of data. Involvement of automation and Artificial Intelligence renders a help in processing the data in larger volume with higher speed to Unveil significant insides for auditing. For paradigm, the past cases of non-compliance, variations in current policies and the fraud became able to be found and utilize to guide the attention of the interior and exterior auditors.

Financial auditors also get helps from Big data to align the reporting process and find fraud. The professionals attain a capability to identify the risk in business in time and perform more focused accurate auditing. All data under the analysis for the mode of auditing first require to be put under verification of accuracy delay less and capacity. In this process, the forth coming decision made by auditors depend upon the reliable and high quality information. This becomes more important in case of auditing due to compliance of decision risk and investment.

Big data also make you to perform automation on the multiple portion of auditing process. Human errors being a main cause for the business to fall out of compliance or expend to much on audit pertaining needs. By converting the manual and repetitive task in automatic ones, auditors are able to establish several controls in advance and supervise how much better a company complies with the set up regulations.

KEY MANAGEMENT

Complete life cycles of cryptographic keys are administered through encryption key management. Generation, usage, storage, achieve and deletion of keys are the processes included in it. To limit the access to the keys physically, logically and via user/ roll access underlay the protection of encrypted keys.

The available solution of Big data security make use of approaches of fragmented cryptographic secure key management deployable in Big data networking to protect user's data and privacy. Three layers cover whole division of our proposed key management scheme. In the layered configuration, lower keys encrypted by upper keys so as to solidify the security of keys. In the put forth scheme, Big data networking servers and others cannot find anything about user's key.

The ciphered information of data owner can be shared to anyone and hierarchical key management being convenient secure and efficient can protect user personal data in the Big Data networking.

DATA SECURITY

A term used for all the measures and tools to safeguard both the data and analytic processes against the assaults, theft or other malicious activities able to damage or affect them adversely is Big data security. Theft of information stored online, ransomware or DDoS assults can crash a server. The issues may be more degraded in case of sensitive and confidential information of the companies.

The incoming data to be intercepted or corrupted in transit, the data stored likely to be stolen or held hostage during resting on cloud or premise servers and the data being outputted appearing to be insignificant able to provide an access point for hackers are the three biggest challenges which come under central role to make the flexible end to end big data security philosophy for some companies.

CONCLUSION AND FUTURE SCOPE

Big data enjoys many merits subject to the condition of a big question mark on the security issue. Managing privacy, trust and transparency of data is a difficult task. Several Big data service providers serve their best but they confront limitations on their own security policies and the diversified policies drive the clients in overwhelming as they all possess various sets of standards and client has no idea of what and how security is being offered by Big data service providers. So it requires a systematic and standard approach to be followed.

A model for big data security is presented to be used as a standard along with a coverage of every zone of security. The type of open standard for big data in SLA has been included and it is certainly beneficial in terms of developing faith and managing privacy by providing variety of security solutions and making service policies transparent. Thus the service provider and clients both will attain benefits and appropriate security technique in accordance with their needs has to be opted by them.

As our proposed work ensues its start in very early stage and in future expansion may be availed in it as a complete framework of big data security by applying some standardizing organization. The framework needs be constructed by experts through merging of service providers and clients at each stage. Other fields like, Host Operating System level security and Hypervisor level security, etc may also be considered to grow.

Conclusion may be drawn for big data security to be at its immaturity level and this endeavor must act as a new ray of expectation.

REFERENCES

Ardagna, C. A., & Damiani, E. (2014). *Business Intelligence meets Big Data: An Overview on Security and Privacy.* NSF Workshop on Big Data Security and Privacy, Dallas, TX.

Big Data, Preliminary Report 2014, ISO/IEC JTC 1. (2014). *Information technology.* www.iso.org/iso/home/about/iso_members.htm,Preliminary

Big Data Security, Challenges and Solutions. (n.d.). https://www.dataversity.net/big-data-security-challenges-and-solutions/

Big Data Security – Issues, Challenges, Tech & Concerns. (n.d.). *Research data alliance*. https://www.rd-alliance.org/group/big-data-ig-data-security-and-trust-wg/wiki/big-data-security-issues-challenges-tech-concerns

Chaudhary & Srivastava. (2016). Big data security issues and challenges. *International Conference on Computing, Communication and Automation (ICCCA2016).*

Moreno, J., Serrano, M. A., Fernandez-Medina, E., & Fernandez, E. B. (2018). *Towards a Security Reference Architecture for Big Data*. http://ceur-ws.org/Vol-2062/paper04.pdf

Yang, H., Park, M., Cho, M., Song, M., & Kim, S. (2014). A System Architecture for Manufacturing Process Analysis based on Big Data and Process Mining Techniques. *2014 IEEE International Conference on Big Data*.

Chapter 6
Machine Learning Approaches for Cardiovascular Disease Prediction:
A Survey

Stuti Pandey
Kamla Nehru Institute of Technology, Sultanpur, India

Abhay Kumar Agarwal
https://orcid.org/0000-0003-2663-8910
Kamla Nehru Institute of Technology, Sultanpur, India

ABSTRACT

In a human body, the heart is the second primary organ after the brain. It causes either a long-term impairment or death of a person if suffering from a cardiovascular disease. In medical science, a proper medical analysis and examination of a cardiovascular disease is very crucial, convincing, and sophisticated task for saving a human life. Data analytics rises because of the absence of sufficient practical tools for exploring the trends and unknown relationships in e-health records. It predicts and achieves information which can ease the diagnosis. This survey examines cardiovascular disease prediction systems developed by different researchers. It also reviews the trend of machine learning approaches used in the past decade with results. Related studies comprise the performance of various classifiers on distinct datasets.

INTRODUCTION

A severe task in medical science is examination and interpretation of a heart disease as heart is an essential organ of a human body. It functions as a pump for various body parts. Through a network arrangement of blood vessels, blood is moved into them and delivers a regular and consistent supply of vital nutritional components and oxygen as well. For the circulatory system, it acts as a centre and in a

DOI: 10.4018/978-1-7998-3327-7.ch006

very short time limit; a person will die if the functioning of heart stops and terminated to pump blood in the different body parts.

For providing medical assistance in specific decision-making activities to doctors, use of information technology is growing day by day. In order to predict a cardiovascular disease risk, the ongoing approaches are failed to identify such people who can receive a medical treatment in prevention. On the other hand, machine learning is offering convenience by utilizing the complex interconnections between the risk factors. It also benefitted the physicians, doctors, scientists, experts and specialist in discovering the hidden patterns, relationships, disease management, medications and analysis for medical analysis.

In order to assist the medical professionals, medical diagnostic based on computer had been developed for analysing the large volumes of the patient data. The system efficacy mainly depends on the features used which must be correlated with some disease state. Based on ECG signals, several signal processing techniques had been implemented successfully. It extracted a set of features and used subsequently by many machine learning classification tools. The main idea of the survey comprises in practice the machine learning assistance in medical and clinical diagnosis as a sign of healthcare benefits.

Many researchers are interested in using classification method for clinical research now days. Correct classifications to find out whether a disease is present or not is very necessary. Thus, it requires the interventions, treatments, prescriptions, analysis, operations and investigations for a medical practitioner. It can be done in an adequate and useful manner, by allowing them to know the root causes of a disease. Thus, the formulation of exact and accurate estimation of a disease, classification is imperative.

In the next section, an overview of machine learning is presented. Subsequently, a literature survey based on machine learning approaches is discussed. Thereafter tabular comparison of algorithms is presented with its trends and results. Lastly, the conclusion of this survey is presented.

OVERVIEW OF MACHINE LEARNING

Machine Learning (ML) helps to gain analytical perception from a massive amount of data which is very inconvenient to humans and sometimes also impractical. Manual analysis of data or building predictive models is almost useless in some scenarios and also a tedious task and less productive. On the other side, Machine learning, produces more stable, reliable, repeatable results which learn from the previous computation tasks. Labelled and unlabeled data records are the primarily two kinds of machine learning data.

Labelled data records are basically used in supervised machine learning as the attributes are provided with them. These data records are further divided into two parts: Categorical data and Numerical data. Here the categorical data values are used for classification purpose while the numerical data values are used for regression purpose. Unlabeled data records are essentially used in unsupervised machine learning so as to determine the presence of structures and patterns in the data set.

Supervised learning involves a mapping among an input variable set X and an output variable set Y by learning. Then the algorithms develop a hypothetical value H for the considered dataset by making generalizations between them. In addition, supervised learning is further classified into two parts: Regression and Classification. In Regression technique, if two or more than two variables are taken into account; a variation in dependent variable depends on a variation in one or more than one independent variables by establishing the statistical relationships between them. Classification on the other hand classifies the objects into absolute categories known as classes.

Compared to supervised learning, unsupervised learning doesn't involve the target output variables. In this, prior biases are carried out for the structural aspects of the inputs that should be contained in the output results. In this learning approach, an input pattern follows the established statistical designs in the total collected input patterns by representation.

LITERATURE SURVEY BASED ON MACHINE LEARNING APPROACHES

This section presents the contribution of researchers made on the diagnosis of a cardiovascular diseases (CVD). To carry out the diagnosis, researchers have utilized numerous ML approaches like Naïve Bayes (NB), Neural Network (NN), Support Vector Machine (SVM), K-Nearest Neighbors (KNN), Decision Tree (DT), J48, Artificial Neural Network (ANN), Random Forest (RF), Logistic Regression (LR), Association rule (AR) etc.

Das *et al.* (2009) gave a method of NN for diagnosing the cardiovascular disease. It was based on version 9.13 of the SAS software. This method used a combination of speculated values obtained from models of multiple predecessors or the posterior probabilities. Using Cleveland heart disease dataset, the accuracy achieved was 89.01% for the classification purpose.

Anbarasi *et al.* (2010) proposed an approach using 13 attributes for cardiovascular disease prediction. The classifiers used in this approach were DT, classification by clustering and NB. Further feature sub selection based on genetic algorithm was applied and almost similar accuracy was obtained. The accuracy attained from NB was 96.5% and classification by clustering was 88.3%. For binary classification, the result achieved by DT was 99.2%.

Vanisree and Jyothi (2011) recommended a Congenital Cardiovascular Disease diagnosis using a Decision Support System (DSS). It was established on a feed forward neural network which is multi layered (i.e. MLFFNN) and also known as back propagation neural network. The used dataset contained 36 attributes over 200 sample data. The attributes represented the hints for the symptoms and the outcomes of the patient's physical assessment. In this system, for training the model, delta learning rule was applied. Training and testing of the model were performed using sample data on the basis of 80-20 rule. Subsequently with an inclusion for the mean square error of 0.016, an accuracy of 90% was gained.

Zhang *et al.* (2012) presented a system using SVM for coronary cardiovascular disease prediction. In this system, different kernel functions were put into account and various important attributes were gathered using Principal Component Analysis (PCA). It was used for a binary class prediction. Grid search method was utilized to find the optimal values and the highest classification accuracy reached was 88.6364% using a Radial Basis Function (RBF).

Vadicherla and Sonawane (2013) predicted a cardiovascular disease diagnosis system based on SVM sequential minimal optimization (SMO) technique. The technique provided the multipliers optimal values during training stage and performance time was also improved, showing better results even on a large amount of dataset.

Elshazly *et al.* (2014) proposed lymph disease diagnosis that uses Genetic Algorithm-SVM for reducing the features of dataset, resulting as 6 features obtained from 18. Initially, 10-fold cross validation was used for the experiment; later on, various kernel functions were utilized. The performance was calculated by different performance metrics. The outcomes for performance metrics like AUC, F-measure, Accuracy and Sensitivity were 84.9%, 82.7%, 83.1% and 82.6% respectively for the GA-linear classifier.

Masethe and Masethe (2014) formulated a cardiovascular attack prediction model by using WEKA tool for comparing numerous data mining algorithms. They performed the experiments using WEKA tool for binary classes and achieved the prediction accuracy of 99.07% with J48, 99.07% with REPTREE, 97.22% with NB, 98.14% with Bayes net and 99.07% with CART.

Boshra and Mirsaeid (2015) developed a model using distinct techniques of data mining to evaluate the cardiovascular disease diagnosis and prediction. The main objective of the developed model was to calculate the performance of KNN, DT, SMO, J48 and NB classification techniques. The performance was evaluated and correlated in terms of accuracy, precision, sensitivity, and specificity. It was analysed that J48 and DT techniques produced the best results for cardiovascular disease prognosis.

Prerana *et al.* (2015) introduced and created a cloud based centralized system for patients and doctors both to view the e-health data by login. They also used the machine learning algorithms like Probabilistic Analysis and classification and NB Classification for predicting a patient's risk level of a cardiovascular disease.

Sharma *et al.* (2015) suggested a system for cardiovascular disease prediction by using partial tree technique and C4.5 rules. Based on the given parameter about health, a prediction of patient's risk level was given by discovering a set of rules. A comparison was done using C4.5 and partial tree. Error classification, accuracy classification and the rules generated measured the performance. The outcome from the comparison achieved a more efficient and potential prediction.

Noura Ajam (2015) ANN was recommended by the author for cardiovascular disease diagnosis. Testing of the model was performed by using feed forward based back propagation learning algorithms. With 20 neurons in hidden layer, the classification accuracy reached to 88%. ANN showed significant result for cardiovascular disease prediction.

Sairabi and Devale (2015) used k-means and NB to figure out a cardiovascular disease. The proposed work used historical heart database to build the system. Author in their research used 13 attributes for constructing the system. The insights from 300 records of Cleveland Heart database were extracted using data mining techniques. The purpose of building the model was diagnosis about the patient whether he/she is having a cardiovascular disease or not.

S. Seema and Kumari (2016) focussed on data mining techniques such as NB, ANN, SVM and DT. These techniques were performed on historical health records for a chronic disease prediction. To measure an accurate rate and improved performance, a comparative study was implemented in which a highest accuracy rate was obtained by SVM. NB yielded highest accuracy for diabetes health records.

Prabhavathi and Chitra (2016) proposed a DNFS technique that stands for DT based Neural Fuzzy System. This approach is used to evaluate and figure out various cardiovascular diseases. They also reviewed the research on cardiovascular disease diagnosis. The proposed research was to create a knowledgeable and profitable system. It was also to enhance the current system results. Data mining approaches were used to improve a cardiovascular disease prediction. From the result obtained in this research it was found that the Support Vector Machine and Neural Networks were conclusive in manner to predict a cardiovascular disease.

Sharan and Sathees (2016) presented an investigation of cardiovascular disease. They suggested a disease prediction using data mining approaches. It was designed to provide the survey of current approaches about extraction of knowledge from the datasets also its usefulness for medical practitioners. The gross time utilized in building the decision tree, was basis of the gained performance. The primary objective of the analysis was using few numbers of attributes for a prediction of disease.

Prajakta Ghadge *et al.* (2016) proposed a method to predict cardiovascular attack using big data. The fundamental objective of the proposed work was to develop a model. Big data and data modelling approaches were to be used for building a model. The research extracts from the database, the various relationships of the patterns which were correlated with cardiovascular disease. The system involves two databases, one the actual big dataset and the second is updated one. To provide a user with reliable system, a java-file system named HDFS was used. This system helped in the healthcare practitioners in making the intelligent decisions.

Patel *et al.* (2016) Data mining and ML techniques were suggested by the authors here to predict a CVD. To predict the cardiovascular system, two objectives were identified. Firstly, the system doesn't know and assume anything about the patient's records in prior. Secondly, the system which was selected should be scalar in running across the huge number of records. WEKA software was used to implement this system. Explorer mode of WEKA and the classification tools were utilized for the testing phase in the experiment.

Dey *et al.* (2016) evaluated a cardiovascular disease prediction method for a binary classification problem. i.e. Both for patients with and without cardiovascular disease. The algorithms analysed were DT, NB and SVM in which the best solution was obtained from SVM by using attribute selection method employing PCA as well.

Purushottam *et al.* (2016) formulated a system using data mining for an enhanced, improved and efficient prediction of a cardiovascular disease. The system was of great help for medical practitioners in making the worthy decision based on the parameter. Based on certain parameter, the accuracy of the system in testing phase and training phase were 86.3% and 87.3% respectively.

K. Gomathi *et al.* (2016) suggested a system based on data mining approaches for multi disease prognosis. In prediction of multiple diseases, data mining played a crucial role. By using data mining approaches, the amount of test data was reduced. The primary focus of this work is prediction of breast cancer, cardiovascular disease and diabetes etc.

Patel *et al.* (2017) recommended a study by applying data mining techniques to extract hidden patterns. The UCI dataset was collected and J48 resulted with the maximum certainty rate in comparison to LMT. The result suggested predictions for a cardiovascular disease.

Marjia and Afrin (2017) developed a system using WEKA tool based on various algorithms such as Multilayer Perception, K Star, Bayes Net, SMO and J48. By using k-fold cross validation method, Bayes Net and SMO outperformed multilayer Perception, K star and J48. But still they didn't achieve a satisfying result. Thus, to get more optimum result for accuracy and improved decisions, an up gradation was done for a disease diagnosis.

Reddy *et al.* (2017) discussed a data mining-based ANN algorithm for cardiovascular disease prediction. The model used distinct criterion such as cholesterol, blood pressure and heart beat rate for evaluation and predicting a patient's health condition. The system accuracy was implemented in Java. This model was developed to reduce the expenses of cardiovascular disease diagnosis.

Polaraju and Prasad (2017) proposed a method containing 13 distinct attributes for training dataset of 3000 instances. This dataset used 70% of that data for training purpose and 30% for testing purpose. This method was used for cardiovascular disease prediction based on Multiple Regression model. Later on, it proved Regression algorithm has better classification accuracy than rest of the algorithms and is more relevant for indicating a chance of getting a cardiovascular disease.

Shahi and Gurm (2017) performed a method to provide quality services to healthcare centres by an automatic diagnosis of a disease based on WEKA software. Based on the comparison, this work suggested SVM is the most accurate and effective data mining algorithm out of the selected algorithms such as KNN, ANN, NB, SVM, AR and DT.

Weng *et al.* (2017) presented a relation of machine learning approaches for the prognosis of a cardiovascular event happened over 10 years. The individual algorithms were NN, LR, RF and Gradient Boosting Machines (GBM). In addition, for parameter optimization; Grid search method was used.

Beyene and Kamat (2018) recommended another prediction for cardiovascular disease by analysis of data mining techniques. This prediction used various medical attributes like age, sex, blood sugar, heart rate etc using WEKA tool for identifying if the person is suffering from a cardiovascular disease or not. The methodology's major objective was diagnosis of a disease in a short time automatically and is also demanding for experts having no more knowledge and skill in healthcare organisations.

Sharmila and Chellammal (2018) proposed a different method which investigates various data mining techniques for cardiovascular disease prediction. It used MapReduce along with SVM and Hadoop Distributed File System (HDFS) as bigdata tools applied on an attribute set which were optimized. SVM was run in side by side mode which provides superior estimation time than sequential SVM. The model proposed the execution of prediction algorithm using SVM in more than one node. As suggested, HDFS was used for storing a huge amount of dataset.

TABULAR COMPARISON OF ALGORITHMS: TRENDS AND RESULTS

Table 1 presents the comparison of all the research that has been performed to predict the cardiovascular disease in past decade. This table shows year-wise progress in the research as well as the trend of used algorithm with their results in form of accuracy.

Table 1. Comparison of literature survey

Year	Authors	Goal/ Topic	Algorithms Used	Result
2009	Das et al.	Diagnosis of heart disease effectively through the use of NN ensembles.	SAS software version 9.13 based NN ensemble method.	Accuracy for classification = 89.01%
2010	Anbarasi et al.	Enhances the prognosis of heart diseases using Genetic algorithm with feature subset selection.	NB, Clustering and DT Feature selection using: Genetic Algorithm	Accuracy: DT = 99.2% Clustering = 88.3% NB = 96.5%.
2011	Vanisree and Jyothi	Based on Signs and Symptoms using NN, Congenital Heart Disease Diagnosis was performed using DSS.	MLFFNN known as back propagation neural network. For training the model, the Delta learning rule is applied.	Overall accuracy = 90% Mean square error = 0.016
2012	Zhang et al.	Diagnosis of coronary heart disease by application of SVM studying.	SVM, PCA were utilized as a classifier to extract the useful features and distinct kernel functions. Optimal parameters values by Grid search method	Highest classification accuracy: Radial Basis Function (RBF). Accuracy = 88.6364%.
2013	Vadicherla and Sonawane	For Heart Disease, DSS based on Sequential Minimal Optimization.	Sequential Minimal Optimization (SMO) technique of SVM	On large dataset SMO shows good results. The time of performance is also enhanced.

continues on following page

Table 1. Continued

Year	Authors	Goal/ Topic	Algorithms Used	Result
2014	Elshazly et al.	Support vector machines with different kernel functions-based Lymph diseases diagnosis approach.	Classification approach GA-SVM Different kernel functions were employed	GA-linear classifier result Accuracy = 83.1% Sensitivity = 82.6% F-measure = 82.7% AUC = 84.9%
	Masethe and Masethe	Using classification algorithms, heart disease prognosis is done.	In contrast of various data mining algorithms like J48, REPTREE, NB, Bayes net and CART using WEKA tool.	Accuracy: J48 = 99.07%, REPTREE = 99.07%, NB = 97.22%, Bayes net = 98.14%, CART = 99.07%
2015	Boshra and Mirsaeid	Diagnosis and Prediction of Heart Disease by Data Mining approaches.	J48, NB, KNN, SMO	The accuracy of J48 was superior than other three approaches.
	Prerana et al.	Prognosis of Heart Disease using Machine Learning Algorithm. Addition to PAC Algorithm. Correlation of Algorithms and HDPS.	NB Classification and Probabilistic Analysis and Classification.	To login and visualize the e-health data on cloud, a unified system established for both doctors and patients.
	Sharma et al.	Cardiovascular Disease prognosis by System Evaluation using C4.5 Rules and Partial Tree.	C4.5 rules and NB algorithm	The accuracy of C4.5 was superior than NB.
	Noura Ajam	Diagnosis of CVD using ANN.	ANN	Accuracy = 88%
	Sairabi and Devale	Cardiovascular Disease prognosis by the use of Modified K-means and NB.	Modified k-means NB.	CVD: Detection = 93%. Undetection = 89%.
2016	S. Seema and Kumari	Lifelong or incurable Disease Prognosis by mining the data.	NB DT SVM	Maximum accuracy In case of heart disease: SVM = 95.56% In case of diabetes: NB = 73.588%
	Prabhavathi and Chitra	Prognosis of distinct Cardiovascular Diseases and their analysis using DNFS Approaches.	NB, DT, C4.5, SVM.	Accuracy: CVD Diagnosis = 85% – 99% CHD Diagnosis = 82% – 92%.
	Sharan and Sathees	Scrutinization and study of CVD Prognosis using Data Mining Approaches.	J48 NB Simple CART	Accuracy = 91.4% Accuracy = 88.5% Accuracy = 92.2%
	Prajakta Ghadge et al.	Using Big Data, an Intelligent CVD Prognosis System was developed.	Hadoop, Mahout, NB.	The mechanization of this System makes greatly advantageous
	Patel et al.	CVD prognosis using ML and Data Mining approaches.	J48, RF algorithm, Logistic model tree algorithm.	J48 Accuracy =56.76% LMT Accuracy = 55.75%
	Dey et al.	Using PCA with Reduced Number of Attributes, CVD prediction is done by speculating Supervised ML Algorithms.	Including or excluding use of PCA for attribute selection on SVM, NB and DT	SVM outperformed the other two.
	Purushottam et al.	Using DT, an efficient CVD prognosis System was developed.	DT classifier	Testing phase=86.3% Training phase=87.3%
	K. Gomathi et al.	Multi Disease prognosis using Data Mining approaches.	NB J48	NB for Heart Disease: 79% NB for Diabetes: 77.6% NB for Breast Cancer: 82.5% J48 for Heart Disease: 77% J48 for Diabetes: 100% J48 for Breast Cancer: 75.5%
2017	Patel et al.	CVD prognosis using ML and Data mining approaches.	LMT, UCI	Accuracy of UCI was better in comparison to LMT.

continues on following page

Table 1. Continued

Year	Authors	Goal/ Topic	Algorithms Used	Result
	Marjia and Afrin	Prognosis of CVD using WEKA tool.	K Star J48 SMO Bayes Net Multilayer Perception	Accuracy = 75% Accuracy = 86% Accuracy = 89% Accuracy = 87% Accuracy = 86%
	Reddy et al.	To develop a CVD prognosis with the help of ANN algorithm in data mining	ANN	Efficiency verified in JAVA.
	Polaraju and Prasad	Prognosis of CVD is performed with the help of Multiple Linear Regression Model.	Multiple Regression Model	Accuracy of Regression algorithm for classification, is improved in comparison to the other algorithms.
	Shahi and Gurm	CVD Prediction System using Data Mining approaches.	WEKA software used. SVM, NB, AR, KNN, ANN, and DT.	For effectiveness and accuracy, SVM was favored.
	Weng et al.	To test using regular clinical data whether ML provides better cardiovascular risk prognosis.	Compared four algorithms: RF, LR, Gradient boosting machines and NN.	Parameter optimization is done using Grid search and it predicted first cardiovascular event happened over 10-years.
2018	Beyene and Kamat	Using data mining approaches, Prediction and Analysis is done for the occurrence of Heart Disease.	J48, NB, SVM.	For individuals it reduced the cost and provided results in short time span for quality services.
	Sharmila and Chellammal	Using data mining techniques, a conceptual method developed for CVD prediction.	SVM runs in parallel fashion.	Parallel SVM provides improved accuracy than sequential one. Also, SVM gives efficient and better accuracy of 85% and 82.35%.

CONCLUSION

This survey has presented the deep observations and understanding of machine learning as well as data mining techniques used in the field of cardiovascular diseases predictions. This survey has also presented the chronological research works performed by various researchers to predict "cardiovascular diseases". The trends of algorithms and their results are presented as well.

It also provides a path to continue with analyzing the discovery of so far hidden relations by using machine learning algorithms for a comparative study.

REFERENCES

Ajam, N. (2015). Heart Disease Diagnoses using Artificial Neural Network. *The International Insitute of Science, Technology and Education*, *5*(4), 7–11.

Anbarasi, M., Anupriya, E., & Iyengar, N. C. S. N. (2010). Enhanced prediction of heart disease with feature subset selection using genetic algorithm. *International Journal of Engineering Science and Technology*, *2*(10), 5370–5376.

Beyene, . (2018). Survey on Prediction and Analysis the Occurrence of Heart Disease Using Data Mining Techniques. *International Journal of Pure and Applied Mathematics*.

Brahmi, B., & Shirvani, M. H. (2015, February). Prediction and Diagnosis of Heart Disease by Data Mining Techniques. *Journals of Multidisciplinary Engineering Science and Technology*, *2*(2), 164–168.

Das, R., Turkoglu, I., & Sengur, A. (2009). Effective diagnosis of heart disease through neural networks ensembles. *Expert Systems with Applications*, *36*(4), 7675–7680. doi:10.1016/j.eswa.2008.09.013

Dey, A., Singh, J., & Singh, N. (2016). Analysis of Supervised Machine Learning Algorithms for Heart Disease Prediction with Reduced Number of Attributes using Principal Component Analysis. *Analysis*, *140*(2), 27–31.

Elshazly, H. I., Elkorany, A. M., & Hassanien, A. E. (2014). Lymph diseases diagnosis approach based on support vector machines with different kernel functions. *Computer Engineering & Systems 9th International Conference (ICCES)*, 198–203. 10.1109/ICCES.2014.7030956

Ghadge, Girme, Kokane, & Deshmukh. (2015). Intelligent Heart Disease Prediction System using Big Data. *International Journal of Recent Research in Mathematics Computer Science and Information Technology*, *2*, 73-77.

Gomathi, K. (2016). Multi Disease Prediction using Data Mining Techniques. *International Journal of System and Software Engineering*, (December), 12–14.

Masethe, H. D., & Masethe, M. A. (2014). Prediction of heart disease using classification algorithms. *World Congress on Engineering and Computer Science 2014, 2*.

Monica, S. (2016, February). Analysis of CardioVasular Disease Prediction using Data Mining Techniques. *International Journal of Modern Computer Science*, *4*(1), 55–58.

Mujawar & Devale. (2015). Prediction of Heart Disease using Modified K-means and by using Naïve Bayes. *International Journal of Innovative Research in Computer and Communication Engineering, 3*, 10265-10273.

Patel, Upadhay, & Patel. (2017). *Heart disease Prediction using Machine Learning and Data mining Technique*. Academic Press.

Patel, Upadhyay, & Patel. (2015). Heart Disease Prediction using Machine Learning and Data Mining Technique. *International Journal of Computer Science and Communication*, 129-137.

Polaraju, K., & Durga Prasad, D. (2017). Prediction of Heart Disease using Multiple Linear Regression Model. *International Journal of Engineering Development and Research Development*.

Prabhavathi & Chitra. (2016). Analysis and Prediction of Various Heart Diseases using DNFS Techniques. *International Journal of Innovations in Scientific and Engineering Research, 2*(1), 1-7.

Prerana, T. H. M., Shivaprakash, N. C., & Swetha, N. (2015). Prediction of Heart Disease Using Machine Learning Algorithms- Naïve Bayes, Introduction to PAC Algorithm, Comparison of Algorithms and HDPS. *International Journal of Science and Engineering, 3*(2), 90-99.

Purushottam, P. (2016). Efficient Heart Disease Prediction System. Academic Press.

Purushottam, S. (2015). Heart Disease Prediction System Evaluation using C4.5 Rules and Partial Tree. *Computational Intelligence in Data Mining*, *2*, 285–294.

Sai, Reddy, Palagi, & Jaya. (2017). Heart Disease Prediction using ANN Algorithm in Data Mining. *International Journal of Computer Science and Mobile Computing*, (April), 168–172.

Shahi, M., & Gurm, R. K. (2017). Heart Disease Prediction System using Data Mining Techniques. *Orient J. Computer Science Technology*, *6*, 457–466.

Sharmila, R., & Chellammal, S. (2018, May). A conceptual method to enhance the prediction of heart diseases using the data techniques. *International Journal on Computer Science and Engineering*.

Shedole & Deepika. (2016). *Predictive analytics to prevent and control chronic disease*. https://www.researchgate.net/punlication/316530782

Sultana & Haider. (2017). *Heart Disease Prediction using WEKA tool and 10-Fold cross-validation*. The Institute of Electrical and Electronics Engineers.

Vadicherla, D., & Sonawane, S. (2013). Decision Support System for Heart Disease Based on Sequential Minimal Optimization in Support. *International Journal of Engineering Sciences and Emerging Technologies*, *4*(2), 19–26.

Vanisree, K. (2011). Decision Support System for Congenital Heart Disease Diagnosis based on Signs and Symptoms using Neural Networks. *International Journal of Computers and Applications*, *19*(6), 6–12. doi:10.5120/2368-3115

Weng, S. F., Reps, J., Kai, J., Garibaldi, J. M., & Qureshi, N. (2017). Can machine-learning improve cardiovascular risk prediction using routine clinical data? *PLoS One*, *12*(4), e0174944. doi:10.1371/journal.pone.0174944 PMID:28376093

Zhang, Y. (2012). Studies on application of Support Vector Machine in diagnose of coronary heart disease. In *Electromagnetic Field Problems and Applications 2012 Sixth International Conference (ICEF)*. IEEE. 10.1109/ICEF.2012.6310380

Chapter 7
Review of an EMG-Controlled Prosthetic Arm

Rishabh Kumar
Sharda University, India

Aditya Kumar Singh
Sharda University, India

Sabyasachi Mukaherjee
Sharda University, India

ABSTRACT

Amputation, especially of the upper limbs, is a condition that exists in almost all parts of the world. There are more than 110 thousand amputees in India itself. It is extremely difficult for amputees to carry out their daily activities and to deal with daily life as normal people do. The purpose of the myoelectric prosthesis is to restore the basic functions of the lost organs in the joint using neural signals produced by the muscles. Unfortunately, the use of such myosignals is complicated. In addition, once detected, it usually requires a computational force strong enough to convert it into a user-controlled signal. Its modification to the actual function of the implant is limited by a number of factors, especially those associated with the fact that each amputee has a different muscle movement. Modified artificial intelligence systems designed for pattern recognition have the potential to improve the size of implants but still fail to provide a system in which artificial arms can be controlled by brain signals.

INTRODUCTION

Electromyography is a medical technique used to measure and calculate electrical activity created by muscle fibers. The tool for making EMG is known as Electromyograph, and the record produced by it is called Electromyogram. Electromyography function is to determine the amount of electrical energy produced by muscle cells when charged electrically or emotionally. It helps physicians to understand how the muscles of the human body communicate with the brain through neurons. Electromyography

DOI: 10.4018/978-1-7998-3327-7.ch007

(EMG) diagnoses and regulates the nervous system of the muscles and nerves. These nerve cells are called motor neurons. They are responsible for transmitting electrical signals that cause synchronization and muscle relaxation. EMG converts these symptoms into useful information (graphs or numbers) that help doctors make a diagnosis. There are basically two ways Electromyography can be done.

SURFACE ELECTROMYOGRAPHY

Surface EMG assesses muscle function by examining muscle function on the surface of the skin. Surface EMG is usually recorded with two electrodes or with a very complex series of multiple electrodes. More than one electrode is required because EMG recording shows the difference in electrical potential between two different electrodes.

NEEDLE ELECTROMYOGRAPHY

Some small needles (also known as electrode needles) are inserted into the surface of the muscles. The electrical activity observed by the electrodes is displayed on the screen in the form of waves. The electrical activity and sensitivity of the muscles is measured by EMG at rest, during periods of minimal reduction and forced reduction. Muscle tissues usually do not show electrical signals at rest but a short period of activity can be recorded when the muscles change from movement to rest or rest to movement.

DETECTION OF EMG SIGNALS

The technology present in artificial arms (i.e. the plastic and polymer arms) does not work as normal human hand operation. A device intended for the purpose of the user dies as a natural flow of muscle. The user's intention in hand control can be traced to sensory-controlled physical signals for example EMG with a combination of technologies such as machine learning, etc. And this technology (sensors + AI) acts as a visual link between human- controlled artificial hand signals. Nowadays, artificial arms contain more electrodes including artificial hand with myoelectric signals produced by human hand. The surface electrode EMG (electromyography) prosthetic arm control signals are felt on the skin of the muscles and is popular because of its simple structure and limited functionality (without cutting human skin). The power of the input organ is very low in the surface electromyography due to the limit of the location of the electrodes to extract signals. By using additional electrodes, it is possible to identify areas with three to four electrodes from the left leg to extract quality control electromyography signals. However, collecting signals for intramuscular electro-myography invasive procedure requires the surgical ability to use an artificial myoelectric sensor. However intramuscular electromyography signals give access to a group of electromyography signals from many locations in the human hand to provide multiple degrees of control in the artificial arm. However, it is possible to achieve simultaneous control with the quality of an artificial arm with intramuscular EMG (electromyography) signals using a surface electrode sensor. In order to make the product available for amputation, we will be using surface EMG because we cannot cut or use skin transplantation in patients. Thus we can easily supply the device to amputees without surgery or operating the skin.

APPLICATION OF ELECTROMYOGRAPHY

Sense and Emulate/Amplify

In this method electro-myography (EMG) is used for recognizing muscle activity in body and the Emg signals are then refined so as to decide the triggering strength as function of time. The triggered strength is then used to decide a resultant force, position or motion of wearable machine like devices, or wearable device for the missing limb. In this application, the electromyography signal is converted so as to decide a digital amplitude for action. This process is used to control HAL exoskeleton designed in 2001(Kasaoka, Sankai).

Figure 1. Flow of EMG signal (Michael Wehner, Wehner Engineering, Application of electromyogrpahy)

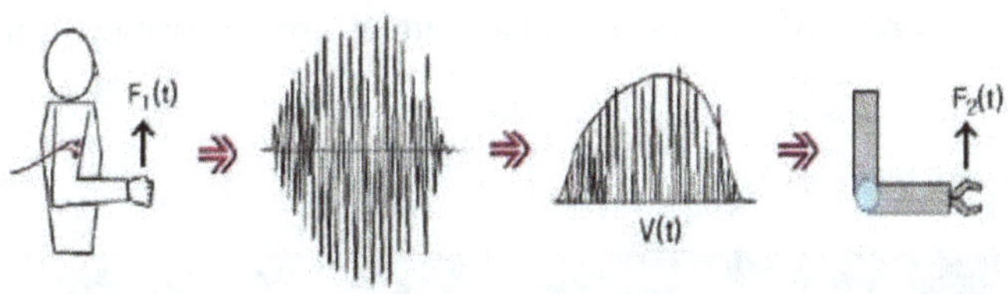

For Checking Muscle Strength

The electromyography signal provides as a signal of the onset of muscle activity and can provide a sequence of shooting of one or more muscles involved in a particular task. Data from the EMG signal is used to show the strength provided by individual muscles and muscle groups.

To Cure Nerve and Muscle Injury Through Clinical Diagnosis

Diagnosis of neuromuscular disease and determining the presence of dysfunction or abnormalities in clinical activities, muscle regeneration using electromyography (EMG) biofeedback, demonstrations kinesiology in anatomy, and use in ergonomics as a tool to study kinesiological muscle function related to posture and other biomarkers of stress, and as an indicator of the pattern of movement and the parameter of the sensory control system of the nervous system.

EMG Signals are used as Control Signals in Prosthetic Instrument

This application is the most common application nowadays and also huge research to be done on this.

PROSTHETIC MARKET IN INDIA AND WORLD

The above artificial limbs go a long way from their origins to the most advanced and experienced architecture in modern times but in our country it does not appear as it comes in countries like US, UK, etc. Today, high-end prosthetics are made of polymer, plastic (which still depends on the sale of artificial limbs) and advanced building materials. The aim is to provide light weight and strong inserts in amputees. Apart from being simple and durable, these devices offer advanced robotic performance. Advanced installation provides basic movements to mimic natural functions. Adding a natural look to silicon arm covers is one of the most used items these days. Research is now underway in India and will come up with these fully automated implants and these developments will accelerate the growth of Indian body parts market in the forecast period. Government support for amputees, the growth of the elderly and growing technology in the health care sector. The features described above will enhance the demand for high- end artificial limbs in the forecast period. So far it is clear that there is no Indian involvement in this area other than conventional plastic and POP prosthetics as shown in fig 4.1, which has 140 million people living in it and an average of 25000- 30000 amputees are performed annually.

Figure 2. Existing plastic prosthesis

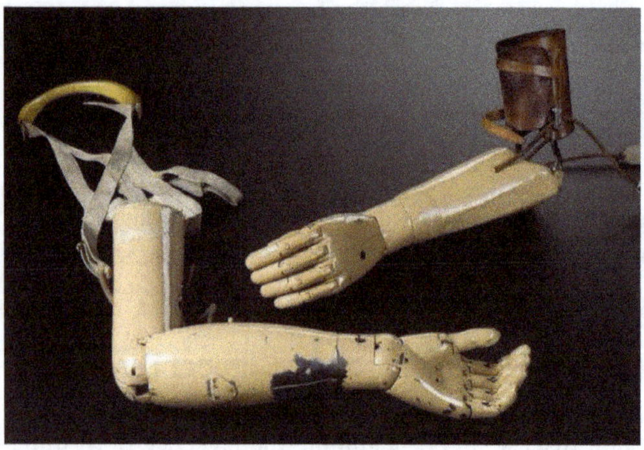

Now looking at the market of prosthesis in world we have so some company and many countries which are doing extremely well in this field that are listed below:

1. John Hopkins laboratory
2. Ottobock, Germany
3. Openbionics
4. Bebionic

And other countries like North America, Germany, UK, Canada, etc. The market size of orthotics and orthotics was estimated at USD 9.2-10 billion by 2020 and ~ 14 billion to 2027 and is expected to reflect the CAGR (Combined Annual Growth Rate) of 4.6% during the weather season. The dynamic increase in the sports caused injuries, road accidents and the increase in the number of diabetes-related amputations, and also the fast growing worldwide Osteosarcoma and this disease can also be a huge share/global market of prosthetics according to Grand View Research. It is therefore a good time for our country to develop in this field to compete with other countries and to make the country disabled and free people. It has also increased the market share and development of our country and is independent of other developed countries in all sectors.

CONTROLLING MECHANISMS

Figure 3. Basic block diagram

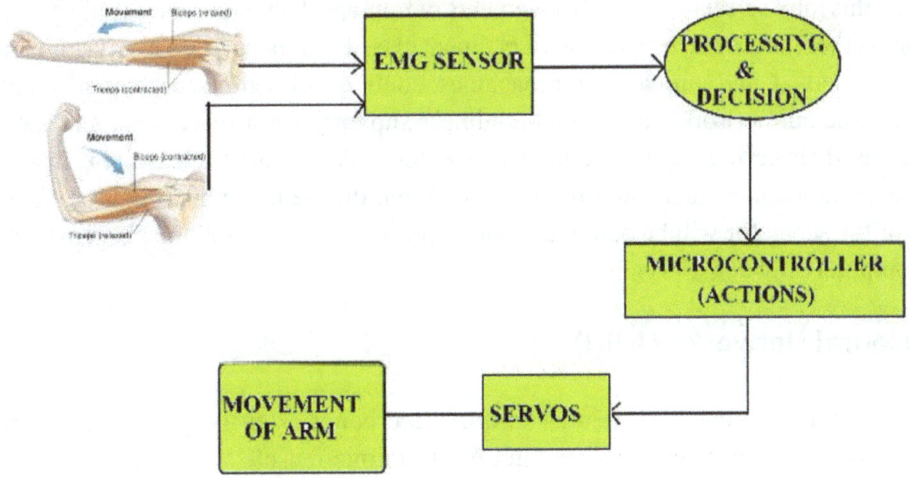

EMG Sensor

The first and the main thing of this research is to extract the electromyography signals that are very small signals in the range of 1mV to 10mV and in 0Hz to 500Hz range. So we will be using EMG surface electrode sensor for extracting and amplification of the electromyography signal. EMG sensor reads the neurological movement of muscles. It is based on a simple analogy that during contraction or expansion of any muscle, electrical activity is produced which spreads through adjoining muscle tissues and bones, and can be recorded from neighboring skin regions where the electrode has been applied.

PROCESSING AND PREDICTION OF EMG SIGNALS/ ARM MOVEMENTS

Processing of Signals Techniques

For the last 10-15 years, the trouble of signal classification have been addressed by numerous students, scientists, professors engineering researchers etc. Advanced classification of signals allow user for several functions of the arm, a fast reaction time, and a more reliable and precise product. While the difficulty of acquiring a sufficient number of degrees of freedom (DOF) is mainly a design issue, signal categorization and prediction is of equal importance. Without a sophisticated signal classification, an EMG based bionic arm would be insusceptible to reliability and performing the movements of grasping patterns and designs. The signal classifier's part can be divided in phases of quality extraction & movement identification, along with both online as well as offline education of arm. A very big research has been committed to amplify quality extraction and realization of pattern, and many methods for all of those classification steps will be discussed in sections below. (Crowder, 1991)

Pattern Identification Methods

The signal segmentation detection section uses extruded features to predict targeted movements, and, when provided this info, select the guided movements of implant. There are many similar ways of pattern identification methods used: Bayesian classifiers, linear classifiers such as linear discriminant functions, the least squares method, and support vector machines, nonlinear classifiers, commonly employing back propagation and neural networks (as well as nonlinear support vector machines), and the fuzzy logic optimization-based clustering algorithm, and so on so forth. As a feature release, it is important to balance the processing load, accuracy and robustness. Often, different segments can be mixed to give a better result or the separator will be combined with a process of optimization as a genetic algorithm to enhance its output. (Crowder, 1991)

Artificial Neural Networks (ANN)

As a pattern learning tool category, neural networks are being extensively used to control artificial prosthetics hands. The neural networks are intended to mimic muscle signal map patterns; according to the activity of neurons, the artificial neural network (ANN) group of nodes (representing neurons) and the relationships between those nodes. Neural networks also include a learning phase that allows differentiation for improving over time. The neural network as shown in Fig. (5.1), is a very well-known patterns for pattern identification of artificial hand control applications. Neural networks can mimic non- linear data also; in return they are often unstable. Often, neural networks are integrated to different algorithm strategies for enhancing stability. An often used combination is the use of NN with AR (Auto-regressive) model, one can use the NN feature combination, and WT weight (Weight) to control a five-finger, non-functional hand designed to enable grip, intermediate grip, finger grip, and grip which is a cylinder. (Tsuji et al., 2000)

3D DESIGN OF THE ARM

The 3D arrangement of the hand consists of a thumb and four fingers, and a palm with forearm structure. All the fingers have 2 joints and thumb will have single joint. The joints are connected with nylon thread in starting phase later on we will be having gears. Nylon thread are then connected to motor and every fingers has its own motor for more accuracy and more number of movements. Below is the image shown of 3d printed fingers, palm and forearm.

Figure 4. Artificial Neural Networks (Design Engineering Technical Conferences and Computers and Information in Engineering Conference IDETC/CIE.)

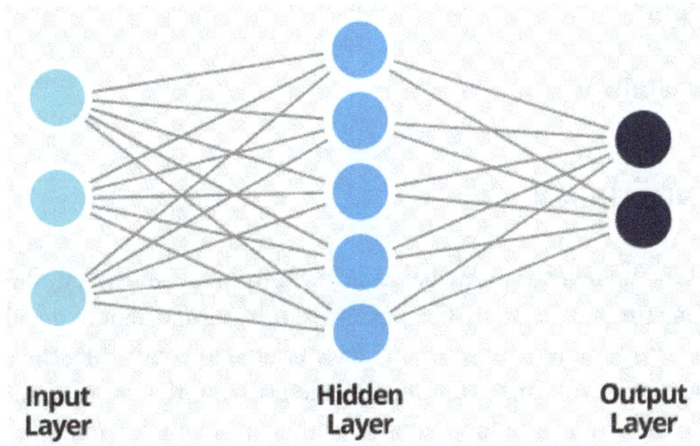

Figure 5. fingers with 2 joints and palm (Nazmus- Design and Implementation of an EMG Controlled)

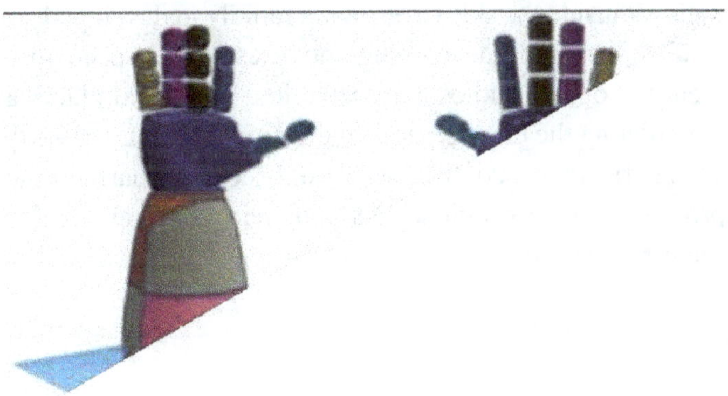

Figure 6. Proposed Model (3dnatives.com)

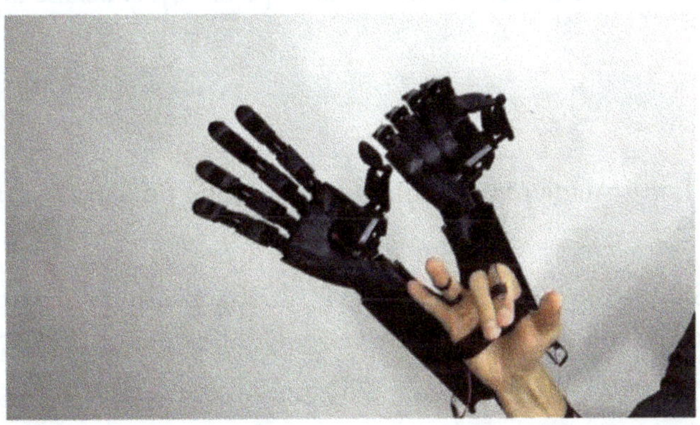

Research Questions

1. Why Myoelctric arm?
2. Need of arm?
3. How the arm is controlled?

RESULT AND CONCLUSION

In this paper, a study of the myoelectric prosthetics arm control mechanism has been performed and has not been developed as the product has only been examined. It is important to develop the product based on these method or methods. One more important factor related to controlling the myo-electric artificial arm is to enable firm grip with the finger numbers made utilizing the car to imitate or control grips of fingers. In addition, the closure of the loop control with sensory motor integration is a fixed field of research, in addition to the corresponding control of quality. Researchers have been experimenting with sensory or sensory stimuli to complete the myoelectric control arm loop. The closed loop control is one of the fields which is most in requirement of attention. In addition, research conducted on myoelectric nerves (implantation of sensor inside the skin) is growing rapidly in developed countries. But nowadays this research is being studied starting in developing countries to keep pace with developing countries and to fill the gap differently. Logical studies have been done in several places across the world and it becomes important for calculating the effectiveness of electromyography treatment vision management techniques. In addition to artificial arm control, we should see other partners such as 3D hand design, 3D arm material to improve efficiency, efficiency and longevity of battery. It also requires combination or incorporation of specialists from different fields to have a product or model work.

FUTURE SCOPE

With the combination of EMG and EEG, then the prosthetic arm will be more precise and accurate. Then we can equip the arm on a wheelchair also in case of paralytic patients. It can also be equipped with touch sensor so that the patients can identify what type of object he/she is touching.

REFERENCES

Crowder, R. M. (1991). Local Actuation of Multijointed Robotic Fingers. *International Conference on Control.*

Tsuji, T., Fukuda, O., Shigeyoshi, H., & Kaneko, M. (2000). Bio-mimetic Impedance Control of an EMG- controlled Prosthetic Hand. *IEEE/RSJ International Conference on Intelligent Robots and Systems*, 377-382. 10.1109/IROS.2000.894634

Chapter 8
Analysis and Design of a Parallel Switched-Inductor DC-DC Converter

Prabhat Kumar
National Institute of Technology, Silchar, India

Amritesh Kumar
National Institute of Technology, Silchar, India

ABSTRACT

This chapter proposes a switched inductor configuration-based non-isolated DC-DC converter with high voltage gain. The proposed converter has two output capacitors instead of a single output capacitor for voltage boosting capabilities. Enhancement of the output voltage with the addition of more number of switched inductor cells is also possible in this configuration. The most advantageous factor of this proposed converter is the use of low-voltage semiconductor devices as they don't require large heat sinks. The converter operation in the steady state is fully analyzed. In addition to that, for the purpose of stability analysis, the small signal model for the proposed converter has also been developed. The frequency response using the small-scale transfer function of the converter has also been done by employing MATLAB. A suitable controller with suitable parameters has also been designed to improve the overall stability of the DC-DC converter in consideration. The results obtained after simulation verifies the feasibility of the converter.

INTRODUCTION

With the massive increase in the population, the demand for the energy is growing each day. To cope up with this massive deficiency of power, the renewable sources of energy like the solar Photovoltaic (PV), Wind, tidal energy are gaining more popularity and their total share in the energy market has been increasing. However, the biggest drawback of such sources of energy is the low voltage output as they are intermittent in nature and their availability largely depends on the condition of the weather as

DOI: 10.4018/978-1-7998-3327-7.ch008

discussed by Shabestary et al. (2015), Salary et al.(2017) and Salary et al.(2015) and Salary et al.(2016) . Owing to these specific drawbacks, there is a need to employ a boost converter having a high voltage conversion capability to meet up the specified voltage requirement of the load as shown by Salary et al.(2016). A conventional boost converter does not have the capability of going towards the extreme values of the duty cycle as presented Salary et al.(2016) and Ghadimi et al. (2007) and Li et al. (2011). The biggest problem encountered during the working of a step-up converter is the issue of EMI while working at high frequency which is not desired as shown by Shu-Kong et al. (2013) and Ismail et al. (2008). Using a transformer operating on high frequency could be another possible solution but would lead to the bulkiness of the system and therefore not feasible. Another solution lies in the cascading of boost converter for obtaining high gain. The recent trend has introduced the switched inductor and the switched capacitor configurations for achieving high voltage gain. There has been some recent development of such converters. However, these configurations have their own set of advantages and drawbacks as shown by Jang et al.(2007), Garcia et al. (2010), Choi et al. (2011), Garcia et al. (2013), Nomura et al. (2006), Zhang et al. (2013), Hwu et al. (2014) and Banaei et al. (2014) and Kerler et al. (2004). The interleaved DC-DC converter can overcome this drawback with the high voltage gain capabilities as presented by Jang et al.(2007) and Garcia et al. (2010) and Choi et al. (2011). The main drawback of this specific topology is encountered in the complexity of control because of the presence of a increased number of the switching devices required. Choi et al. (2011) have presented a novel circuit configuration of interleaved boost converter with the purpose of obtaining a high voltage gain as compared to a conventional boost DC-DC converter. How two intermediary capacitors can be used in order to double the voltage at the output terminals of the boost converter has been beautifully shown in Nomura et al. (2006). A diode-assisted DC-DC converter has been presented by Zhang et al. (2013). This topology has a enhanced voltage gain. In recent times, some DC-DC converters have been presented that make use of the KY converters. These KY converters have also been employed in constructing a converter with a high output voltage as shown in Hwu et al. (2014). A new topology of buck-boost DC-DC converter working in voltage boost mode is proposed in Banaei et al. (2014). The Optimum sizing of battery for EV applications requiring 200-800 volts has been proposed by authors in Kerler et al. (2004).

For the past some years, the Distributed Energy Systems (DES) have been gaining large popularity owing to their various advantages like islanding mode of operation. These DES have various households having PV panels installed at their premises. So, in such a distributed network, the consumers have the advantage of supplying surplus power back to the grid. This paper presents a modified boost DC-DC converter topology utilizing a switched inductor configuration and a combination of two output capacitors instead of a single capacitor for obtaining high output voltage. This topology consists of 3 switched-inductors. One can increase the number of inductors to achieve higher voltage levels as per the requirements. This paper also focuses on the stability analysis of the proposed converter using MATLAB. Bode plots for the uncompensated converter is presented. A PID controller having suitable values of K_p, K_i and K_d is chosen and the system is made immune to the disturbances in the parameters of the system.

THE PROPOSED TOPOLOGY

Figure 1 shows the conventional boost converter topology and Figure 2 represents the constructional details of the presented topology to obtain high gain. The proposed topology consists of three inductors L_1, L_2 and L_3 which are identical to each other. The inductors are arranged in a switched inductor configuration for providing the high output voltage. In addition to the inductors, there are two identical output capacitors C_1 and C_2, two switches and various diodes to facilitate various operating modes of the proposed topology. The voltage at load terminals is the sum of the capacitor voltages.

There are three working modes of the proposed topology based on the different switching combinations within one a time period. The whole circuit is supplied with a DC voltage source V_S. The diodes remains in forward or reverse bias depending on the switching conditions of both the switches.

Figure 1. A conventional boost DC-DC converter

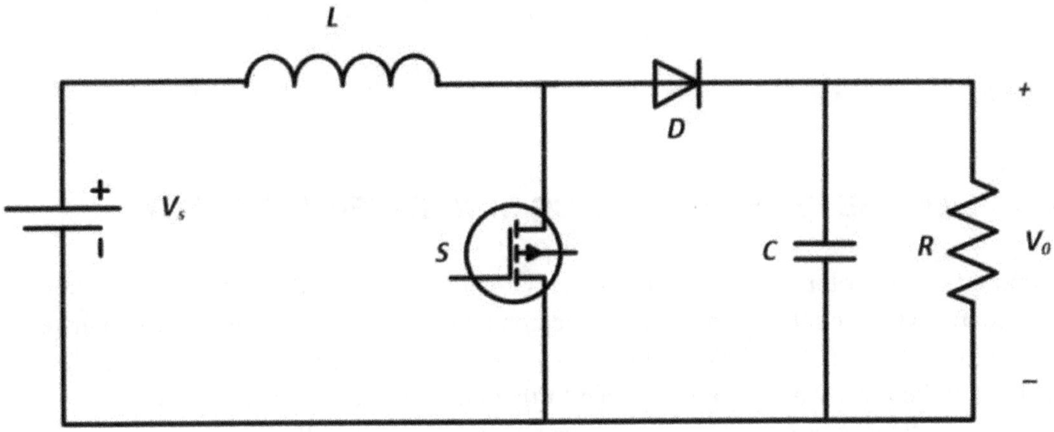

The voltage conversion ratio for a boost converter is given as:

$$\frac{V_0}{V_s} = \frac{1}{1-D} \tag{1}$$

Figure 2. The presented DC-DC converter topology

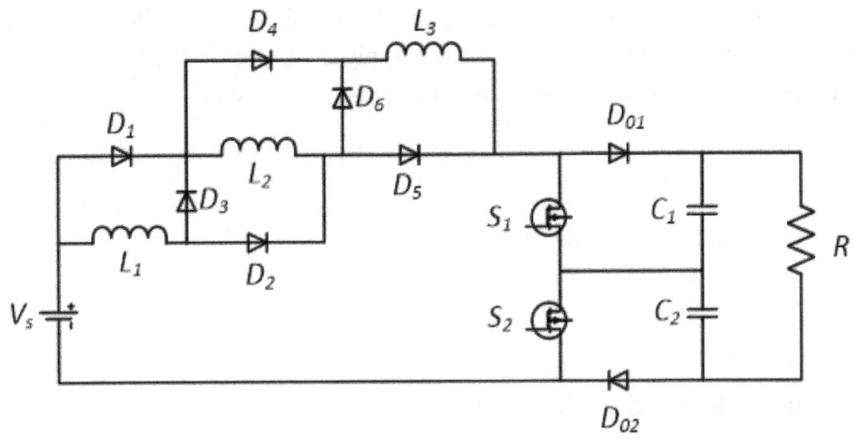

The output voltage is given as:

$$V_o = V_{C1} + V_{C2} \tag{2}$$

Where,

$$V_{C1} = V_{C2} = \frac{V_o}{2} \tag{3}$$

also

$$L_1 = L_2 = L_3 \tag{4}$$

ANALYSIS OF THE PRESENTED DC-DC CONVERTER TOPOLOGY

The proposed converter topology has three operating modes. In the mode 1, the inductors store energy from the source when both the switches are turned on simultaneously. The energy accumulated by the inductor in mode 1 is then transferred to the capacitors C_1 and C_2 in mode 2 and mode 3. These modes of operation of the proposed topology has been further described and fully analyzed.

Some assumptions were considered during analysis which are as follows:

1. The values of capacitors C_1 and C_2 are large enough and same.
2. The values of all the inductors are same and are of large value.
3. Inductor currents and the capacitor voltages are same.
4. All the components are ideal.

Mode 1: The switches S_1 and S_2 in this mode of operation, are closed and the diodes D_3, D_6, D_{01} and D_{02} remains in reverse biased condition whereas other diodes are forward biased. The inductors get connected in parallel to each other and the whole arrangement of the inductor gets connected to the DC supply V_S. With the connection of the inductors across the supply as shown in Figure 3, they start storing energy during this period. The output capacitors supply the power to the load during this switching period. The voltages of all the inductors working in this mode is given as:

$$v_{L1} = v_{L2} = v_{L3} = v_L = V_S \tag{5}$$

The duration of this mode is given by

$$t = DT_s \tag{6}$$

Where, $T_s = \frac{1}{f}$, f being the switching frequency in Hz.

Figure 3. Mode 1 of the proposed topology

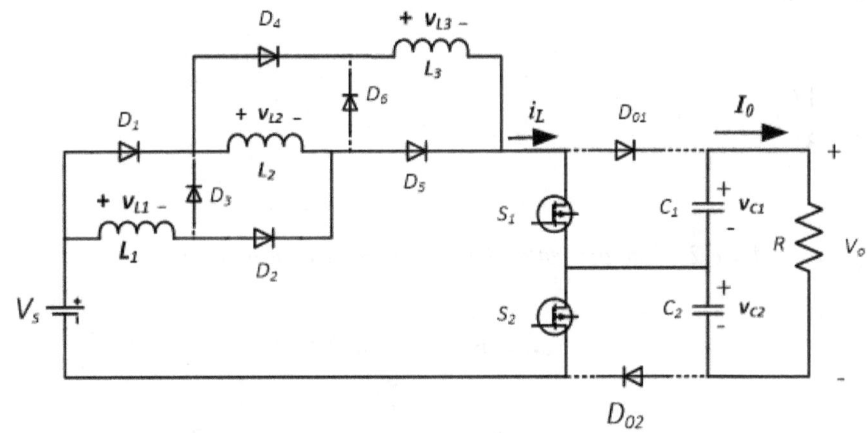

V_s

The values of the capacitor current in this mode is found to be:

$$i_{C1} = i_{C2} = -I_0 = -\frac{V_o}{R} \tag{7}$$

Mode 2: This mode starts with the switch S_1 in conducting state but the switch S_2 is in non-conducting state. The output capacitor C_1 starts discharging and supplies the load during this period. The diodes D_3, D_6 and D_{02} are working in forward biased whereas rest are reverse biased. The inductors that were previously in parallel to each other in mode 1 now gets connected in series fashion. The energy accumulated by the inductor working in the mode 1 is supplied to the capacitor C_2 and it starts charging. The duration of the mode 2 is given by:

$$t = (1-D)T_s = \frac{(1-D)}{f} \tag{8}$$

The connection diagram of the proposed topology working in mode 2 has been shown in Figure 4. From Figure 4, using KVL in the closed loop yields following relations:

$$v_{L1} + v_{L2} + v_{L3} = (V_s - v_{C2}) = \left(V_s - \frac{v_0}{2}\right) \tag{9}$$

Since the inductors are of equal value, it can be implied that

$$v_{L1} = v_{L2} = v_{L3} = v_L$$

Using this value, the equation (ix) becomes,

$$3v_L = (V_s - \frac{V_0}{2})$$

$$v_L = \frac{1}{3}\left(V_s - \frac{V_0}{2}\right) \tag{10}$$

Figure 4. Mode 2 of the presented DC-DC converter topology

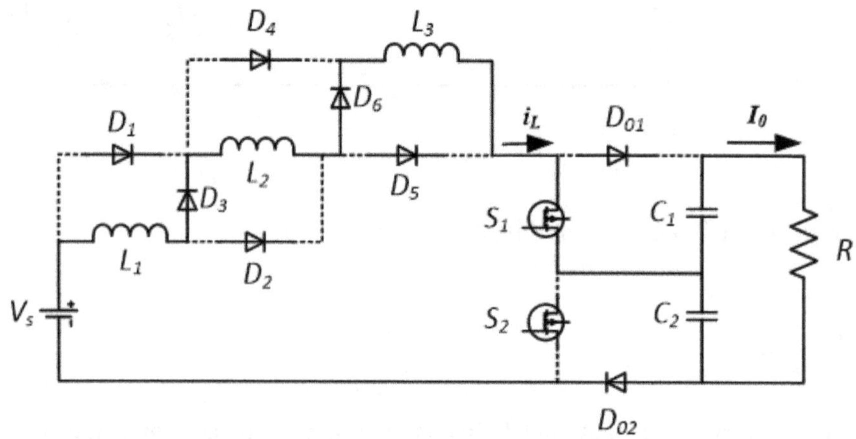

Current through the capacitors are given as:

$$i_{C1} = -I_0$$

$$i_{C2} = i_L - I_0 \tag{11}$$

Mode 3: The switch S_1 is in OFF position while the switch S_2 is in ON position in this mode as shown in Figure 5. With such switching configuration, the path of energy flow towards the capacitor C_2 is directed towards the capacitor C_1 and it starts charging. Meanwhile, the capacitor C_2 feeds the load during this time. The diodes D_3 and D_6 continue to remain forward biased while the diode D_{01} that was reverse biased in the mode 2 now starts conducting whereas diode D_{02} does not conduct. The inductor current i_L continues to decrease. The inductors in this mode continues to remain in series with the voltage across all of them being the same.

Application of KVL in Figure 5 yields,

$$v_{L1} + v_{L2} + v_{L3} = (V_s - v_{C1}) = \left(V_s - \frac{V_0}{2}\right) \tag{12}$$

As

$$v_{L1} = v_{L2} = v_{L3} = v_L$$

$$\therefore 3v_L = \left(V_s - \frac{V_0}{2}\right)$$

and

$$v_L = \frac{1}{3}\left(V_s - \frac{V_0}{2}\right) \tag{13}$$

Figure 5. Mode 3 of the proposed topology

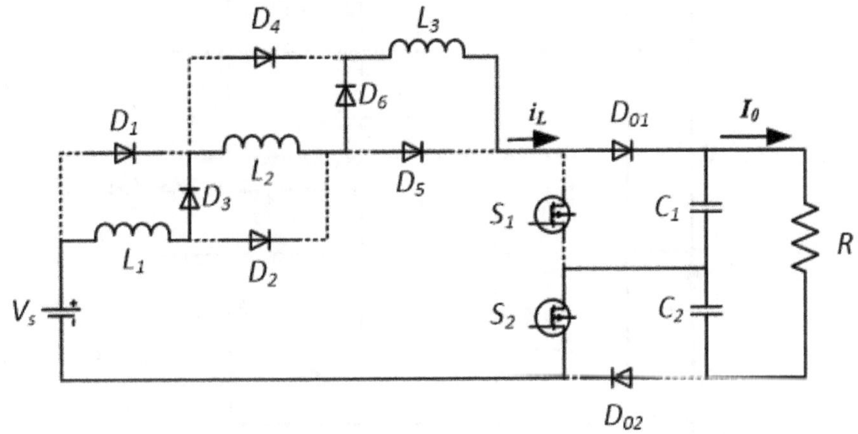

The capacitor currents during this mode are:

$$i_{C1} = i_L - I_0$$

$$i_{C2} = -I_0 \tag{14}$$

Applying voltage-second balancing on the inductor in all the three modes gives:

$$V_s D T_s + \left(\frac{1}{3}\left(V_s - \frac{V_0}{2}\right)\right)\left(\frac{1-D}{2}\right)T_s + \left(\frac{1}{3}\left(V_s - \frac{V_0}{2}\right)\right)\left(\frac{1-D}{2}\right)T_s \tag{15}$$

Simplifying equation (15), we get,

$$\frac{V_0}{V_s} = \frac{2(2D+1)}{1-D} \tag{16}$$

The switching pulses of both the switches and the values of voltages across different components during all the three modes has been shown in Figure 6.

Figure 6. Switching pulse and voltages of different components

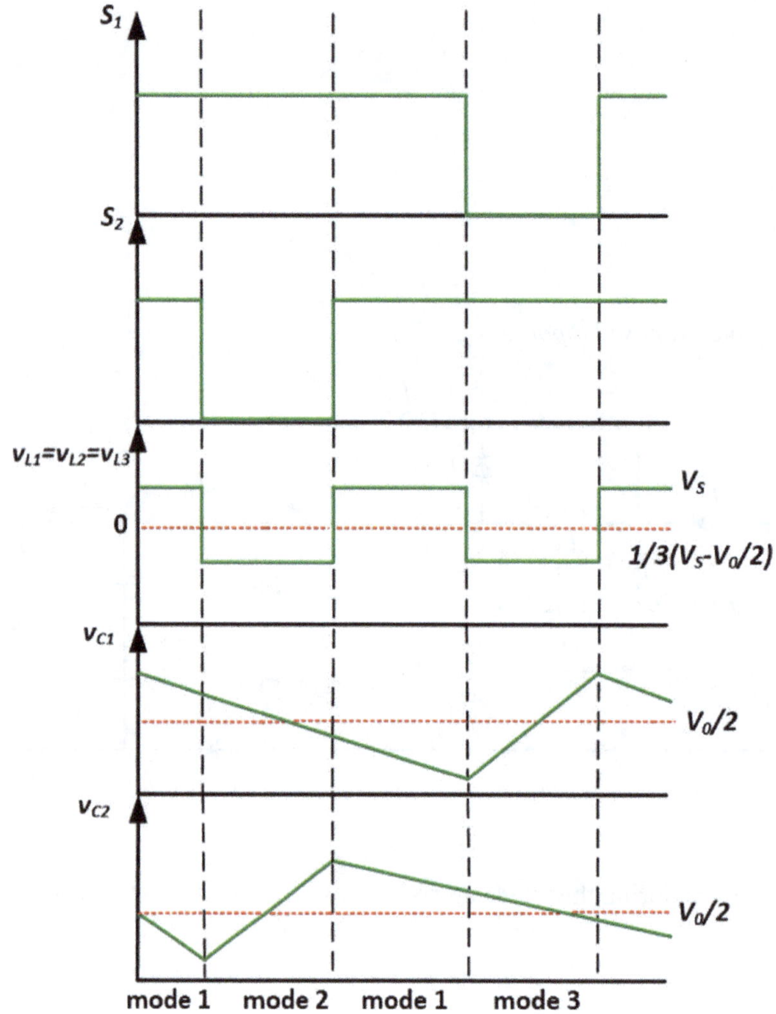

Figure 7 shows the variation of voltage conversion ratio for the presented converter in comparison to a boost converter.

AVERAGE LARGE-SIGNAL MODELLING OF THE PROPOSED TOPOLOGY

With the value of inductor voltage and capacitor currents obtained during all the three modes, the Average large-signal model can be derived for yielding the small-signal model of the proposed converter topology. The small signal representation is very necessary for the purpose of stability analysis of the topology in consideration. The equations governing the inductor voltages and capacitor currents for all the three mode are given by equation (5), (6), (10), (11), (13), (14). The average signal model is obtained for each mode as:

Figure 7. variation of voltage gain with duty cycle

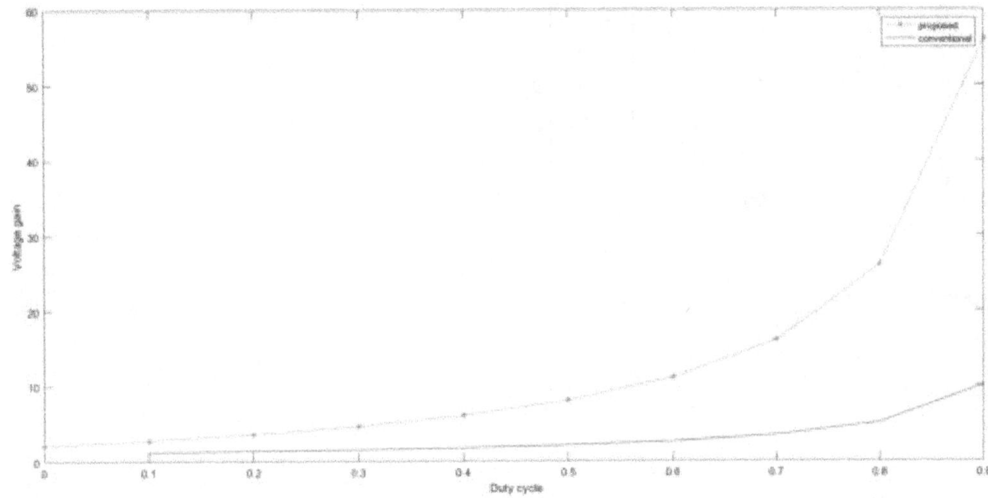

For mode 1:

$$\frac{di_L}{dt} = \frac{v_s}{L} \tag{17}$$

Also,

$$\frac{dv_{C1}}{dt} = -\frac{v_0}{RC_1} \tag{18}$$

$$\frac{dv_{C2}}{dt} = -\frac{v_0}{RC_2} \tag{19}$$

The standard form of state-space representation is given as

$$\dot{x} = Ax + Bu$$

$$y = Cx + Du$$

The equations (17), (18) and (19) can also be written in state-space representation as:

$$\begin{bmatrix} \frac{di_L}{dt} \\ \frac{dv_{C_1}}{dt} \\ \frac{dv_{C_2}}{dt} \end{bmatrix} = \begin{bmatrix} 0 & 0 & 0 \\ 0 & -\frac{2}{RC_1} & 0 \\ 0 & 0 & -\frac{2}{RC_2} \end{bmatrix} \begin{bmatrix} i_L \\ v_{C_1} \\ v_{C_2} \end{bmatrix} + \begin{bmatrix} \frac{1}{L} \\ 0 \\ 0 \end{bmatrix} v_s \tag{20}$$

For mode 2:

$$\frac{di_L}{dt} = \frac{v_s}{3L} - \frac{v_{C_2}}{3L} \tag{21}$$

From the equation of capacitor currents in mode 2,

$$\frac{dv_{C_1}}{dt} = -\frac{2v_{C_1}}{RC_1} \tag{22}$$

$$\frac{dv_{C_2}}{dt} = \frac{i_L}{C_2} - \frac{2v_{C_2}}{RC_2} \tag{23}$$

The equations (21), (22), (23) in state-space representation form can be written as:

$$\begin{bmatrix} \frac{di_L}{dt} \\ \frac{dv_{C_1}}{dt} \\ \frac{dv_{C_2}}{dt} \end{bmatrix} = \begin{bmatrix} 0 & 0 & -\frac{1}{3L} \\ 0 & -\frac{2}{RC_1} & 0 \\ \frac{1}{C_2} & 0 & -\frac{2}{RC_2} \end{bmatrix} \begin{bmatrix} i_L \\ v_{C_1} \\ v_{C_2} \end{bmatrix} + \begin{bmatrix} \frac{1}{3L} \\ 0 \\ 0 \end{bmatrix} v_s \tag{24}$$

Similarly, for mode 3,

$$\frac{di_L}{dt} = \frac{v_s}{3L} - \frac{v_{C_1}}{3L} \tag{25}$$

and from the capacitor currents,

$$\frac{dv_{C1}}{dt}=\frac{i_L}{C_1}-\frac{2v_{C_1}}{RC_1} \tag{26}$$

$$\frac{dv_{C_2}}{dt}=-\frac{2v_{C_2}}{RC_2} \tag{27}$$

Representing the equations (25), (26), (27) in state-space form,

$$\begin{bmatrix}\frac{di_L}{dt}\\ \frac{dv_{C_1}}{dt}\\ \frac{dv_{C_2}}{dt}\end{bmatrix}=\begin{bmatrix}0 & -\frac{1}{3L} & 0\\ \frac{1}{C_2} & -\frac{2}{RC_1} & 0\\ 0 & 0 & -\frac{2}{RC_2}\end{bmatrix}\begin{bmatrix}i_L\\ v_{C_1}\\ v_{C_2}\end{bmatrix}+\begin{bmatrix}\frac{1}{3L}\\ 0\\ 0\end{bmatrix}v_s \tag{28}$$

In state-space representation for the three modes, the average-large signal model is obtained as:

$$A=\begin{bmatrix}0 & 0 & 0\\ 0 & -\frac{2}{RC_1} & 0\\ 0 & 0 & -\frac{2}{RC_2}\end{bmatrix}d+\begin{bmatrix}0 & 0 & -\frac{1}{3L}\\ 0 & -\frac{2}{RC_1} & 0\\ \frac{1}{C_2} & 0 & -\frac{2}{RC_2}\end{bmatrix}\frac{(1-d)}{2}+\begin{bmatrix}0 & -\frac{1}{3L} & 0\\ \frac{1}{C_2} & -\frac{2}{RC_1} & 0\\ 0 & 0 & -\frac{2}{RC_2}\end{bmatrix}\frac{(1-d)}{2}$$

$$B=\begin{bmatrix}\frac{1}{L}\\ 0\\ 0\end{bmatrix}d+\begin{bmatrix}\frac{1}{3L}\\ 0\\ 0\end{bmatrix}\frac{(1-d)}{2}+\begin{bmatrix}\frac{1}{3L}\\ 0\\ 0\end{bmatrix}\frac{(1-d)}{2}$$

Upon simplification,

$$A = \begin{bmatrix} 0 & -\frac{(1-d)}{6L} & -\frac{(1-d)}{6L} \\ \frac{(1-d)}{2C_1} & -\frac{2}{RC_1} & 0 \\ \frac{(1-d)}{2C_2} & 0 & -\frac{2}{RC_2} \end{bmatrix}, B = \begin{bmatrix} \frac{(1+2d)}{3L} \\ 0 \\ 0 \end{bmatrix}$$

Using the values of matrices A and B, the average Large scale model can be written as:

$$\begin{bmatrix} \frac{di_L}{dt} \\ \frac{dv_{C_1}}{dt} \\ \frac{dv_{C_2}}{dt} \end{bmatrix} = \begin{bmatrix} 0 & -\frac{(1-d)}{6L} & -\frac{(1-d)}{6L} \\ \frac{(1-d)}{2C_1} & -\frac{2}{RC_1} & 0 \\ \frac{(1-d)}{2C_2} & 0 & -\frac{2}{RC_2} \end{bmatrix} \begin{bmatrix} i_L \\ v_{C_1} \\ v_{C_2} \end{bmatrix} + \begin{bmatrix} \frac{(1+2d)}{3L} \\ 0 \\ 0 \end{bmatrix} v_s \tag{29}$$

SMALL-SIGNAL MODELLING OF THE PRESENTED TOPOLOGY

The small-signal model of the presented topology is achieved by representing the variables with a steady-state part and a small change across the steady state point as:

$$x = X + \hat{x}$$

The different variables can thus be represented as:

$$i_L = I_L + \widehat{i_L}$$

$$v_{C1} = V_{C1} + \widehat{v_{C1}}$$

$$v_{C_2} = V_{C_2} + \widehat{v_{C_2}}$$

$$v_s = V_s + \widehat{v_s}$$

$$d = D + \hat{d}$$

Small-signal model is represented as:

$$\begin{bmatrix} \dfrac{d(I_L+\hat{i}_L)}{dt} \\ \dfrac{d(V_{C1}+\widehat{v_{C1}})}{dt} \\ \dfrac{d\left(V_{C2}+\widehat{v_{C2}}\right)}{dt} \end{bmatrix} = \begin{bmatrix} 0 & -\dfrac{\left(1-\left(D+\hat{d}\right)\right)}{6L} & -\dfrac{\left(1-\left(D+\hat{d}\right)\right)}{6L} \\ \dfrac{\left(1-\left(D+\hat{d}\right)\right)}{2C_1} & -\dfrac{2}{RC_1} & 0 \\ \dfrac{\left(1-\left(D+\hat{d}\right)\right)}{2C_2} & 0 & -\dfrac{2}{RC_2} \end{bmatrix} \begin{bmatrix} I_L+\hat{i}_L \\ V_{C1}+\widehat{v_{C1}} \\ V_{C_2}+\widehat{v_{C_2}} \end{bmatrix} + \begin{bmatrix} \dfrac{\left(1+2\left(D+\hat{d}\right)\right)}{3L} \\ 0 \\ 0 \end{bmatrix} = (V_s+\widehat{V_s})$$

Separating the steady-state terms and eliminating the higher-order terms, we obtain the following set of equations:

$$\begin{bmatrix} \dfrac{d\widehat{i_L}}{dt} \\ \dfrac{d\widehat{v_{C_1}}}{dt} \\ \dfrac{d\widehat{v_{C_2}}}{dt} \end{bmatrix} = \begin{bmatrix} 0 & -\dfrac{(1-D)}{6L} & -\dfrac{(1-D)}{6L} \\ \dfrac{(1-D)}{2C_1} & -\dfrac{2}{RC_1} & 0 \\ \dfrac{(1-D)}{2C_2} & 0 & -\dfrac{2}{RC_2} \end{bmatrix} \begin{bmatrix} \widehat{i_L} \\ \widehat{v_{C_1}} \\ \widehat{v_{C_2}} \end{bmatrix} + \begin{bmatrix} \dfrac{2V_s}{3L} \\ 0 \\ 0 \end{bmatrix} \hat{d} \tag{30}$$

The equation (30) represents the small-signal model of the presented topology. The transfer function is derived employing the small-signal model by putting the values of different parameters of the circuit from Table 1 as:

$$\frac{v_0(s)}{d(s)} = \frac{\left(16\times10^6\right)S+\left(8\times10^7\right)}{S^3+5S^2+51541.66S+207708.33} \tag{31}$$

Table 1. Different parameters of the system

Parameters	Value
L_1, L_2, L_3	500 μH
C_1, C_2	1000 μF
D	0.5
R	400 Ω
V_S	25V

FREQUENCY RESPONSE OF THE PRESENTED CONVERTER TOPOLOGY

Frequency response can be conveniently obtained using the small signal transfer function represented by (31) using the Bode plots. MATLAB has been used to obtain the bode plot for the above transfer function as shown in the Figure 8.

As seen from Figure 8, the phase margin for the converter is a very small which is an indication of instability of the system even to the slight variation in the duty cycle. Thus, this overall system requires a closed loop control to ensure better stability and reliability.

Figure 8. Bode plot of the small signal model transfer function

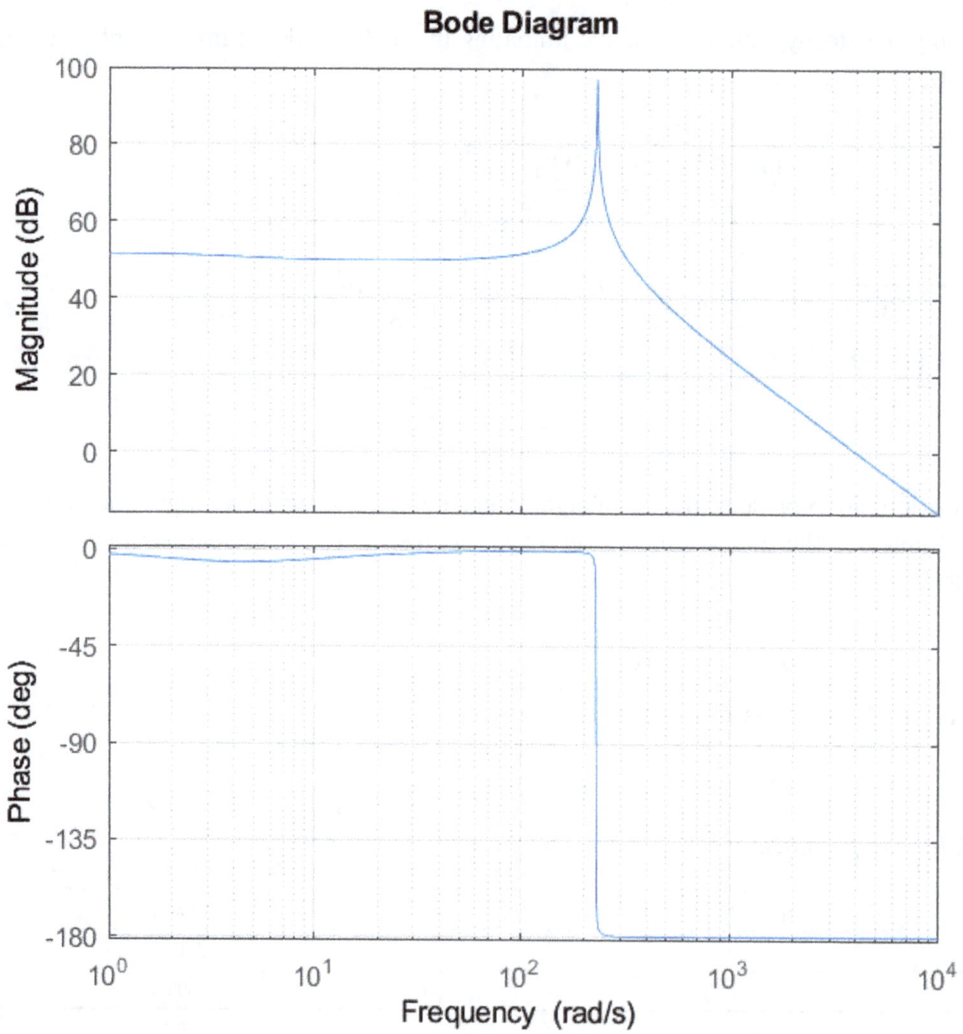

CONTROLLER FOR THE PROPOSED SYSTEM

For increasing the overall stability and compensation, a PID controller can be incorporated in the system is represented in Figure 9. This controller parameters can be suitably adjusted as per the requirement of the system. A value of K_i= 0.001, K_p=0.2 and K_d=0.00025 has been used for improving the stability of the whole system.

Figure 9. Control Scheme for the presented topology

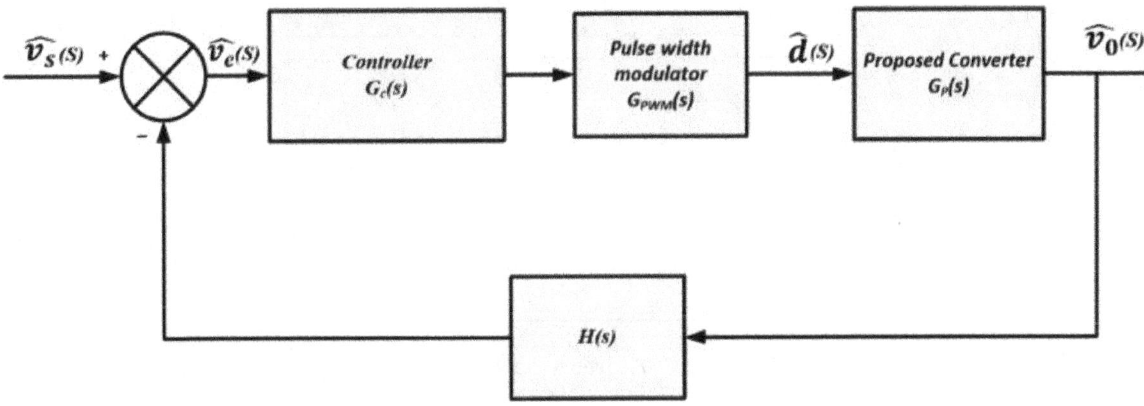

The frequency response obtained for the compensated topology has been shown in the Figure 10. It is observed that the phase margin of the compensated system is now around 54° which shows that the system has been made immune to the variation in the system parameters and an overall closed loop stability.

RESULTS

The proposed topology has been simulated with the help of MATLAB/Simulink with the values of component given in Table 1.

The output voltage of around 206 V was obtained with a supply voltage of 25 V. The output voltage, switching pulses of two switches, inductor current, voltage across any of the inductors and the voltage across the two switches has been shown in Figure 11.

CONCLUSION

The switched inductor step-up DC-DC converter has been discussed having high voltage conversion ratio. The steady-state analysis was carried out for the presented DC-DC converter along with average large-signal and small-signal model of the converter. Employing the small-signal transfer function, converter stability was analyzed by employing MATLAB. Finally, the stability of the converter was improved and the system was made immune to the disturbances in the system parameters by designing a controller with suitable values of K_i, K_p and K_d.

Figure 10. Bode plot of the compensated system

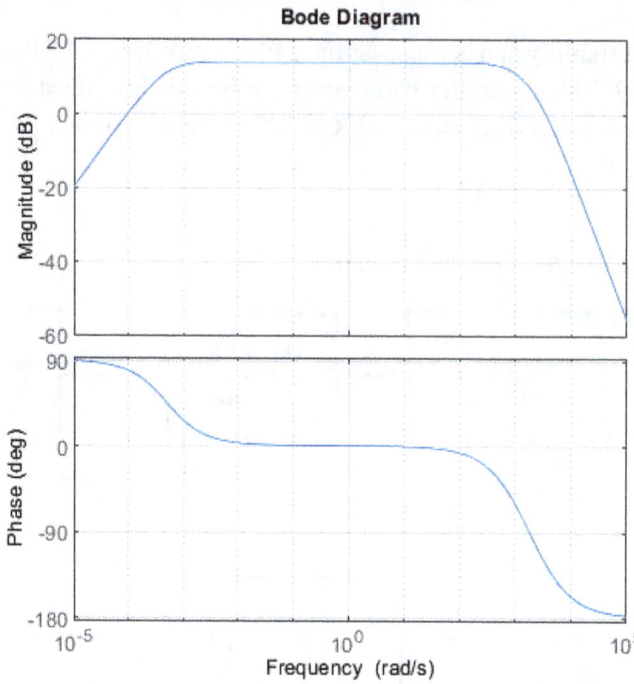

Figure 11. (a) Output voltage (b) Switching pulse for S_1 (c) Switching pulse for S_2 (d) Inductor current (e) Inductor voltage (f) Voltage across switch S_1 (g) Voltage across switch S_2

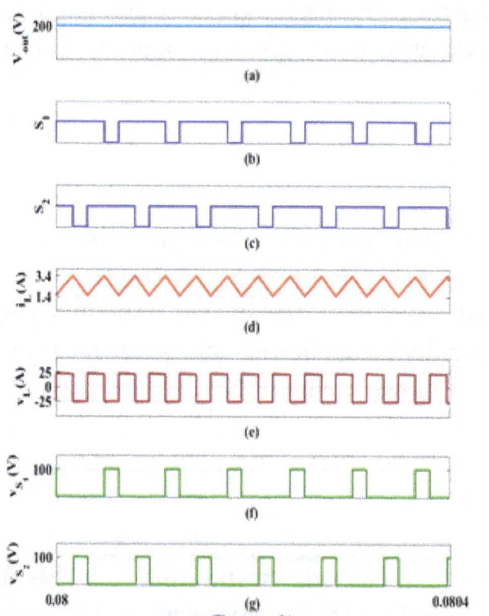

REFERENCES

Alizadeh Shabestary, S. M., Saeedmanesh, M., Rahimi Kian, A., & Jalalabadi, E. (2015). Real-time frequency and voltage control of an islanded mode microgrid. *J. Iran. Assoc. Electr. Electron. Eng.*, *12*(3), 9–14.

Asghar Ghadimi, A., Rastegar, H., & Keyhani, A. (2007). Development of average model for control of a full bridge PWM DC-DC converter. *J. Iran. Assoc. Electr. Electron. Eng.*, *4*(2), 52–59.

Banaei, M. R., Ardi, H., & Farakhor, A. (2014). Analysis and implementation of a new single-switch buck–boost DC-DC converter. *IET Power Electronics*, *7*(7), 1906–1914. doi:10.1049/iet-pel.2013.0762

Choi, S., Agelidis, V. G., Yang, J., Coutellier, D., & Marabeas, P. (2011). Analysis, design and experimental results of a floating-output interleaved-input boost-derived dc–dc high-gain transformer-less converter. *IET Power Electronics*, *4*(1), 168–180. doi:10.1049/iet-pel.2009.0339

Garcia, F. S., Pomilio, J. A., & Spiazzi, G. (2010). Modeling and control design of the six-phase interleaved double dual boost. *Proc. 9th IEEE Int. Conf. Ind. Appl., 2010*, 1–6.

Garcia, F. S., Pomilio, J. A., & Spiazzi, G. (2013). Modeling and control design of the interleaved double dual boost converter. *IEEE Transactions on Industrial Electronics*, *60*(8), 3283–3290. doi:10.1109/TIE.2012.2203770

Hwu, K. I., & Jiang, W. Z. (2014). Voltage gain enhancement for a step-up converter constructed by KY and buckboost converters. *IEEE Transactions on Industrial Electronics*, *61*(4), 1758–1768. doi:10.1109/TIE.2013.2263779

Ismail, E. H., Al-Saffar, M. A., & Sabzali, A. J. (2008). High conversion ratio DC–DC converters with reduced switch stress. *IEEE Trans. Circuits Syst. I*, *55*(7), 2139–2151. doi:10.1109/TCSI.2008.918195

Jang, Y., & Jovanovic, M. M. (2007). Interleaved boost converter with intrinsic voltage-doubler characteristic for universal-line PFC front end. *IEEE Transactions on Power Electronics*, *22*(4), 1394–1401. doi:10.1109/TPEL.2007.900502

Kerler, M., Burda, P., Baumann, M., & Lienkamp, M. (2004). A Concept of a High-Energy, Low-Voltage EV Battery Pack. *2004 IEEE International Electric Vehicle Conference (IEVC).*

Li, W., & He, X. (2011). Review of non-isolated high-step-up DC-DC converters in photovoltaic grid-connected applications. *IEEE Transactions on Industrial Electronics*, *58*(4), 1239–1250. doi:10.1109/TIE.2010.2049715

Nomura, H., Fujiwara, K., & Yoshida, M. (2006). A new DCDC converter circuit with larger step-up/down ratio. *Proc. 37th IEEE Power Electron. Spec. Conf.*, 1 - 7.

Salary, Banaei, & Ajami. (2017). Design of novel stepup boost DC-DC converter. *Iran. J. Sci. Technol. Trans. Electr. Eng., 41,* 13-22.

Salary, E., Banaei, M. R., & Ajami, A. (2015). Step-up DC-DC converter based on partial power processing. *Gazi Univ. J. Sci.*, *28*(4), 599–607.

Salary, E., Banaei, M. R., & Ajami, A. (2016). Multi-stage DC-AC converter based on new DC-DC converter for energy conversion. *J. Oper. Autom. Power Eng.*, *4*, 42–53.

Shu-Kong, K., & Lu, D. D. C. (2013). A high step-down transformer-less single-stage single-switch AC/DC converter. *IEEE Transactions on Power Electronics*, *28*(4), 36–45.

Zhang, Y., Liu, J., & Zhang, C. (2013). Improved pulse-width modulation of diode-assisted buck-boost voltage source inverter. *IEEE Transactions on Power Electronics*, *28*(8), 3675–3699. doi:10.1109/TPEL.2012.2227816

Chapter 9
Comparison of Machine Learning Algorithms for Cardiovascular Disease Prediction

Stuti Pandey
Kamla Nehru Institute of Technology, Sultanpur, India

Abhay Kumar Agarwal
https://orcid.org/0000-0003-2663-8910
Kamla Nehru Institute of Technology, Sultanpur, India

ABSTRACT

Cardiovascular disease prediction is a research field of healthcare which depends on a large volume of data for making effective and accurate predictions. These predictions can be more effective and accurate when used with machine learning algorithms because it can disclose all the concealed facts which are helpful in making decisions. The processing capabilities of machine learning algorithms are also very fast which is almost infeasible for human beings. Therefore, the work presented in this research focuses on identifying the best machine learning algorithm by comparing their performances for predicting cardiovascular diseases in a reasonable time. The machine learning algorithms which have been used in the presented work are naïve Bayes, support vector machine, k-nearest neighbors, and random forest. The dataset which has been utilized for this comparison is taken from the University of California, Irvine (UCI) machine learning repository named "Heart Disease Data Set."

INTRODUCTION

The cause of death is mostly due to cardiovascular diseases worldwide as compared to any other disease. 80% of these deaths are caused by heart attacks and strokes. A cardiovascular disease death in people, less than 70 years of age is about thirty four percent. Manual building of predictive models or

DOI: 10.4018/978-1-7998-3327-7.ch009

data analysis for cardiovascular disease prediction is impractical, less productive as well as tedious task to be performed by a human being. On the other side, machine learning provides a convenient way of observing a massive volume of data (Ghadge et al., 2016) (Prerana et al., 2015). It also learns from past experiences and computation for producing a better and precise result in the future.

From the past decade, machine learning has provided some major research inputs and subsequent improvements in the realm of medical science. Numerous machine learning algorithms have been used to develop the model of cardiovascular disease prediction (Weng et al., 2017). These models can learn from training data and examined when tested by new untrained data. Most of the approaches that have been utilized to develop the model of cardiovascular disease prediction are based on some standard dataset (Patel et al., 2017). One of the most widely used dataset is Cleveland Clinic Foundation dataset available at UCI repository.

Cleveland Clinic Foundation Dataset

UCI repository was started at University of California, Irvine in 1987. It was a venture of David Aha and his assistant scholars. This repository contains huge number of datasets from miscellaneous realms. These datasets are utilized from time to time by many researchers and experts to perform their empirical studies and analysis. Besides Cleveland Clinic Foundation Dataset, there are three more sources of "Heart Disease Data Set".

The Cleveland Clinic Foundation dataset contains minimum missing values among all four datasets as well as it is the most widely used dataset by experts. Robert Detrano, (M.D, Ph.D) provided this dataset. There are 76 attributes in the dataset but due to scope of better analysis, a subset of 14 attributes is utilized in all published experiments (Patel et al., 2017). Among fourteen attributes, the dataset contains 'thirteen' dependent and 'one' independent variable to be anticipated. This anticipated variable ranges from 0 to 4 which describe the heart conditions. For the comparison presented in this paper, class labels '1 to 4' are merged into one class and labelled as '1'. Therefore, anticipated class can be either '0' representing healthy heart or '1' representing unhealthy heart. These fourteen attributes are mentioned in table 1.

Objective

The objective of this research work is to compare the machine learning algorithms for cardiovascular disease prediction on the basis of performance metrics (Masethe & Masethe, 2014). The performance metrics which have been used for evaluation of machine learning algorithms are accuracy, precision, recall, F1-score and ROC. The four machine learning algorithms which have been compared on the basis of performance metrics are Naïve Bayes (NB) (Anbarasi et al., 2010) (Boshra & Mirsaeid, 2015) (Dey et al., 2016) (S. Seema & Kumari, 2016) (Shahi and Gurm, 2017) (Beyene & Kamat, 2018) (Prabhavathi & Chitra, 2016) (Sairabi & Devale, 2015) (Sharan & Sathees, 2016) (Sharma et al., 2015), Support Vector Machine (SVM) (Vadicherla & Sonawane, 2013) (Elshazly et al., 2014) (Sharmila & Chellammal, 2018) (Zhang et al., 2012), K-Nearest Neighbors (KNN) (Boshra & Mirsaeid, 2015) and Random Forest (RF) (Patel et al., 2016). This paper presents the contribution of optimal attributes in the evaluation of machine learning algorithms. Hence in the later part of this paper, it is concluded how attribute selection impacts the performance metrics.

Table 1. Attributes of dataset

Sr. No.	Attributes	Type	Description
1	age	Numeric	Represent the 'age' of an instance. The value is represented in 'years';
2	sex	Binary	Defines the 'gender' of instances. Here, '0' represent 'female' and '1' represent 'male' instances;
3	cp	Categorical	Describes the 'chest pain type' for an instance. It has three values. The value '1' represents 'typical angina' and value '2' represents 'atypical angina' while value '3' represent asymptomatic;
4	trestbps	Numeric	Explains the 'resting blood pressure' of an instance in mm Hg;
5	chol	Numeric	Used to describe the 'serum cholesterol' of an instance in mg/dl;
6	fbs	Binary	Explains the 'fasting blood sugar' of an instance. It is a binary attribute which has two values '0' representing 'below threshold' and '1' representing 'above threshold'. The threshold value is '120 mg/dl';
7	restecg	Categorical	Describe the 'resting ECG results'. It is expressed with three values. The value '0' represents the 'normal' and '1' represents 'ST-T wave abnormality' while '2' represents 'LV hypertrophy';
8	thalach	Numeric	Used to describe the 'maximum heart rate achieved' by an instance;
9	exang	Binary	Describes the 'exercised induced angina'. It has two values namely '0' for 'No' and '1' for 'Yes'.
10	oldpeak	Numeric	Used to describe 'ST depression induced by exercise relative to rest'.
11	slope	Categorical	Describes 'slope of the peak exercise ST segment'. It has three values. The value '1' represents 'up-sloping' and value '2' represents 'flat' while value '3' represent 'down-sloping'.
12	ca	Categorical	Used to describe 'number of major vessels colored by fluoroscopy'. It has four different values ranges from '0 to 3'.
13	thal	Categorical	Explains the 'defect types'. It has three different values. The value '3' represents the 'normal' and '6' represents 'fixed defect' while '7' represents 'irreversible defect'.
14	num	Categorical	Explains the 'diagnosis of heart disease'. It has five different values ranging from 0 to 4. The value '0' represents the 'healthy' heart while values '1 to 4' describes the severity of 'unhealthy' heart.

Related Work

A lot of research works have been performed to compare the machine learning algorithms for cardiovascular disease prediction. But still, there exist the limitations in these works. These limitations involve in terms of accurate analysis and comparison of models for cardiovascular disease prediction. Since, the comparison approaches of machine learning algorithms are based on performance metrics. Most of the researchers have compared these algorithms on the basis of either accuracy or precision. Few of them have used recall, while ROC is used by a smaller number of researchers. F1-score is rarely used by any experts. All these performance metrics are not used simultaneously for comparison of machine learning algorithm. Another type of limitation is based on the method of attribute selection on the dataset used for predicting the cardiovascular diseases.

Next section presents the research methodology including the tool, algorithms used in comparison, attribute evaluators, search methods and performance metrics. After this, proposed method for attribute selection is presented. Following the sequence, all the experiments which have been performed for this comparison is presented with its experimental results. Last section presents the conclusion of this comparison.

EXPERIMENTAL SETUP

Research Methodology

The steps pursued as component of research methodology are as follows:

- Cleveland Clinic Foundation dataset is selected;
- Weka tool is selected for experiments;
- Preprocessing of dataset to divide it into training and testing dataset files;
- Normalization of training and testing dataset files;
- NB, SVM, KNN and RF are used for comparison on Weka to classify the instances of dataset as healthy or unhealthy;
- Correlation based Feature Selection (CFS) and Classifier Subset Evaluation (CSE) are used as attribute evaluator during attribute selection;
- The search methods used with attribute evaluator during attribute selection are Best First Search (BFS), Linear Forward Selection (LFS) and Genetic Search (GS).

It must be noted that Coefficient of Variation (CV) Parameter Selection is used for performance optimization while attribute selection is performed on the basis of proposed method.

Weka

Waikato Environment for Knowledge Analysis (Weka) is a suite of machine learning software written in Java and runs on almost any platform. It was developed at the University of Waikato, New Zealand and is free software licensed under the GNU General Public License. To solve the real-world data mining problem on a given dataset, they can be enforced straight or called from your own Java code. The version used in this research work is 3.6.9. Weka receives dataset files in two formats. These formats are comma separated value (csv) and attribute-relation file format (arff). To perform experiments, arff files are created with Cleveland Clinic Foundation dataset. Files with lesser attributes are created by preprocessing tab of the tool.

Algorithms used in Comparison

Machine learning is a powerful craft based on artificial intelligence and comprises a lot of algorithms to develop a model. These models are trained with input data known as training dataset and examined by unknown input termed as testing dataset. Training of these models also depends on algorithms. Although numerous algorithms are present in machine learning, but NB, SVM, KNN and RF produce better accuracy in results for cardiovascular disease prediction. In this paper, NB, SVM, KNN and RF is selected for comparison from past experience.

NB classifiers utilizes the operation of Bayes theorem. These are simply a group of "probabilistic classifiers". In a learning problem, they require numerous linear parameters for the number of features and establish assumptions among the features with strong (naïve) independence.

SVM classifier uses a hyper-plane for classifying the data into different classes. After modeling with training data, SVM produces a hyper-plane in the test data as an output. By drawing a hyper-plane, discrete data classes are separated. This can be done by finding the space in the data matrix by SVM model.

KNN is a non-parametric machine learning algorithm. Primarily, it is based on Euclidian Distance Formula for classifying the new training instances. The problem of classification as well as regression can be solved by KNN algorithm. In the feature space, k-closest training examples are comprised in the input for both kinds of problem.

A RF classifier includes a group of decision trees. Specifically, for the construction of each tree, an algorithm 'A' is applied on a given training set 'S'. Further a random vector θ is sampled independently and identically using some kind of distribution on the training set S. For the predictions of individual trees, a majority vote is applied over them in Random Forest classifier (Shai & Shai, 2014).

Attribute Evaluators

While operating with the model prediction, it is important to know the features which are useful, consistent, compatible and relevant in predicting a model. For a particular problem domain, it also requires a deep understanding and knowledge in automatic selection of attributes. Choosing a smaller number of attributes is necessary as it makes the complexity of the model to a minimum. A simpler model is easy to understand, explain and interpret. In this paper, two attribute selection methods are used namely: Correlation based Feature Selection (CFS) and Classifier Subset Evaluation (CSE).

In Correlation based Feature Selection (CFS), for the purpose of selecting the most relevant and important attributes in the given dataset, a prominent technique known as correlation is used. In statistics, it is known as Pearson's correlation coefficient. A correlation between an output variable and each of the attribute can be calculated for the selection of attributes.

In Classifier Subset Evaluation, in order to evaluate the 'merit' out of a given set of attributes, a classifier is used. It either separately holds out the testing set or estimates the attribute subsets on the given training data. It further uses the various internal machine learning algorithms for the attribute selection which includes Logistic Regression, Naïve Bayes etc.

Search Methods

Attribute Evaluators always demands a search method to perform attribute selection. Some of these can only be utilized with particular search methods. One example is correlation-based feature selection. It works only with best first search or ranker search methods. In the analysis of this study, search methods used for attribute selection are BFS, LFS and GS.

BFS uses a backtracking facility. Basically, it utilizes greedy hill climbing augmentation for searching the space of attribute subsets. The level of done backtracking can be controlled by allowing the consecutive non-improving nodes to be fixed. A BFS can be started in two modes. First mode performs searching in backward direction by considering entire set of attributes. Second mode starts with an empty attribute set and then search in a forward direction. One more method is available by adding and deleting all single possible attributes at a particular point and then start searching in both the directions

LFS is basically an expansion of the BFS method. It considers a reduced k number of attributes. A fixed number k of attributes is selected by a fixed-set while in each of the step; k is increased with selected fixed-width. This search method either performs ranking of the k-attributes or an initial ordering for selecting the top most k-attributes. Directions to be used in this search method can be forward or floating forward with an optional backward search step.

GS method is based on the principle given by Charles Darwin "survival of the fittest" (Khare and Burse, 2016). It stimulates generally for the evolutions in natural systems. The composition of the algorithm is based on three different operators: mutation, crossover and the reproduction. In machine learning, it is also a search method based on the theory and for attribute selection; it is competent definitely in examining large search spaces.

Performance Metrics

The comparison of algorithms is evaluated by performance metrics like accuracy, precision, recall, F1-score and ROC (Bramer, 2013). All these performance metrics are fundamentally calculated from the four key elements of the confusion metrics. These key elements depict the actual and predicted classes. The elements are:

- True Positive (TP): TP is a value which is true positive and truly classifies a positive value.
- False Positive (FP): FP is a value which is false positive and falsely classifies a positive value.
- False Negative (FN): FN is a value which is false negative and falsely classifies a negative value.
- True Negative (TN): TN is a value which is true negative and truly classifies a negative value.

Accuracy (Bramer, 2013) can be defined as the ratio of truly classified instances to all the instances. Basically, it is a fraction of all the instances which are correctly predicted or classified. Accuracy can be calculated as:

$$Accuracy = \frac{\mathrm{TP} + \mathrm{TN}}{\mathrm{TP} + \mathrm{FP} + \mathrm{FN} + \mathrm{TN}} \tag{1}$$

Precision (Bramer, 2013) can be defined as the ratio of true positive to the sum of true positive and false positive. Basically, it is a fraction of instances which are predicted as true positive from all the instances which are predicted as positive instances. Precision can be calculated as:

$$Precision = \frac{\mathrm{TP}}{\mathrm{TP} + \mathrm{FP}} \tag{2}$$

Recall (Bramer, 2013) can be defined as the ratio of true positive to the sum of true positive and false negative. Basically, it is a fraction of instances which are predicted as true positive from all the instances which are exactly positive instances. It is also known as sensitivity. Recall can be calculated as:

$$Recall = \frac{\mathrm{TP}}{\mathrm{TP} + \mathrm{FN}} \tag{3}$$

F1-score (Bramer, 2013) is used to combine the precision and recall. It is expressed as twice of the ratio of precision and recall multiplication to the sum of precision and recall. Basically, it is used to express the quality of classifier. F1–score can be calculated as:

$$F1 - score = \frac{2\text{*Precision*Recall}}{\text{Precision} + \text{Recall}} \tag{4}$$

Receiver Operating Characteristic (ROC) Curve is a graph which is plotted for true positive rate vs false positive rate. It is used to express the performance of a classifier.

PROPOSED METHOD FOR ATTRIBUTE SELECTION

A method is proposed for selecting the attributes of normalized training dataset to compare the machine learning algorithms. The method uses Attribute Evaluator for evaluating the attributes of dataset based on some classifier and search method. Following table 2 shows the proposed method for attribute selection:

Table 2. Point Table for Attribute Selection

Attribute Evaluator	Classifier	Search Method	Attribute Serial Number mentioned in Table 1												
			1	2	3	4	5	6	7	8	9	10	11	12	13
CFS	–	BFS			1				1		1	1		1	1
		GS		1	1				1		1	1		1	1
CSE	KNN	GS				1			1	1		1			1
	Logistic	BFS	1		1			1	1			1	1	1	1
	Logistic	LFS	1		1	1		1	1			1	1	1	1
	NB	GS			1		1	1	1		1		1	1	1
	RF	GS	1	1		1			1	1	1		1		
	SVM	BFS			1				1				1	1	1
	SVM	GS	1	1	1	1	1		1	1			1	1	1
Total Points			4	3	7	4	2	3	9	3	4	5	6	7	8

Here, CFS (Correlation based Feature Selection) and CSE (Classifier Subset Evaluation) are used as attribute evaluator. CFS does not need any classifier but CSE needs classifier to evaluate the features. Classifiers used with the attribute evaluator are Logistic, Naïve Bayes, SVM, Random Forest and KNN. The search methods used with attribute evaluator are BFS, LFS and GS.

In above Table 2, first column represents the "Attribute Evaluator". Similarly, second and third columns represent the "Classifier" and "Search method" used with attribute evaluator respectively.

Remaining columns represents the "attributes (attribute number) of dataset". We have used only 13 attributes because last attribute is used to predict cardiovascular disease based on these 13 attributes. So, there is no need to include it.

Each row represents a feature selection method, so '1' is put in the cells for the selected attribute number which have been selected by used combinations. For example, CFS uses BFS method, the selected attributes are 3, 7, 9, 10, 12 and 13. So, '1' is put in the corresponding cells of attribute number. In the same way, CSE uses Logistic classifier with LFS method and selected attributes are 1, 3, 4, 6, 7, 10, 11, 12 and 13. Therefore, cells corresponding to these attribute numbers contain '1'. In the last row, the total points are calculated corresponding to each attribute number by summing up each column values i.e. '1'. For example, attribute number 7 is selected by each combination of attribute evaluator, classifier and search method; so, its total point is '9'. It means this attribute is very crucial to predict the cardiovascular disease.

EXPERIMENTS

Three different approaches have been used to perform the experiments in this comparison. The first approach is to compare the performance of machine learning algorithms on simple training set by using 10-fold cross validation. The second approach is based on test dataset using normalization and without normalization. In second approach, performance optimization is also tested by CV Parameter Selection.

The third approach includes the attribute selection. The attribute selection is performed with the help of 'proposed method for attribute selection'. K-fold cross validation and normalization are used in all the experiments. The purpose of k-fold cross validation is to avert over-fitting and under-fitting of the model. Furthermore, the purpose of normalization is to use a common scale in the dataset for the values of numeric columns without deforming values of range differences or information loss. The dataset is partitioned into two parts to perform all the experiments. These two parts are training dataset and test dataset. Before partitioning, data pre-processing (cleaning and scaling) is applied on the dataset. Basically, data pre-processing applies a set of methods to change raw data into intelligible format. The purpose of data pre-processing is to detect outliers, missing values and inconsistencies in the data. As a remedy, data pre-processing decides the actions taken with outliers, filling up missing values and removing inconsistencies.

The used dataset Cleveland comprises 303 rows. There are 14 attributes in which 5 are numerical and 9 are categorical. There are 6 missing values in the dataset. These missing values are from 'Ca' and 'Thal' columns which are categorical attributes. For categorical attributes, missing values are replaced by mode value of the respective columns. Number of outliers in the dataset are 43, but it is not noise. These outliers are from Chol and Trestbps columns because the values in these columns are above the normal scale. But, in medical scenario these are not outliers because the patients who are identified with cardiovascular disease are considered as outliers. So, these are kept as they are.

Training of model is the main step in machine learning which is performed by using the training set. Training set is generated by partitioning the main dataset. In general, it is considered as standard to use 60 to 70 percent of data as a training set. But it can vary depending on the need and purpose of the experiments. Proposed work has used 70% of the dataset as training dataset.

After training the model with training set, testing of the model is performed. To test the model, test dataset is desired. Test dataset is another part of main dataset which is not the part of training of the model. It is treated as dependent variable of the data. Proposed work has used 30% of the dataset as testing dataset. When data is tested using cross validation, performance can be better or worse. The performance depends on the model used. Therefore, performance optimization is used to guarantee that every model is working in its highest. Coefficient of Variation Parameter Selection is utilized in this comparison for performance optimization.

EXPERIMENTAL RESULTS

This section provides the experimental results, analysis and observations gathered from the simulation performed on Weka tool for the comparison of machine learning algorithms to predict cardiovascular disease. Since, the objective of the proposed work is to test which algorithm classifies diseases the best. Initially by applying 10-fold cross validation method on the training dataset, performance is well evaluated. Then the trained model performance is calculated against the test dataset without using normalization.

After this, performances of the same algorithms are obtained on normalized dataset. Followed by results obtained on normalized dataset, the CVParameterSelection is applied for performance optimization and results are analyzed. In the last, performances of machine learning algorithms are evaluated on the dataset with normalized and reduced features.

All the results are presented in tables and results are analyzed on the basis of performances metrics.

Performance Result on Training Set

Initially, all four algorithms are examined on training set. The performance is evaluated over 10-fold cross validation because it produces almost optimal results. The results are as follows:

Table 3. Performance results on training set

Algorithms	Accuracy (%)	Precision (%)	Recall (%)	F1-score (%)	ROC (%)
NB	82.5	82.5	82.5	82.5	90.0
SVM	53.3	59.5	53.3	38.4	50.6
KNN	79.7	79.7	79.7	79.7	79.1
RF	81.6	81.7	81.6	81.5	89.0

Referring to Table 3, it can be seen that NB outperforms all the algorithms. RF is very close to NB in performance. KNN is at third position. The performance of SVM is lowest among all the algorithms and also very poor.

Performance Result on Test Set without Normalization

After the performance analysis on training set, the algorithms are evaluated on test set without normalization. The models which have been built by these algorithms are examined by test set. Training of models are performed over 10-fold cross validation. The results are as follows:

Table 4. Performance results on test set

Algorithms	Accuracy (%)	Precision (%)	Recall (%)	F1-score (%)	ROC (%)
NB	86.8	86.8	86.8	86.8	93.3
SVM	58.2	75.9	58.2	44.0	51.3
KNN	75.8	75.8	75.8	75.8	75.3
RF	78.0	78.1	78.0	77.7	87.4

Referring to Table 4, it can be observed that the performance of NB has increased on test dataset and it is best among all the algorithms used. Performances of RF and KNN are decreased against the test dataset. RF and KNN are at second and third positions respectively. The performance of SVM is lowest among all the algorithms and also very poor against the test dataset.

Performance Result on Test Set using Normalization

Since, most of the algorithms perform only on normalized dataset, so the performance analysis of all the algorithms are evaluated on normalized test set. The training of models is also performed with normalized training dataset by using 10-fold cross validation. The performance results are tabulated in following table 5.

Table 5. Performance results on normalized test set

Algorithms	Accuracy (%)	Precision (%)	Recall (%)	F1-score (%)	ROC (%)
NB	84.6	84.6	84.6	84.6	92.7
SVM	80.2	80.4	80.2	79.9	78.8
KNN	75.8	76.3	75.8	75.9	76.0
RF	79.1	79.1	79.1	78.9	83.3

Referring to Table 5, it can be seen that the performance of NB remains the same. Performance of SVM has increased significantly on normalized training and test datasets. RF and KNN have also shown the same performances; as without normalized test dataset.

SVM is now at second position in performance. Third is RF and KNN is at fourth position.

CVParameterSelection (Performance Optimization) Result

It selects the best value for a parameter to optimize performances using cross validation and also optimizes accuracy (classification) or root mean–squared error (regression). All the algorithms are evaluated by using CVParameterSelection on normalized test set. The results are as follows:

Referring to Table 6, it can be seen that the performances of all the algorithms remains the same as the results mentioned in table 5. CVParameterSelection has not produced significant changes in the performance of algorithms.

Table 6. CVParameterSelection results of algorithms on normalized test set

Algorithms	Accuracy (%)	Precision (%)	Recall (%)	F1-score (%)	ROC (%)
NB	84.6	84.6	84.6	84.6	92.7
SVM	80.2	80.4	80.2	79.9	78.8
KNN	75.8	76.3	75.8	75.9	76.0
RF	78.0	78.0	78.0	78.0	88.1

Performance Result on Nine Selected Attributes

The proposed attribute selection method has given points to all the 13 attributes which will decide the 14th attribute (num) i.e. whether the instance is affected from cardiovascular disease or not. On the basis of point table 2, those attribute which scores '4 or more than 4 points' have been selected for prediction of cardiovascular disease. Referring to point table 2, the attributes are: age, cp, trestbps, restecg, exang, oldpeak, slope, ca and thal. Performance results of the same algorithms with these reduced attributes are tabulated in following table 7.

Table 7. Performance Result on Nine Selected Attributes

Algorithms	Accuracy (%)	Precision (%)	Recall (%)	F1-score (%)	ROC (%)
NB	84.6	84.6	84.6	84.6	90.9
SVM	80.2	80.4	80.2	79.9	78.8
KNN	73.6	73.8	73.6	73.7	73.4
RF	81.3	81.4	81.3	81.1	85.0

Referring to Table 7, it can be seen that the performance of RF is increased with these attributes. RF is at second position now. NB shows the best performance on reduced attribute datasets. SVM also achieved good result, followed by RF. KNN is at last position with very less satisfactory performance.

Performance Result on Six Selected Attributes

Similar to previous procedure for selection of attributes, this time only those attributes have been selected which scores '5 or more than 5 points' in the point table 2. The selected attributes are: cp, restecg, oldpeak, slope, ca and thal. Performance results of the same algorithms with these reduced attributes are tabulated in following table 8.

Referring to Table 8, it can be observed that the performance of NB is increased with only six attributes but performance of RF is decreased with these attributes. Performance of KNN is improved to some extent with six attributes, but SVM shows great fall in the performance with these six attributes comparatively to those nine attributes selected earlier.

Table 8. Performance Result on Six Selected Attributes

Algorithms	Accuracy (%)	Precision (%)	Recall (%)	F1-score (%)	ROC (%)
NB	85.7	85.8	85.7	85.7	90.3
SVM	73.6	73.5	73.6	73.2	72.1
KNN	76.9	76.9	76.9	76.9	77.3
RF	75.8	75.7	75.8	75.6	82.1

Results Comparison

Following figure 1(a) and 1(b) compares accuracy and precision of NB, SVM, KNN and RF on the basis of all the results tabulated in table 3, table 4, table 5, table 6, table 7 and table 8.

Figure 1a. Overall comparison of accuracy

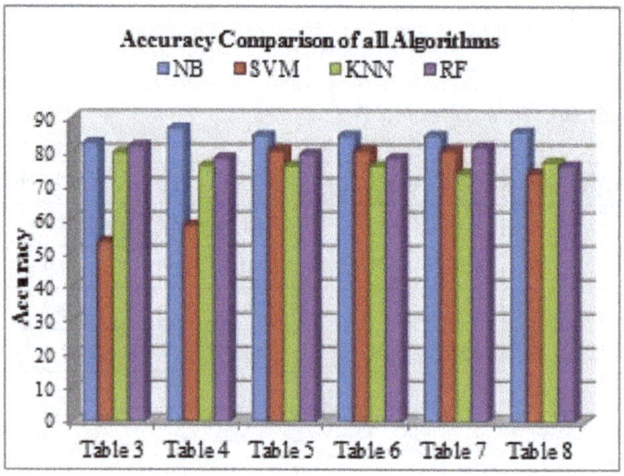

Figure 1b. Overall comparison of precision

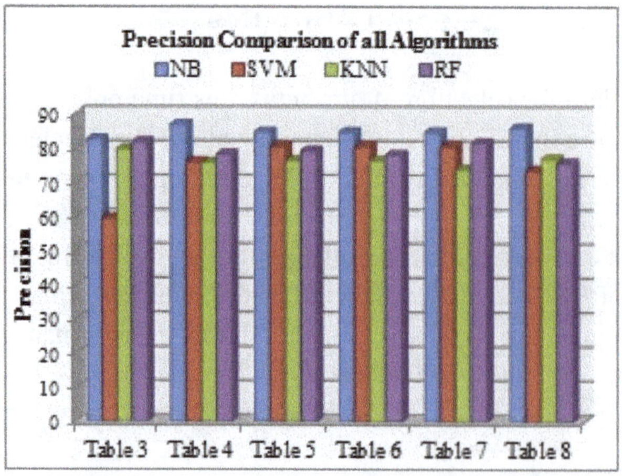

The key observations from figure 1(a) and 1(b) are as follows:

- Accuracy of NB is above 80% in all the tables and outperforms all the algorithms.
- Accuracy of SVM is low in table 3 and table 4.
- Precision of NB is also above 80% in all the tables.
- Precision of SVM is lowest in table 3.

Following figure 2(a), 2(b) and 2(c) compares recall, f1-score and ROC of NB, SVM, KNN and RF on the basis of all the results tabulated in table 3, table 4, table 5, table 6, table 7 and table 8.

Figure 2a. Overall comparison of recall

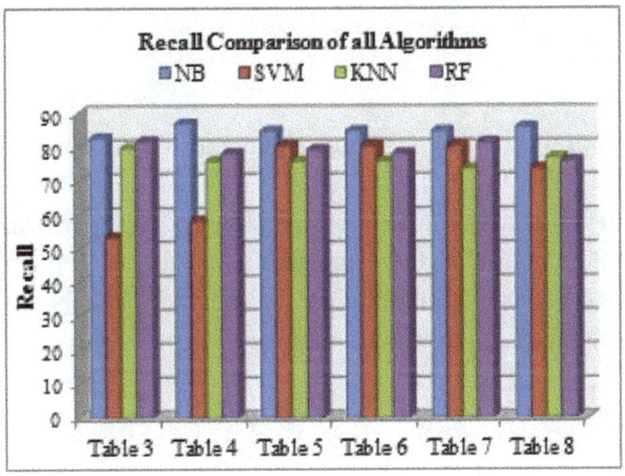

Figure 2b. Overall comparison of f1-score

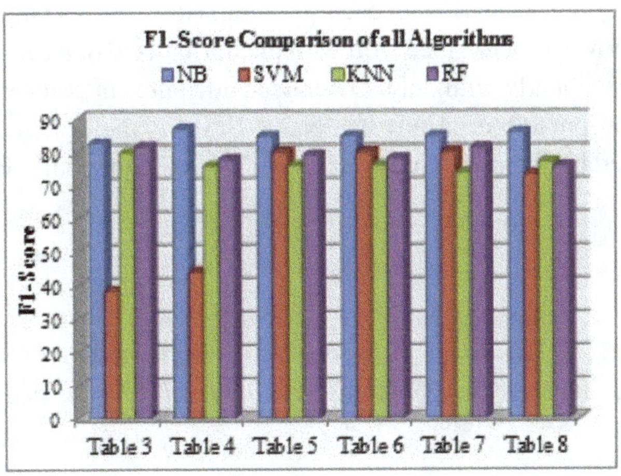

Figure 2c. Overall comparison of ROC

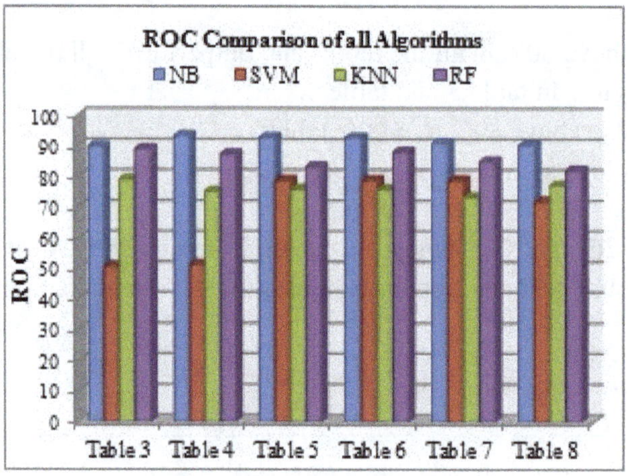

The key observations from figure 2(a), 2(b) and 2(c) are as follows:

- Recall and f1-score of NB is above 80% in all the tables and outperforms all the algorithms.
- ROC of NB is above 90% in all the tables and also outperforms all the algorithms.
- Recall and f1-score of SVM is low in table 3. F1-score of SVM is below 40% in table 3.
- ROC of SVM is almost equal in table 3 and table 4.

CONCLUSION

In this research work, all the results obtained through different methods are tabulated in tables and compared against performance metrics. Each table has explained the accuracy, precision, recall, F1-score and ROC of all the algorithms in different scenarios. In each situation, Naïve Bayes has performed well to predict cardiovascular disease. SVM has performed well with the normalized dataset. After attributes selection, Naïve Bayes performed excellently with only six reduced attributes in comparison to other algorithms.

In all scenarios, this proposed work finds that Naïve Bayes is the best classifier for cardiovascular disease prediction followed by Random Forest, KNN and SVM sequentially for the used dataset.

REFERENCES

Anbarasi, M., Anupriya, E., & Iyengar, N. C. S. N. (2010). Enhanced prediction of heart disease with feature subset selection using genetic algorithm. *International Journal of Engineering Science and Technology*, *2*(10), 5370–5376.

Best First Search. (n.d.). Available at: http://weka.sourceforge.net/doc.dev/weka/attributeSelection/BestFirst.html

Beyene, . (2018). Survey on Prediction and Analysis the Occurrence of Heart Disease Using Data Mining Techniques. *International Journal of Pure and Applied Mathematics.*

Brahmi, B., & Shirvani, M. H. (2015, February). Prediction and Diagnosis of Heart Disease by Data Mining Techniques. *Journals of Multidisciplinary Engineering Science and Technology*, *2*(2), 164–168.

Cardiovascular disease. (n.d.). Available at: https://www.who.int/cardiovascular_diseases/en/

Classifier Subset Evaluation. (n.d.). Available at: http://weka.sourceforge.net/doc.dev/weka/attributeSelection/ClassifierSubsetEval.html

Correlation Based Feature Selection. (n.d.). Available at: https://machinelearningmastery.com/perform-feature-selection-machine-learning-data-weka/

Dey, A., Singh, J., & Singh, N. (2016). Analysis of Supervised Machine Learning Algorithms for Heart Disease Prediction with Reduced Number of Attributes using Principal Component Analysis. *Analysis*, *140*(2), 27–31.

Elshazly, H. I., Elkorany, A. M., & Hassanien, A. E. (2014). Lymph diseases diagnosis approach based on support vector machines with different kernel functions. *Computer Engineering & Systems 9th International Conference (ICCES),* 198–203.

Ghadge, Girme, Kokane, & Deshmukh. (2015). Intelligent Heart Disease Prediction System using Big Data. *International Journal of Recent Research in Mathematics Computer Science and Information Technology, 2*, 73-77.

K-Nearest Neighbour. (n.d.). Available at: https://en.wikipedia.org/wiki/K-nearest_neighbors_algorithm

Linear Forward Selection. (n.d.). Available at: http://weka.sourceforge.net/doc.stable/weka/attributeSelection/LinearForwardSelection.html

Machine Learning RepositoryU. C. I. (n.d.). Available at: https://archive.ics.uci.edu/ml/machine-learning-databases/heart-disease/

Masethe, H. D., & Masethe, M. A. (2014). Prediction of heart disease using classification algorithms. *World Congress on Engineering and Computer Science 2014, 2.*

Max, B. (2013). *Principles of Data Mining* (2nd ed.). Springer.

Monica, S. (2016, February). Analysis of CardioVasular Disease Prediction using Data Mining Techniques. *International Journal of Modern Computer Science*, *4*(1), 55–58.

Mujawar & Devale. (2015). Prediction of Heart Disease using Modified K-means and by using Naïve Bayes. *International Journal of Innovative research in Computer and Communication Engineering, 3*, 10265-10273.

Naïve Bayes Classifier. (n.d.). Available at: https://en.wikipedia.org/wiki/Naive_Bayes_classifier

Patel, Upadhay, & Patel. (2017). *Heart disease Prediction using Machine Learning and Data mining Technique.* UCI Machine Learning Repository. Available at: http://archive.ics.uci.edu/ml/about.html

Patel, Upadhyay, & Patel. (2015). Heart Disease Prediction using Machine Learning and Data Mining Technique. *International Journal of Computer Science and Communication*, 129-137.

Prabhavathi & Chitra. (2016). Analysis and Prediction of Various Heart Diseases using DNFS Techniques. *International Journal of Innovations in Scientific and Engineering Research, 2*(1), 1-7.

Prerana, T. H. M., Shivaprakash, N. C., & Swetha, N. (2015). Prediction of Heart Disease Using Machine Learning Algorithms- Naïve Bayes, Introduction to PAC Algorithm, Comparison of Algorithms and HDPS. *International Journal of Science and Engineering, 3*(2), 90-99.

Priyanka & Kavita. (2016). Feature Selection Using Genetic Algorithm and Classification using Weka for Ovarian Cancer. *International Journal of Computer Science and Information Technologies, 7*(1), 194–196.

Purushottam, S. (2015). Heart Disease Prediction System Evaluation using C4.5 Rules and Partial Tree. *Computational Intelligence in Data Mining, 2*, 285–294.

Shahi, M., & Gurm, R. K. (2017). Heart Disease Prediction System using Data Mining Techniques. *Orient J. Computer Science Technology, 6*, 457–466.

Shalev-Shwartz & Ben-David. (2014). *Understanding Machine Learning*. Cambridge University Press.

Sharmila, R., & Chellammal, S. (2018, May). A conceptual method to enhance the prediction of heart diseases using the data techniques. *International Journal on Computer Science and Engineering*.

Shedole & Deepika. (2016). *Predictive analytics to prevent and control chronic disease*. https://www.researchgate.net/punlication/316530782

Support Vector Machine. (n.d.). Available at: https://docs.opencv.org/2.4.13.7/doc/tutorials/ml/introduction_to_svm/introduction_to_svm.html

Vadicherla, D., & Sonawane, S. (2013). Decision Support System for Heart Disease Based on Sequential Minimal Optimization in Support. *International Journal of Engineering Sciences and Emerging Technologies, 4*(2), 19–26.

Waikato environment for knowledge analysis (weka) version 3.6.9. (2013). Available at: https://download.cnet.com/Weka-32-bit/3000-10254_4-75852610.html

Weng, S. F., Reps, J., Kai, J., Garibaldi, J. M., & Qureshi, N. (2017). Can machine-learning improve cardiovascular risk prediction using routine clinical data? *PLoS One, 12*(4), e0174944.

Zhang, Y. (2012). Studies on application of Support Vector Machine in diagnose of coronary heart disease. In *Electromagnetic Field Problems and Applications 2012 Sixth International Conference (ICEF)*. IEEE.

Chapter 10
IoT-Based Dynamic Traffic Management and Control for Smart City in India

Deepali Kothari
Shri G.S. Institute of Technology and Science, India

Anjana Jain
Shri G.S. Institute of Technology and Science, India

Arun Parakh
Shri G.S. Institute of Technology and Science, India

ABSTRACT

The world is becoming smart as IoT is now an integral part of individuals' routine lives. To control any devices at any place and at anytime from anywhere is now just a matter of access. The goal of this work is to provide simple, efficient, cost-effective, and reliable communication system for traffic management. Keeping in view the aim of smart city, after cleanliness, traffic is the major concern nowadays. A case study is presented through proposed model in this work that will help in improving traffic condition of the city. The available data is analyzed and processed through Raspberry-pi. This data is simultaneously being updated at the web server through cloud. Based on the data available in real time, the system enables controlling traffic system dynamically. This helps in reducing congestion and provides fast going way for heavy vehicular traffic. The system can be clubbed with existing centralized traffic control system in the Indore city to manage traffic conditions in a better way.

INTRODUCTION

India is a democratic republic country having parliamentary form of government (National Portal of India, 2019). The country stood on second position with a population exceeding 1.38 billion residents (India's Population, 2020). Therefore, it has become a global management issue. Indore is the most

DOI: 10.4018/978-1-7998-3327-7.ch010

populous and largest city in west-central India. Indore is the commercial capital of Madhya Pradesh, having a population of around 5.12 million approximately (Indore Population Estimation, 2020). The most crowded city of M.P. is known for its food. Also it is a research hub having IIT, IIM and RRCAT, an education hub, religious hub, medical hub, tourist hub and shopping hub. The city is surrounded by Industry areas of Pithampur and Mhow.

According to the World's largest cleanliness survey, Swachh Survekshan 2020, Indore is declared as the cleanest city for the fourth time (Swachh Survekshan, 2020). After achieving first step towards smart city project, government's next mission is to deal with Traffic. The system proposed here involves the use of IoT with current running systems. This will be easy to implement and handle traffic in better way. IoT is a broader term which inculcates the use of online network to make things easier. The advent in technology enables to build a smart environment for communicating devices that can find applications in almost every field including automation, security, monitoring, surveillance, detection or control. The domain may include industry, educational institutions, agriculture, domestic applications etc. The market of IoT in India is anticipated to increase with Compound annual growth rate of more than 28% in course of 2015-2020 according to "India Internet of Things Market Forecast & Opportunities, 2020" (Internet of Things, 2019). In past recent years, the paradigm of IoT has shown significant growth. Ericsson Mobility Report tells that number of IoT devices is projected to be quadrupled by 2021 (Information About Growth of IoT, 2019). Figure 1 represents the six elements of IoT. Table 1 gives the description of all these elements.

Figure 1. Internet of Things: from connecting devices to human value.

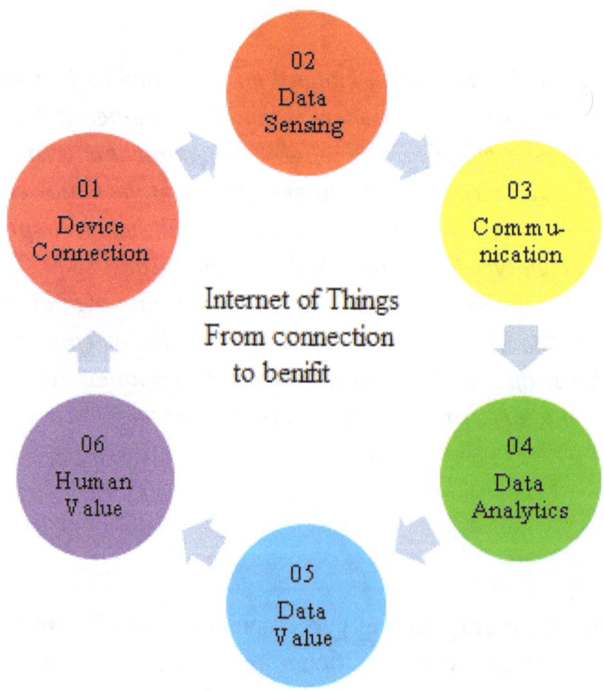

Table 1. Six Elements of IoT

Elements	Description
01 Device connection	Includes IoT devices and their connectivity with Intelligence Embedded in them.
02 Data sensing	Allows reading sensor data & storage.
03 Communication	Enables transportation of data & can be accessed through cloud or network.
04 Data analytics	Provides the cognitive analysis of data, analysis of Big data & AI.
05 Data value	Analyzes actions, APIs & processes data, provides actionable Intelligence.
06 Human value	Provides tangible benefits to stakeholder, Smart Applications.

With the rapid growth in population in recent years, the problem of traffic is even more severe here. Mobility is inversely proportional to population growth. This even leads to congestion, theft and accidents. Air pollution and noise pollution is also increasing which adversely affects health of humans. Smoke from the vehicles emits harmful gases that create breathing problems. Therefore, to solve these issues of day-to-day life, a possible solution could be if we combine existing centralized traffic light control system with a module that controls traffic system dynamically through IoT. The rest of the paper is described in four sections. Section II discusses the related work in the concerned domain. Section III explains about present scenario and challenges. The proposed system model is presented in Section IV. Section V concludes the research and Section VI mentions the references.

Background

A valuable amount of research has been carried out on IoT in concerned domain:

1. IoT Platforms and Technologies

The work presented by Nikola Petrov et al. (Petrov et al., 2016), gives the overview of available platforms for IoT technologies. It explains the applicability of Raspberry Pi for IoT environments. The system is low cost, efficient, flexible system. A survey on Internet of Things by Luigi Atzori et al. (Atzori et al., 2010) explores IoT in different domains. A Comparative survey of RFID systems, WSN and RFID sensor networks is shown. The current aspect, issues which are addressed and which needs to be further accessed with regard to IoT is explained. A case study that describes protocols used and realization of architectures in Padova, Italy is depicted by A. P. Castellani et al. (Castellani et al., 2010). This case study has been validated through the guidelines of SENSEI. WSN testbed has been deployed in Information Engineering Department of Padova University, Italy. IPv6 is used to reach nodes and realizes using IoT technology.

The work done by Nils Walravens et al.(Walravens & Ballon, 2013) shows an approach for making business models. It develops mobile apps for different cities for expanding existing mobile network in business models. Pooja Yadav et al. (Yadav et al., 2018) show the growth of IoT in India. The paper reviews challenges and risk already existing in India; possible solutions to be done by government. The work presented by Maxim Chernyshev et al. (Chernyshev et al., 2018), gives the idea of the increasing use of IoT and depicts a comparison with existing technologies like WSN, networking and cloud computing. The work done by A. P. Castellani et al. (Castellani et al., 2012) formulates a Google based web toolkit, WebIoT to enhance communication between human and things and amongst things. The design has been validated through health care applications and web resources communication.

2. IoT Based Detection and Monitoring Systems

The research done by Umera Anjum (Umera, 2017), uses image processing on live video to detect theft. A camera is used to capture video, a LCD display, IR for night vision and USB for storage. The captured image is transmitted over IoT and can be viewed by user online. This paper presented by Dr. B. Sivasankari et al. (Sivasankari, 2017), shows a system which monitors the air quality using IoT and Raspberry Pi. The system uses various types of gas sensors. Graph is also plotted periodically using thing speak platform. The paper presented by Nazeem Basha (Nazeem Basha et al., 2016), presents an Intrusion Detection system made using Raspberry pi. Door motion is sensed by sensor, this data is sent to AWS IOT console through Raspberry pi. This invokes AWN SNS module to send notification (push mail) to the owner. Data is updated in Google spread sheet. The research done by Hyeongrae Kim et al. (Kim et al., 2018), proposes a fault detection model. Entire system is configured with multiple IoT's. Look-up table method proves to be an effective method to monitor the nodes of IoT that are freezed. Bluetooth and Wi-Fi technologies are also used. Sangwon Lee developed an intelligent parking system (Lee et al., 2008) using wireless sensor networks. Ultrasonic sensors with magneto motors are used for accurate detection of vehicle in parking lots.

3. IoT with Raspberry Pi and Wireless Sensor Networks

This article presented by Cheah Wai Zhao et al. (Zhao et al., 2015) explores use of Raspberry pi as a server. Three cases are studied: connection over wi-fi for smart applications, communication over zigbee and Pi as server by using 'Samba', open source software which provides services like: file & print services, authentication, authorization, name resolution and service announcement. The paper presented by Dhiraj Sunehra & B. Harish Kumar (Harish Kumar, 2017), explains communication and transfer between two nodes; pi and Arduino. The water pump is turned on; and correspondingly values are up-dated on the webpage through Pi web server. The work proposed by Fatma Salih et al. (Salih & Omer, 2018), depicts the video server made with the help of Raspberry pi. Stream of video or audio is created in real-time through LAN using pi. This data is processed through HTTP and sent with RTP protocol.

Omid Avatefipour et al. (Avatefipour & Sadry, 2018) compare four different methods for managing traffic system namely Traffic Light System, RFID, static & dynamic traffic system, traffic system using IoT are compared for traffic management. The implementation of smart monitoring system by Nicola Bressan et al. (Bressan et al., 2010) deployed using WSN and actuator network. A binary web service is providing portable user friendly app development. The work presented by D. Wang et al. (Wang et al., 2018), explains a method to handle Big data intelligently using 5G Intelligent Internet of Things (5G I-IOT). It consists of three parts; cloud for intelligent computation, a processing center to generate results and link between the two i.e. object processor to process data and sends data to previous block and as an acknowledgement and takes feedback to make decisions. Paolo Casari et al. (Casari et al., 2009) give an overview of WISE-WAI project in Padova, Italy. A large testbed is deployed for large scale wireless sensor networks. It incorporates environmental sensors such as temperature, humidity, light intensity, a transceiver along with a microcontroller all deployed in building of Information Engineering Department in University of Padova, Italy.

4. **Smart city concept with IoT:**

The research done by Mahesh Lakshminarasimhan (Lakshminarasimhan, 2016) describes a traffic control system in big cities which uses RFID and analyzes big data through IoT. The system would help in reducing congestion in traffic and also provides a mechanism for security of vehicles. The research work of Andrea Zanella et al. (Zanella et al., 2014) shows how urban IOT contributes to smart city vision with value added services for administration and for citizens. Smart city project of Padova city, Italy is depicted. With the conjunction of city municipality, a successful deployment of an IoT island got possible in Padova. The work presented by C. E. A. Mulligan et al. (Mulligan & Olsson, 2013) proposes a business model for smart cities. Evolution of the architecture for smart city models is proposed. Existing system architectures and business models are depicted. It also depicts the role of human being for making a city smart.

Mischa Dohler et al. (Dohler et al., 2011) depict an action plan for smart cities. The paper suggests creating smart city departments to facilitate smart city development. The plan is divided into three phases. First phase includes revenue generation and utilization. Supported services are described in phase two; whereas in third phase deals with permissible leisure and fun oriented activities. The smart traffic management system is shown by Sabeen Javaid et al. (Javaid et al., 2018). It also provides a way for fire brigade and ambulances at the time of emergency. It dynamically controls the traffic by putting green light ON for particular direction according to density of traffic.

A dynamic road traffic management using IoT is presented by Syed Misbahuddin et al. (Misbahuddin et al., 2015). This model is developed to manage traffic in city Makkah, Saudi Arabia. Onsite traffic police officer maintains traffic to allow pilgrims to reach the mosque. RPi is used for dynamic control of traffic through GPIO pins. The research done by Ch V A Satwik et al. (Satwik et al., 2018) shows a system for daily transit of vehicles. In place of traditional road divider, smart pole system is developed which reduces traffic congestion and uses road efficiently. It shows implementation of smart divider system in manual and automatic system as well. Mauro De Sanctis et al. (De Sanctis et al., 2016) presents the communication through satellite for Internet of Remote Things. Service management, IPv6 support, communication between sensors or actuator and satellite are discussed. Scenarios where satellite plays an importance are also described.

PRESENT CHALLENGES

Indore is a city of fixed infrastructure. Bridges, old structures, buildings constitute major portion of city. Indore is known for business center for food, jewellery and cloth. Areas become so clumsy which leads to traffic congestion. So, there is less scope of changing model of city. Therefore, the only option left is to deal with available space and manage traffic congestion. Challenges includes supervision of traffic, parking system in building, shopping malls, traffic light system at squares etc. needs to be managed. Physical constraints always pose challenges in the system. Many over bridges are being built in the city and some are under construction as well. To manage traffic in such situation, the work is bit tedious. The number of road accidents has seriously been increased in recent years. There are minimum twelve cameras implanted at majority squares in Indore. Violation of traffic rules leads to generate E-challan for the vehicle (Smart Mobility Solution in Indore, 2019).

Presently in Indore city, the traffic light controlling systems are microprocessor based which are usually pre-programmed. Depending on the density of traffic, the traffic light controlling patterns are modified. In present scenario, centralized traffic light system is used to control the traffic; where red, yellow and green signal light is switched ON for fixed time duration. This time depends on density of traffic at different times. For example, during rush hours in evening from 5:00pm-7:00pm IST, the green signal light is probably ON for less duration say 15-20 seconds, which may be increased in afternoon due to light traffic and during moderate traffic hours in early morning. But since this condition of traffic is not fixed all the time therefore, a system has to be designed which automatically senses vehicular traffic density and accordingly manages the traffic. The system can also provide a way for emergency vehicle. A system which gives result on its own and reduces human effort efficiently is everyone's demand today.

Proposed IoT Based Traffic Management Model

Any system works on certain input, processes it and gives desired result. The system works more effectively with the use of new and upcoming technologies even without human intervention. Figure 2 shows the proposed model for dynamic control of traffic light system.

Figure 2. Proposed system for traffic management.

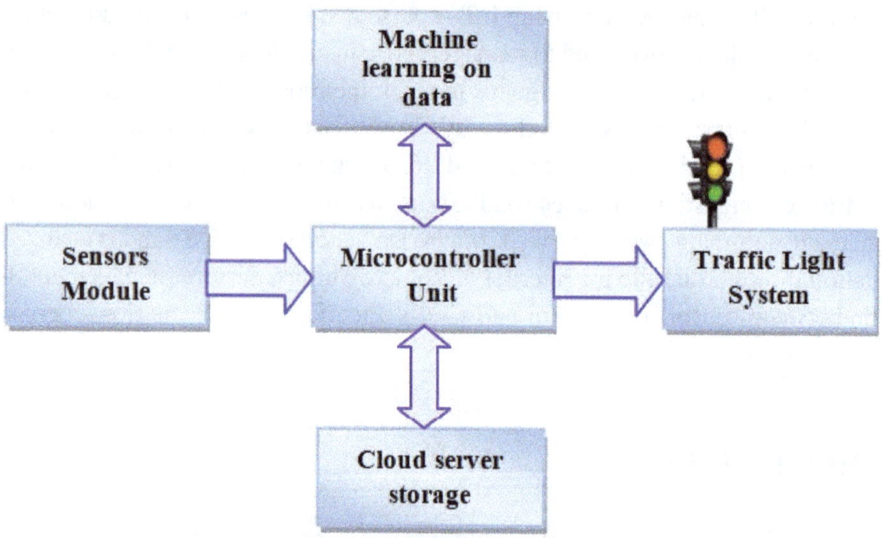

Different sections are explained as follows:

1. **Input Section**

Input is taken through some sensors. These sensors are implemented in each direction will sense the number of vehicles in an on-going real-time traffic. Four sets of IR sensor are implemented at the square; one set for each direction. Sensors are used to sense the density of traffic at square. This helps in counting the number of vehicles at each side. This data is sent to microcontroller unit for further processing. If any vehicle crosses the stop line while the red signal is ON then, it can also find the person who is violating the traffic rules.

2. **Microcontroller Unit**

The data available from the input section i.e. through sensors is given to microcontroller unit for processing. This processing may involves the use a single board computer such as Raspberry pi or the type with additionally interfaced modules to handle the incoming data. Here pi, an on-board computer is used. Data is made available continuously in real-time which need to be processed. The unit takes the data, processes it and sends it for further storage and necessary actions.

3. **Cloud Server Storage**

The input processed data is made available for storage at the cloud. This data is continuously stored on the web. If data is stored on the web it can be accessed anytime from anywhere.

4. **Machine Learning on Data**

Large amount of data is made available through sensors. This helps in creating database for handling the traffic in better way. Once training of this data set is done, supervised learning algorithms on this dataset will be applied for further analysis. Based on the results of learning methods, necessary action is performed. Due to which density of traffic is controlled automatically.

5. **Traffic Light System**

Based on analysis of traffic data, database and signals from sensors, it can be easily predicted which traffic route is to be opened first. So, if there is no traffic to move towards north say, then the model should show red light indication i.e. stop, continuously (till vehicle arrives) for the vehicles coming towards north direction. It should put green signal light ON only when there is a vehicle waiting to go on to other side.

A CASE STUDY

To assure feasibility of any model or project, it has to be tried and tested first. For this purpose, it has to be implemented and tested in real time. A demo model of the system is developed for the traffic light control system. The set of two IR sensors is implanted at each side of the square. One of the sensors will sense the vehicles and increase the counter. The other one will check fraudulent if happens like if signal is not green and vehicle crosses the stop line or zebra crossing line by breaking traffic rules. This way it also checks for person who breaks traffic rules. A threshold value say ten is set for indicating low, medium and high density of traffic. Accordingly, data is checked continuously and periodically updated in database in real-time. The data sensed by sensors helps to create database and it is also displayed on the webpage.

Existing technologies implements controlling mechanism for traffic light with fixed duration timers generated by microcontrollers. Management is done by RFID or some wireless technologies but it was human dependent. The proposed system can be combined with the existing centralized control system for traffic. The scope of this system is not just to reduce the traffic but also to reduce human effort with cost-effective solutions. It is being observed that there is huge traffic during working hours of office, college, industries etc. i.e. morning and evening hours; while during day time traffic is minimal. So, during peak hours traffic can be controlled dynamically. By the term dynamic, it is meant that, the side which is having high density of traffic should be allowed to go first compared to the side having only two to three vehicles. To meet any emergency, signal of that side should be opened at priority basis. This model also saves time of an individual where they keep waiting for timer to be ON for their turn in periodical fashion. Since, the data is maintained at cloud so can be accessed at any instant of time from anywhere. Dynamic management of traffic according to density of vehicles while working on real-time data proves a landmark in making a city smart.

Figure 3. Sensor interface with microcontroller unit.

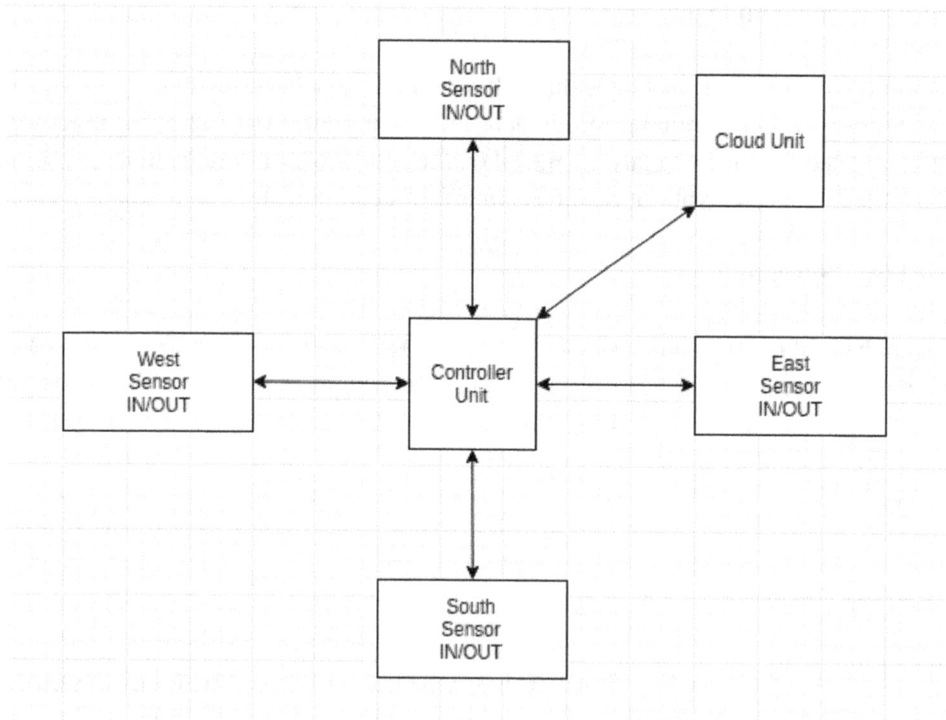

RESULTS

Each sensor is a set of two sensors which works together to count the vehicle.

Example 1: South Sensor Unit

The sensor at the South side unit consists of two IR sensors S1 and S2.

- S1 Counts for Incoming vehicle and S2 make sure that the Incoming vehicle is counted without any error and fail.
- Initially, the counter for the vehicle is assumed as zero, when sensor S1 gets High it is incremented.
- S1 is high and S2 is low means the vehicle is detected and in a steady position which makes sure our count is valid.

Once all the sensors have posted the data to the control unit, it gets compared to the previous data and density can be measured.

Figure 4. Snapshot of south direction sensor data at real-time

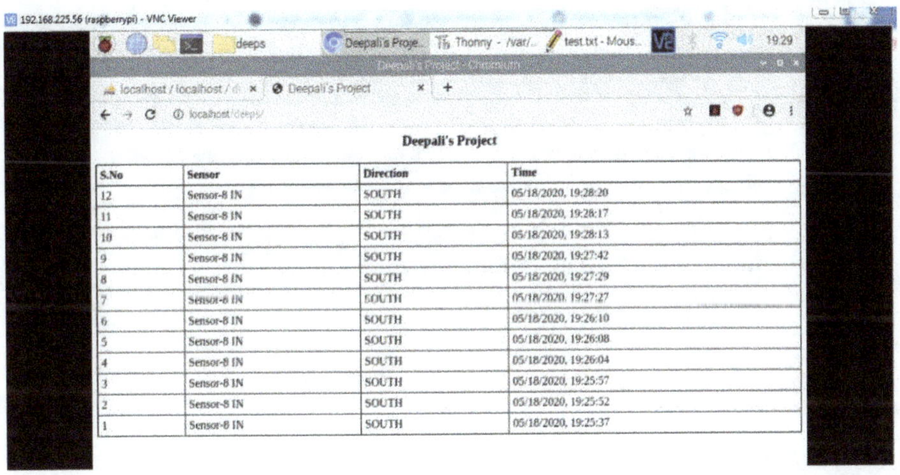

Deepali's Project

S.No	Sensor	Direction	Time
12	Sensor-8 IN	SOUTH	05/18/2020, 19:28:20
11	Sensor-8 IN	SOUTH	05/18/2020, 19:28:17
10	Sensor-8 IN	SOUTH	05/18/2020, 19:28:13
9	Sensor-8 IN	SOUTH	05/18/2020, 19:27:42
8	Sensor-8 IN	SOUTH	05/18/2020, 19:27:29
7	Sensor-8 IN	SOUTH	05/18/2020, 19:27:27
6	Sensor-8 IN	SOUTH	05/18/2020, 19:26:10
5	Sensor-8 IN	SOUTH	05/18/2020, 19:26:08
4	Sensor-8 IN	SOUTH	05/18/2020, 19:26:04
3	Sensor-8 IN	SOUTH	05/18/2020, 19:25:57
2	Sensor-8 IN	SOUTH	05/18/2020, 19:25:52
1	Sensor-8 IN	SOUTH	05/18/2020, 19:25:37

Density calculation:

Example 2: South Sensor Unit

Suppose South signal has a density of 50 vehicles at once and system have a count of 40 vehicles. South signal can occupy more vehicles based on this data hence signal timer can be increased.

CONCLUSION AND FUTURE SCOPE

With the advent of IoT paradigm in past recent years, human intervention has been greatly reduced. The paper reviews different domains where IoT was implemented. Keeping in view the concept of Smart city, traffic was the most suitable target to be achieved. Research tells that work is going on worldwide for handling traffic. The paper discussed present challenges and possible solution to control the problem of traffic for making a city smart in India. Current centralized system is not fully able to manage traffic in better ways. Hence, for the sake of need, a model for dynamic control of traffic is presented. This will help to reduce traffic congestion mainly at densely populated squares. Raspberry Pi along with IR sensors is used in the proposed model. Data is also maintained at cloud which provides easy access at any time.

Figure 5. Snapshot of screen while code is running and showing status of sensor with increase in vehicle count in real time

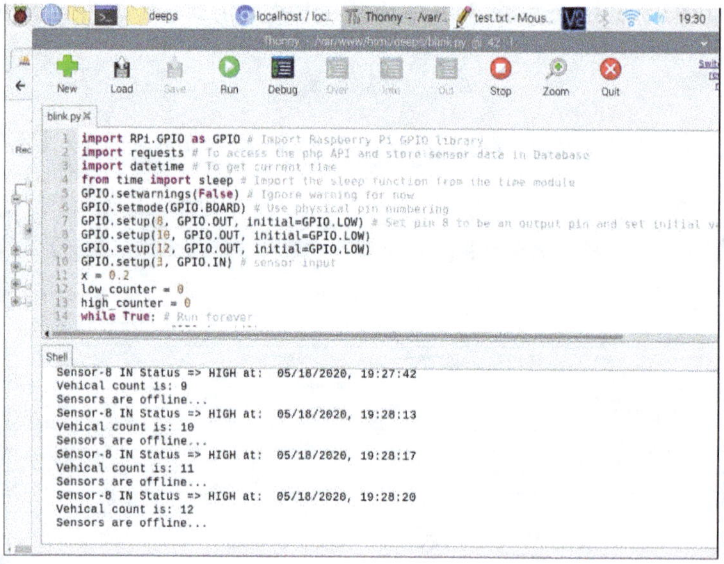

Figure 6. Snapshot of php page

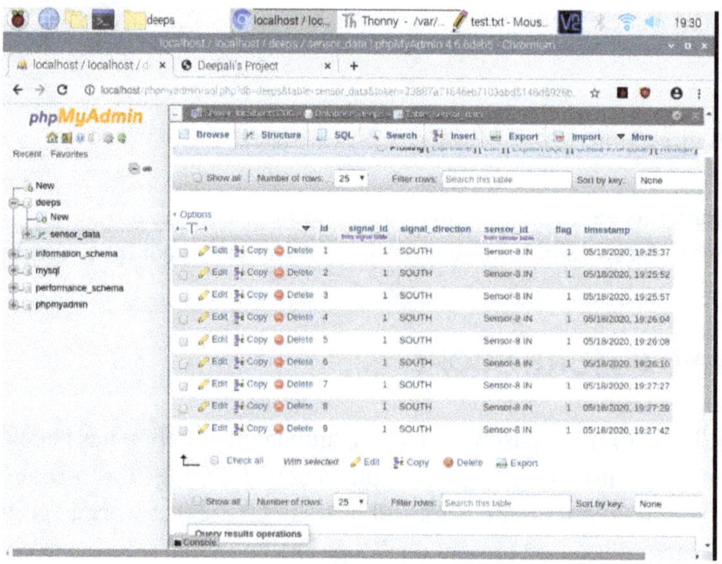

Dynamic traffic management is one way to manage traffic conditions in the city. Furthermore, green way for the vehicles throughout the road can be done wherever possible. If a vehicle will find green signal at every square then it can move smoothly and continuously. This will reduce the traffic greatly at squares. More efficient circuit and sensors can be implemented at squares to read real-time data therefore, making system more efficient. Machine learning and IoT together can help to achieve milestone for traffic conditions of the city. After successful implementation of this model, it can be further implanted at other most populous part of the city to reach the goal of smart city in India. The goal is to make Traffic Smart of Smart city and hence, serve the government to achieve its mission.

REFERENCES

Atzori, A. L., Iera, A., & Morabito, G. (2010). The internet of things: A survey. *Computer Networks*, *54*(15), 2787–2805. doi:10.1016/j.comnet.2010.05.010

Avatefipour, O., & Sadry, F. (2018). Traffic Management System Using IoT Technology - A Comparative Review. *IEEE International Conference on Electro/Information Technology (EIT)*, 1041-1047. 10.1109/EIT.2018.8500246

Bressan, N., Bazzaco, L., Bui, N., Casari, P., Vangelista, L., & Zorzi, M. (2010). The deployment of a smart monitoring system using wireless sensor and actuator networks. *Proc. IEEE Smart Grid. Comm.*, 49–54.

Casari, P., Castellani, A., Cenedese, A., Lora, C., Rossi, M., Schenato, L., & Zorzi, M. (2009, June). The WIreless SEnsor networks for city-Wide Ambient Intelligence (WISE-WAI) project. *MDPI J. Sensors*, *9*(6), 4056–4082. doi:10.339090604056 PMID:22408513

Castellani, A. P., Bui, N., Casari, P., Rossi, M., Shelby, Z., & Zorzi, M. (2010). Architecture and protocols for the Internet of Things: A case study. *Proc. 8th IEEE Int. Conf. Pervasive comput. Commun. WorkshopS (PERCOM Workshops)*, 678–683.

Castellani, A. P., Dissegna, M., Bui, N., & Zorzi, M. (2012). WebIoT: A web application framework for the internet of things. *Proc. IEEE Wireless Commun. Netw. Conf. Workshops*. 10.1109/WCNCW.2012.6215491

Chernyshev, M., Baig, Z., Bello, O., & Zeadally, S. (2018, June). Internet of Things (IoT): Research, Simulators, and Testbeds. *IEEE Internet of Things Journal*, *5*(3), 1637–1647. doi:10.1109/JIOT.2017.2786639

De Sanctis, M., Cianca, E., Araniti, G., Bisio, I., & Prasad, R. (2016, February). Satellite Communications Supporting Internet of Remote Things. *IEEE Internet of Things Journal*, *3*(1), 113–123. doi:10.1109/JIOT.2015.2487046

Dohler, M., Vilajosana, I., Vilajosana, X., & Llosa, J. (2011). Smart Cities: An action plan. *Proc. Barcelona Smart Cities Congress*, 1–6.

Harish Kumar, B. (2017). WSN based Automatic Irrigation and Security System using Raspberry Pi Board. *International Conference on Current Trends in Computer, Electrical, Electronics and Communication (CTCEEC)*, 1097-1103. 10.1109/CTCEEC.2017.8455140

India's Population. (2020). http://worldpopulationreview.com/countries/india-population/

Indore Population Estimation. (2020). https://indiapopulation2020.in/population-of-indore-2020.html

Information About Growth of IoT. (2019). https://www.thalesgroup.com/en/markets/digital-identity-and-security/iot/magazine/infographic-growth-iot

Internet of Things, Market Forecast. (2019). https://www.researchandmarkets.com/reports/3395405/india-internet-of-things-iot-market-forecast

Javaid, S., Sufian, A., Pervaiz, S., & Tanveer, M. (2018). Smart traffic management system using Internet of Things. *20th International Conference on Advanced Communication Technology (ICACT)*, 393-398.

Kim, H., Lee, D., Cho, J., & Park, D. (2018). Software execution freeze-safe microcontroller using power profile tracking for IoT-driven connected services. *IEEE 4th World Forum on Internet of Things*, 237-240.

Lakshminarasimhan, M. (2016). *IoT Based Traffic Management System*. Academic Press.

Lee, S., Yoon, D., & Ghosh, A. (2008). Intelligent parking lot application using wireless sensor networks. *Proc. Int. Symp. Collab. Technol. Syst.* 10.1109/CTS.2008.4543911

Misbahuddin, S., Zubairi, J. A., Saggaf, A., & Basuni, J. (2015). IoT based dynamic road traffic management for smart cities. *12th International Conference on High-capacity Optical Networks and Enabling/Emerging Technologies (HONET),* 1-5. 10.1109/HONET.2015.7395434

Moreno, M. V., Terroso-Saenz, F., Gonzalez-Vidal, A., Valdes-Vela, M., Skarmeta, A. F., Zamora, M. A., & Chang, V. (2017, April). Applicability of Big Data Techniques to Smart Cities Deployments. *IEEE Transactions on Industrial Informatics*, *13*(2), 800–809. doi:10.1109/TII.2016.2605581

Mulligan, C. E. A., & Olsson, M. (2013, June). Architectural implications of smart city business models: An evolutionary perspective. *IEEE Communications Magazine*, *51*(6), 80–85. doi:10.1109/MCOM.2013.6525599

National Portal of India. (2019). https://www.india.gov.in/my-government

Nazeem Basha, Jilani, & Arun. (2016). An Intelligent Door System using Raspberry Pi and Amazon Web Services IoT. *International Journal of Engineering Trends and Technology, 33*, 84-89.). doi:10.14445/22315381/IJETT-V33P217

Petrov, N., Dobrilovic, D., Kavalić, M., & Stanisavljev, S. (2016). *Examples of Raspberry Pi usage in Internet of Things.*). doi:10.20544/AIIT2016.15

Rath, M. (2018). Smart Traffic Management System for Traffic Control using Automated Mechanical and Electronic Devices. *IOP Conference Series. Materials Science and Engineering*, *377*, 012201. doi:10.1088/1757-899X/377/1/012201

Salih & Omer. (2018). *Raspberry pi as a Video Server*. .8515817 doi:10.1109/ICCCEEE.2018

Satwik, C. V. A., Kumar, L. P., Vineeth, K., & Pillai, K. N. (2018). Intelligent Road Management System for Daily Transit. International Conference on Communication and Signal Processing (ICCSP), 523-526.

Sivasankari, B. (2017). IOT based Indoor Air Pollution Monitoring using Raspberry PI. *International Journal of Innovations in Engineering and Technology*, *9*(2), 16–2.

Smart Mobility Solution in Indore. (2019). http://urbanmobilityindia.in/Upload/Conference/16d93b11-5067-48da-b854-1bff507a99c2.pdf

Swachh Survekshan. (2020). http://www.swachhsurvekshan2020.org/

Umera, A. (2017). *IOT Based Theft Detection using Raspberry Pi*. Academic Press.

Walravens, N., & Ballon, P. (2013, June). Platform business models for smart cities: From control and value to governance and public value. *IEEE Communications Magazine, 51*(6), 72–79. doi:10.1109/MCOM.2013.6525598

Wang, D., Chen, D., Song, B., Guizani, N., Yu, X., & Du, X. (2018, October). From IoT to 5G I-IoT: The Next Generation IoT-Based Intelligent Algorithms and 5G Technologies. *IEEE Communications Magazine, 56*(10), 114–120. doi:10.1109/MCOM.2018.1701310

Yadav, E. P., Mittal, E. A., & Yadav, D. H. (2018). IoT: Challenges and Issues in Indian Perspective. *2018 3rd International Conference on Internet Of Things: Smart Innovation And Usages (IoT-SIU)*, 1-5. 10.1109/IoT-SIU.2018.8519869

Zanella, A., Bui, N., Castellani, A., Vangelista, L., & Zorzi, M. (2014, February). Internet of Things for Smart Cities. *IEEE Internet of Things Journal, 1*(1), 22–32. doi:10.1109/JIOT.2014.2306328

Zhao, C.W., Jegatheesan, J., & Loon, S.C. (2015). *Exploring IOT Application Using Raspberry Pi*. Academic Press.

Chapter 11
Machine Learning Based Intrusion Detection System:
A Survey

Ashish Pandey
Kamla Nehru Institute of Technology, Sultanpur, India

Neelendra Badal
Kamla Nehru Institute of Technology, Sultanpur, India

ABSTRACT

Security is one of the fundamental issues for both computer systems and computer networks. Intrusion detection system (IDS) is a crucial tool in the field of network security. There are a lot of scopes for research in this pervasive field. Intrusion detection systems are designed to uncover both known and unknown attacks. There are many methods used in intrusion detection system to guard computers and networks from attacks. These attacks can be active or passive, network based or host based, or any combination of it. Current research uses machine learning techniques to make intrusion detection systems more effective against any kind of attack. This survey examines designing methodology of intrusion detection system and its classification types. It also reviews the trend of machine learning techniques used from past decade. Related studies comprise performance of various classifiers on KDDCUP99 and NSL-KDD dataset.

INTRODUCTION

The growth of technology has made the world more dependent on communication devices and sensors. Moreover, sensor data is increasing at the rate of around 13%. By the 2020, it is expected to reach 35% of whole data communication. The Internet traffic also increases 30 GB on average and it was near about 10 GB in 2016 (Cisco, 2017). This increase in devices, collected data and associated services are becoming the catalyst to accelerate the attack rate in the network. Therefore, it becomes mandatory to design robust tools which can defend networks against vulnerability and security threats. In this way

DOI: 10.4018/978-1-7998-3327-7.ch011

importance of network security is gaining attention. Latest tools are still based on datasets which try to make the Intrusion Detection System (IDS) powerful against new attacks.

Intrusion Detection Systems are defined as a system that monitors network traffic in order to analyze that any security violations have taken place. Result of monitoring is detection of anomalies and intrusions. Most of the present-day IDS classifications are based on only one feature, like: the machine learning algorithm that can be probably used by researchers (Hamed et al., 2018) (Hodo et al., 2017), the distinctive attributes of IDS (Debar et al., 1999) (Amer & Hamilton, 2010) or the designing features of IDS which can be used by researchers during design (Varma et al., 2018). This survey presents the intrusion detection system in three folds:

- A survey of IDS Design Classification
- A survey of IDS Implementation
- A chronological survey of IDS.

In the next section, survey of IDS design classification is presented. Subsequently, a survey of IDS implementation is described. Thereafter, a chronological survey on research works that have been performed in the field of IDS is presented. Last section presents the conclusion of this survey.

A SURVEY OF IDS DESIGN CLASSIFICATION

A detailed classification is presented in (Hindy et al., 2018) which is devoted to the design of IDS and also describes the various features that IDS possessed. The general attributes that characterize IDS are: Role of IDS, Output Information provided by IDS, Requirements and Usage. When an intrusion occurs, the IDS logs the information related to the intrusion. Basically, this log results are used by network experts for further analysis to prevent the breach and for the training process of IDS itself. At the same time when it logs the information of intrusion; it should also trigger alerts of threat detection for the particular system affected. These alerts help the authorized user to perform corrective actions and mitigate the threat. Effective IDS must include mitigation features to take corrective actions against threats (Butun et al., 2014).

The output information is very demanding for analysis which is given by intrusion detection system to the end user. The information collected by IDS must comprise intruder identification, their location, time of intrusion or timestamp, intrusion layer (i.e. OSI layer), intrusion activity (i.e. type of attack) and lastly intrusion type (Butun et al., 2014). For the identification of intruder, user credentials and internet protocol addresses are used.

An IDS is considered effective, if its detection rate is high and false alarm rate is low which are key aspect during design. Other important factors include low resource consumption, throughput, transparency and safety of the overall system. An evaluation method must be applied before the deployment of IDS; therefore, it cannot perform unusual behavior when becomes functional. Evaluation methods can be fixing of bugs in the IDS and detecting anomalies by fuzzing. The intruders can exploit these anomalies to render the IDS worthless or start a DoS attack (Butun et al., 2014).

A SURVEY OF IDS IMPLEMENTATION

Intrusion Detection System can be scattered over numerous nodes in the network. In this scenario, decision is either independent or collaborative for intrusion detection. Multiple nodes share a single decision in collaborative mode by using statistical technique. But individual node takes all decision on the network in independent mode.

Furthermore, in this scattered manner, a flat infrastructure is that where each node has same working capacity. But, when all the nodes correspond to clusters having contrasting capabilities and also contribute in decisions differently, it is treated as clustered infrastructure. The location for the computation is another aspect of distributed intrusion detection system. The computation location which is centralized focuses to work on data collected from the whole network. But disregarding decisions from other nodes, the stand-alone computation location focuses to work on local data. The combination of both centralized and stand-alone is Cooperative computation. In this mode each node detects an intrusion for its own and also takes part in overall decision. Lastly, intrusion detection system can also be operated in hierarchal computational mode. In this mode a cluster delivers all the information related to intrusion detection to center node for decision making (Butun et al., 2014).

The gross accuracy of the system depends on the location of Intrusion Detection System because IDS location on the network can extremely affect the detection process of the threat or attack. To respond the threats in real time, the placement of IDS should be on a host computer i.e. host-based. It can also be inline. But if inline IDS is placed on a busy network; its detection rate often degrades. To implement a hybrid system, hosts can be used as sensors for swarm intelligence. It can be scattered both on the host and through the network.

The detection method is also a crucial feature of all IDS. Signature-based IDS are based on database of previous attacks or threat detection to create correct signatures. For the known attacks, this method delivers high accuracy rate. However, new and diversified threats cannot be detected by this method (Bellekens et al., 2014). Misuse Detection is another name of Signature-based. Anomaly-based IDS focuses on pattern identification and comparing these patterns to usual traffic patterns. In this type of IDS, the system is trained first before the deployment. Anomaly-based IDS delivers better accuracy rate for new and diversified threats in comparison to signature-based IDS. Yet, it also produces huge false alarm rate.

A CHRONOLOGICAL SURVEY OF IDS

By using published available datasets, numerous IDS models have been proposed and evaluated in the last ten years. This section presents the efforts given by different researchers in the area of intrusion detection in past decade. All the research works which are presented here have used KDDCUP99 and NSL-KDD datasets. Table 1 presents year-wise published work by different researchers for IDS. Moreover, this table also describes progression of the algorithms used in the last decade. The shift in the use of particular algorithm can be seen by this table which is as follows:

Table 1. IDS research works from past decade

Year 2008	
Giacinto et al. (2008) "Intrusion Detection in Computer Networks by a Modular Ensemble of one-class Classifiers"	
Used Algorithm:	Parzen Classifier, v-SVC, K-means
To provide the solution for Unsupervised Anomaly Detection problem, authors have proposed a modular solution by one-class classifier for designing of each module. They proposed a mapping technique to map the results of one-class classifier into density function. For given threshold false alarm rate, they computed the false alarm rate for each module using heuristic design. KDDCUP99 dataset was used to perform all the experiments. Experimental results have concluded that v-SVC algorithm outperforms Parzen classifier and K-means on average.	
Hu et al. (2008) "AdaBoost-based Algorithm for Network Intrusion Detection"	
Used Algorithm:	AdaBoost
Authors have proposed AdaBoost-based algorithm to detect the intrusion. Decision stumps have been used as weak classifiers in the algorithm. All the experiments were performed on KDDCUP99 dataset. Decision rules have been designed for both continuous and categorical features of dataset. The experimental results have shown that the detection rate and false alarm rate of proposed algorithm was best among Genetic Clustering, Hierarchical SOM, SVM, Bagged C5 and RSS-DSS. The running speed of proposed algorithm was also faster in learning stage than another algorithm.	
Year 2009	
Shafi and Abbass (2009) "An Adaptive Genetic-based Signature Learning System for Intrusion Detection"	
Used Algorithm:	Genetic-based
Authors have presented UCS (supervised learning classifier system) with real time Signature Extraction system framework which was named as UCSSE. They have used merge and contact generalization operators to modify rule boundaries which was helpful in minimizing the overlap and conflict among signatures. All experiments were performed on KDD Cup dataset and results of performance metrics showed that extended UCS performed better than baseline UCS.	
Wu and Yen (2009) "Data mining-based Intrusion Detectors"	
Used Algorithm:	C4.5
Authors have measured the accuracy, FAR and DR of different attacks by using KDD cup 99 dataset. They have used C4.5 and SVM as a classifier for comparison of results. The results calculated after experiments have shown that C4.5 was better than Support Vector Machine in detection rate and accuracy of DoS, User to Root and Probe attacks. In the case of FAR, SVM performance was better than C4.5.	
Year 2010	
Mok et al. (2010) "Random Effects Logistic Regression Model for Anomaly Detection"	
Used Algorithm:	Logistic Regression
Authors have proposed a method to forecast anomaly detection by using random effects logistic regression. They applied this method to accommodate risk elements of the disclosures and changeableness which was not described by such elements. The proposed method included the particular elements of the risk class like employed 'protocol type' and 'logged in'. Classification accuracy of proposed model was 98.94% and 98.68% for the training and validation data set respectively.	
Jawhar and Mehrotra (2010) "Design Network Intrusion Detection System using hybrid Fuzzy-Neural Network"	
Used Algorithm:	NN, FCM Clustering
Authors have proposed a classification model using FCM and Neural Network. It was designed in two stages in which FCM clustering was first stage and its classification rate was 99.99%. Neural Network was used as second stage of the model. Resilient back propagation with sigmoid function was used in proposed model. Experimental results have shown that the classification rate of proposed model was highest than previous work. False negative and false positive rates were 0.01% which was also lowest in comparison to previous works.	

continues on following page

Table 1. Continued

Year 2011	
Rahman et al. (2011) "Adaptive Intrusion Detection based on Boosting and Naive Bayesian Classifier"	
Used Algorithm:	AdaBoost, NB
Authors have used an ensemble approach of boosting and Naïve Bayes classifier to design an adaptive IDS. The main goal of the model was to improve the performance of Naïve Bayesian classifier in detection of intrusions. They used KDD99 dataset to perform experiments. It has been revealed by experimental results that the adaptive IDS algorithm outperformed k-NN, C4.5, SVM, NN and GA in detection of DoS, Probe, R2L and U2R attacks and "Normal" as well.	
Su (2011) "Real-time Anomaly Detection Systems for Denial-of-Service Attacks by Weighted k-Nearest-Neighbor Classifiers"	
Used Algorithm:	Genetic Algorithm, Weighted k-NN
Author has proposed weighted KNN classifier to detect denial of service attacks (large-scale attacks) in real time. For feature selection and weighting, the combination of GA and KNN was used. To test the implemented NIDS, only top most attributes have been selected after weighting all primary 35 attributes in training phase. The gross accuracy rate was 97.42% for known attacks when 19 attributes were selected. But, in case of unknown attacks, the gross accuracy rate was 78% when 28 attributes were selected.	
Year 2012	
Aneetha and Bose (2012) "The Combined Approach for Anomaly Detection using Neural Networks and Clustering Techniques"	
Used Algorithm:	Modified SOM, K-means
Authors have proposed a method for anomaly detection which occurs in networks. They have used clustering algorithm and neural network for the proposed method. Distance threshold, connection strength and neighborhoods functions assisted to create network by using the modified SOM (Self Organizing Map). Nodes are grouped by K-means clustering algorithms in the network on the basis of similarity measures. Detection rate was improved by 2% by modified SOM. But after deploying K-means; it was again increased by 1.5%. Experiments performed manifested that proposed method was very effective with 98.5% detection rate against denial-of-service attack.	
Lin et al. (2012) "An Intelligent Algorithm with Feature Selection and Decision Rules Applied to Anomaly Intrusion Detection"	
Used Algorithm:	SVM, DT, SA
Authors have proposed an intelligent algorithm which was able to detect anomalies. The proposed algorithm also involved decision rules and features selection. Proposed algorithm was tested with the help of KDD'99 dataset against anomaly intrusion detection. Feature selection was performed by SA (simulated annealing) and SVM. For the detection of new attacks, SA and DT had generated decision rules and proven their efficiency. Experimental results have manifested that the performance of proposed algorithm was efficient in detection of Denial of Service, Probe, Remote to Local, User to Root and Normal classes of KDD'99 dataset in comparison to other existing approaches.	
Year 2013	
Zainaddin and Hanapi (2013) "Hybrid of Fuzzy Clustering Neural Network over NSL Dataset for Intrusion Detection System"	
Used Algorithm:	Fuzzy Clustering NN
Author has introduced the framework experiment to test the reliability and stability of existing technique of Fuzzy Clustering and Artificial Neural Network over NSL KDD dataset. This technique was earlier used with random selection of KDDCUP99 dataset. They attained detection of precision more than 90% with maintaining the recall value more than 80%. NSL KDD dataset has shown increment over 90% for harmonic mean value in comparison to KDDCUP99 dataset. Especially for low frequent attacks, the existing hybrid approach performed better detection rates over NSL KDD dataset. The reason behind this was removal of redundant and uncertain data from original KDD dataset.	
Lisehroodi et al. (2013) "A Hybrid Framework based on Neural Network MLP and K-means Clustering for Intrusion Detection System"	
Used Algorithm:	K-means, NN MLP
Authors have performed two different experiments for comparison of single classifier and hybrid approach. Neural network Multi-Layer Perceptron was selected as single classifier for comparison. To produce the hybrid approach, K-means clustering and Multi-Layer Perceptron was combined and resulted as Kmeans+NeuralNetwork(KM-NEU). Single ANN and proposed approach were evaluated in terms of accuracy, detection rate and false alarm rate. At the same time proposed approach was also compared with other related approaches. KDDCUP99 dataset was used to perform experiments. Proposed approach had shown better classification due to neural network capability to produce less false positives and negatives with the ability of K-means clustering for grouping the data which was used as pre-classification algorithm.	

continues on following page

Table 1. Continued

Year 2014	
Shrivas and Dewangan (2014) "An Ensemble Model for Classification of Attacks with Feature Selection based on KDD99 and NSL- KDD Data Set"	
Used Algorithm:	ANN-Bayesian Net-GR ensemble: ANN, Bayesian Net with GR feature selection
Authors have proposed an ensemble approach by using Artificial Neural Network with Bayesian Net classifier for intrusion detection system. Gain Ratio was used for feature selection. All experiments were performed with KDD99 and NSL-KDD dataset. Accuracy of proposed model was 99.42% with 31 features of KDD99 dataset. While the accuracy was 98.07% with 35 features of NSL-KDD dataset.	
Ranjan and Sahoo (2014) "A New Clustering Approach for Anomaly Intrusion Detection"	
Used Algorithm:	K-medoids
Authors have proposed a method based on k-medoids clustering algorithm with fixed alterations of it. The proposed method described a procedure to select initial medoids for IDS. The advantages of the proposed method were that it had removed the need of primary centroids, as well as reliance on the irrelevant clusters and number of clusters. Experimental results have manifested that the proposed method outperformed FCM, Y-means and K-means in detection of DoS, Probe, R2L and U2R attacks	
Year 2015	
Eesa et al. (2015) "A Novel Feature-selection Approach based on the Cuttlefish Optimization Algorithm for Intrusion Detection Systems"	
Used Algorithm:	DT, Cuttlefish Optimization Algorithm (Feature Selection)
Authors have proposed a model for IDS which was the combination of Cuttlefish Algorithm (CFA) and Decision Tree (DT). The basic purpose of the model was selection of features from KDDCUP99 dataset to evaluate its performance. CFA was modified before using as an attribute selection apparatus. After this, Decision Tree classifier was used as evaluation apparatus for the features produced by CFA. Experimental results have shown that the rate of detection and accuracy went on increasing with the number of features less than or equal to 20.	
Lin et al. (2015) "CANN: An Intrusion Detection System based on Combining Cluster Centers and Nearest Neighbors"	
Used Algorithm:	K-means, K-NN
Authors have proposed an approach named as Cluster Center and Nearest Neighbors (CANN) for powerful IDS. The experimental results have revealed that the proposed approach outperformed K-NN and SVM over 6 selected features dataset. The accuracy and DR were higher with low FAR for CANN. On the other hand, CANN, K-Nearest Neighbor and Support Vector Machine performed equally over 19 selected features dataset but computational time of CANN was less than both K-NN and SVM over training and test dataset. The drawback of CANN was that it could not detect U2R and R2L attacks effectively.	
Year 2016	
Hadri et al. (2016) "Intrusion Detection System using PCA and Fuzzy PCA Techniques"	
Used Algorithm:	PCA and Fuzzy PCA, K-NN
Authors have proposed a method for reduction of large number of input data features from KDDCUP99 dataset for effective detection of intrusions. For this purpose, authors have used principal component analysis (PCA) and Fuzzy PCA algorithms. Proposed method kept just meaningful and relevant information from used dataset. All the experiments performed have resulted that Fuzzy PCA was better than PCA in detection of Denial of Service and U2R attacks.	
Nskh et al. (2016) "Principal Component Analysis based Intrusion Detection System Using Support Vector Machine"	
Used Algorithm:	SVM, PCA
Authors have compared Intrusion Detection Systems built by using different kernels of Support Vector Machines. All the models were implemented with Principal Component Analysis (PCA) and without PCA. By support of PCA, both higher and lower dimensional dataset were used for classification. Experimental results have shown that the detection rate kept on increasing with the increase in dimension size. But detection time was reduced when PCA was implemented. Performance of RBF Kernel was better in terms of detection rate among all the kernels implemented. But in case of detection speed, Polynomial kernel based SVM was faster than the others.	

continues on following page

Table 1. Continued

Year 2017	
Syarif and Gata (2017) "Intrusion Detection System using Hybrid Binary PSO and K-Nearest Neighborhood Algorithm"	
Used Algorithm:	Binary PSO, K-NN
Authors have proposed an approach which was the combination of K- Nearest Neighbor and Particle Swarm Optimization (PSO). A comparison was performed between proposed approach (KNN+PSO) and K- NN for classification accuracy of KDDCUP99 dataset. The accuracy rate of the proposed approach was better than K-NN in detection of each type of attack class even after increasing the value of k. For the Normal class, increasing rate of accuracy value has exceeded up to 2% for n = k = 5, 10 and 15. Experimental results have shown that proposed approach has achieved accuracy rate 99.91% for Denial-of-Service attack class which was the best among all attack classes when the value of k = 5.	
Effendy et al. (2017) "Classification of Intrusion Detection System (IDS) Based on Computer Network"	
Used Algorithm:	K-means, NB, K-means, Information Gain
Authors have proposed an Intrusion Detection System model by using Naïve Bayes algorithm. All the experiments were performed on NSL-KDD dataset. The experimental results concluded that the performance of NB algorithm for classifying intrusion types can be optimized. For the optimized results of NB, continuous variable discretization using k-means clustering and feature selection using Information Gain algorithms have been applied.	
Year 2018	
He et al. (2018) "An Improved Kernel Clustering Algorithm Used in Computer Network Intrusion Detection"	
Used Algorithm:	Kernel Clustering
Authors have proposed a refined kernel-based clustering algorithm. The proposed algorithm was used for intrusion detection system with KDDCUP99 dataset. In proposed method, they mapped original data objects into high dimensional space. After this, analysis of mapped data was accomplished with clustering algorithm. After applying the clustering, numbers of representative points were increased. All the experiments performed have manifested that the proposed method produced better result than SVM and traditional kernel clustering algorithm in terms of DR and FAR. The proposed method was also suitable for most of the attack types at the same time.	
Shone et al. (2018) "A Deep Learning Approach to Network Intrusion Detection"	
Used Algorithm:	DL, NDAE, Stacked NDAEs
Authors have proposed Non-symmetric Deep Auto-Encoder (NDAE) method for unsupervised feature learning to solve the problem of existing Network Intrusion Detection System approaches. For this purpose, they have constructed a classification model by using stacked NDAE and Random Forest algorithm. The proposed model was implemented in Tensor-Flow. All the experiments performed have manifested that the proposed method achieved higher accuracy, precision and recall in minimum training time. They have also compared the proposed model against mainstream Deep Belief Networks technique. The comparison result showed that the proposed model has provided up to 5% improvement in accuracy with reduction in training time up to 98.81%.	

CONCLUSION

This survey provides the deep insight about Intrusion Detection System. The classification methods used to explain general features of IDS, attributes related to decision making and infrastructure as well as location on the network have demonstrated the IDS very well. The survey about the most pre-eminent IDS research of past decade has shown the growth in technology and algorithms to implement robust IDS. Finally, this survey has provided the base to design an efficient model of IDS.

REFERENCES

Amer, S. H., & Hamilton, J. (2010). Intrusion detection systems (IDS) taxonomy-a short review. *Defense Cyber Security, 13*(2), 23–30.

Aneetha, A. S., & Bose, S. (2012). The combined approach for anomaly detection using neural networks and clustering techniques. *Computing in Science & Engineering*, *2*(4), 37–46.

Bellekens, Tachtatzis, Atkinson, Renfrew, & Kirkham. (2014). Glop: Enabling massively parallel incident response through gpu log processing. In *Proceedings of the 7th International Conference on Security of Information and Networks*. ACM.

Butun, Morgera, & Sankar. (2014). A Survey of Intrusion Detection Systems in Wireless Sensor Networks. *IEEE Communications Surveys & Tutorials, 16*(1), 266–282. .2013.050113.00191 doi:10.1109/SURV

Cisco. (2017). *Cisco Visual Networking Index: Forecast and Methodology, 2016-2021*. https://www.cisco.com/c/en/us/solutions/collateral/service-provider/visual-networking-index-vni/complete-white-paper-c11-481360.html

Dahlia, A. A. Z., & Hanapi, Z. M. (2013). Hybrid of fuzzy clustering neural network over NSL dataset for intrusion detection system. *Journal of Computational Science, 9*(3), 391–403. doi:10.3844/jcssp.2013.391.403

Debar, H., Dacier, M., & Wespi, A. (1999). Towards a taxonomy of intrusion-detection systems. *Computer Networks, 31*(8), 805–822.

Di He, X. C., Zou, D., Pei, L., & Jiang, L. (2018). *An Improved Kernel Clustering Algorithm Used in Computer Network Intrusion Detection*. Advance online publication. doi:10.1109/ISCAS.2018.8350994

Eesa, A. S., Orman, Z., & Adnan, M. A. B. (2015). A novel feature-selection approach based on the cuttlefish optimization algorithm for intrusion detection systems. Academic Press.

Effendy, D. A., Kusrini, K., & Sudarmawan, S. (2017). Classification of intrusion detection system (IDS) based on computer network. In *Proceedings of 2017 2nd International conferences on Information Technology, Information Systems and Electrical Engineering (ICITISEE)*. IEEE.

Giacinto, G., Perdisci, R., Del Rio, M., & Roli, F. (2008). *Intrusion detection in computer networks by a modular ensemble of one-class classifiers*. Academic Press.

Hadri, A., Chougdali, K., & Touahni, R. (2016). Intrusion detection system using PCA and Fuzzy PCA techniques. In *Advanced Communication Systems and Information Security (ACOSIS), International Conference on*. IEEE. 10.1109/ACOSIS.2016.7843930

Hamed, Ernst, & Kremer. (2018). A Survey and Taxonomy of Classifiers of Intrusion Detection Systems. In Computer and Network Security Essentials. Springer.

Hindy, Brosset, Bayne, Seeam, Tachtatzis, Atkinson, & Bellekens. (2018). A Taxonomy and Survey of Intrusion Detection System Design Techniques. *Network Threats and Datasets, 1*, 1. doi:10.1145/nnnnnnn.nnnnnnn

Hodo, E., Bellekens, X., Hamilton, A., Tachtatzis, C., & Atkinson, R. (2017). *Shallow and deep networks intrusion detection system: A taxonomy and survey*. arXiv preprint arXiv:1701.02145.

Hu, Hu, & Maybank. (2008). AdaBoost-based algorithm for network intrusion detection. *IEEE Transactions on Systems, Man, and Cybernetics, Part B: Cybernetics, 38*(2), 577–583. doi:10.1109/TSMCB.2007.914695

Lin, S.-W., Ying, K.-C., Lee, C.-Y., & Lee, Z.-J. (2012). *An intelligent algorithm with feature selection and decision rules applied to anomaly intrusion detection*. Academic Press.

Lin, W.-C., Ke, S.-W., & Tsai, C.-F. (2015). *CANN: An intrusion detection system based on combining cluster centers and nearest neighbors*. Academic Press.

Lisehroodi, M. M., Muda, Z., & Yassin, W. (2013). A hybrid framework based on neural network MLP and K-means Clustering for Intrusion Detection System. In *Proceedings of 4th International Conference on Computing and Informatics*. ICOCI.

Mok, M. S., Sohn, S. Y., & Yong, H. J. (2010). *Random effects logistic regression model for anomaly detection*. Academic Press.

Muna Mhammad, T. J., & Mehrotra, M. (2010). Design network intrusion detection system using hybrid fuzzy-neural network. *International Journal of Computer Science and Security, 4*(3), 285–294.

Nskh, P., Varma, M. N., & Naik, R. R. (2016). Principle component analysis based intrusion detection system using support vector machine. In *Recent Trends in Electronics, Information and Communication Technology (RTEICT), IEEE International Conference on*. IEEE. 10.1109/RTEICT.2016.7808050

Rahman, C. M., Farid, D. M., & Rahman, M. Z. (2011). Adaptive intrusion detection based on boosting and naive bayesian classifier. *International Journal of Computers and Applications, 24*(3), 11–19.

Ranjan, R., & Sahoo, G. (2014). *A new clustering approach for anomaly intrusion detection*. arXiv preprint arXiv:1404.2772.

Ravi Kiran Varma, P., Valli Kumari, V., & Srinivas Kumar, S. (2018). A Survey of Feature Selection Techniques in Intrusion Detection System: A Soft Computing Perspective. In Progress in Computing, Analytics and Networking. Springer.

Shafi, K., & Abbass, H. A. (2009). *An adaptive genetic-based signature learning system for intrusion detection*. Academic Press.

Shone, N., Ngoc, T. N., Phai, V. D., & Shi, Q. (2018). A deep learning approach to network intrusion detection. *IEEE Transactions on Emerging Topics in Computational Intelligence, 2*(1), 41–50. doi:10.1109/TETCI.2017.2772792

Shrivas, A. K., & Dewangan, A. K. (2014). An ensemble model for classification of attacks with feature selection based on KDD99 and NSL-KDD data set. *International Journal of Computers and Applications, 99*(15), 8–13. doi:10.5120/17447-5392

Su, M.-Y. (2011). *Real-time anomaly detection systems for Denial-of-Service attacks by weighted k-nearest-neighbor classifiers*. Academic Press.

Syarif, A. R., & Gata, W. (2017). Intrusion detection system using hybrid binary PSO and K-nearest neighborhood algorithm. In *Information and Communication Technology and System (ICTS), 2017 11th International Conference on*. IEEE.

Wu, S.-Y., & Yen, E. (2009). *Data mining-based intrusion detectors*. Academic Press.

Chapter 12
Low-Cost Wireless Speed Control and Fault Mitigation of Three-Phase Induction Motor

Priya Vijayvargiya
Shri G.S. Institute of Technology and Science, India

Arun Parakh
Shri G.S. Institute of Technology and Science, India

ABSTRACT

This chapter presents a design proposal for low-cost speed control and electrical fault mitigation of three-phase induction motors. The proposed system can control and monitor TIMs (three-phase induction motors) from far-flung areas. Here authors have proposed a relay-free system for fast fault clearance. IoT technology and low-cost microcontrollers have helped in achieving a system that is more reliable, economical, user friendly, and fast. It can be controlled by mobile application at the comfort of home. Data related to fault occurrence can be stored and analyzed for preventive maintenance. V/f scalar control method is used for speed control of TIM and able to control it in a wide range. Electrical faults such as over-current, over-temperature, over-voltage, and under-voltage are considered in this chapter. Simulation of the proposed design is done using Proteus 8 software. ESP32 is used to runs a web server that connects the mobile app with simulation.

INTRODUCTION

In this industrial era, every type of industry requires and uses a three-phase induction motor (TIMs). These motors are predominant to other electrical motors due to their robustness, reliability, and economical availability. Some of its most predominant features such as self-starting and constant speed operation have made it industries most praise electrical machine. TIM's various applications in the industry require speed controlling either in manufacturing unit or in-service unit. Some of its application is found in conveyors, refrigerators, pumps, wind tunnels, etc. A current belief indicates that the internet of things

DOI: 10.4018/978-1-7998-3327-7.ch012

technology is leading the world towards the fourth industrial revolution. If smart-phones can control an electric bulb, then industrial motors can definitely be controlled with it besides monitoring its running condition. Protection of TIMs from various faults is another target of the proposed system. Since buying or repairing industrial workhorses i.e. TIMs further adds the industrial operational cost which is neither good nor feasible for any industry as it halts the manufacturing process. Losses due to the halting of TIMs include additional labor cost, service cost, per day goods manufacturing loss, additional monetary loss due to repairing, etc. In a study, it is found that out of all failures motor failure occurs 33% due to electrical fault, 31% due to mechanical fault, and 36% due to other factors such as environmental maintenance, etc.(Thomson, 2001). An Induction motor's health can be overseen by the health of its main components such as stator, rotor, and bearing. Here the authors have considered electrical faults. Faults like over-current, over-voltage, under-voltage, phase failure, over temperature are general electrical faults. These types of faults are commonly occurring in all industries which have variable load demand. Speed control and fault mitigation are two very important aspects with respect to industrial demand. The authors tried to amalgamate two systems into one and that too with minimizing the complexity of circuits. Another challenge was to make it cost-effective, technically advanced and operator friendly. For these purposes, IoT technology is used along with a low-cost microcontroller. It has converted the whole controlling process into a smart one. The fault mitigation system is relay-free as the inverter trips the motor when the microcontroller gave a fault occurrence signal to IGBT drivers and the gate pulse turned off. The whole system is simulated using software Proteus 8. The whole system is connected to the internet and controlled by a mobile application with the help of an ESP32 microcontroller.

Background

Several studies have been conducted on speed control of three-phase induction motor (Lüdtke, 1995). The primary goal is to control the rotational speed and maintain it too, at the instance of disturbances. Some of the high-performance speed control strategies are based on direct torque control and field orientation method (Lüdtke, 1995; Negm, 2006 ; Peña, 2016) but the V/f control method is also capable of high performances at low-speeds (Garcia,1998). Along with it, scalar V/f provides an easy as well as a practical way while controlling real-time TIMs (Seutake, 2011; Peña, 2016). This approach is best suitable for steady states. There are two types of the scheme in V/f control, i.e. Open-loop and closed-loop. The open-loop approach is commonly used due to simplicity, but closed-loop provides higher performance characteristics (Peña, 2016). Besides speed control, another important aspect for TIM is to give emphasis on fault condition too. Any faults if not addressed on time may lead to overall motor failure. Various studies show the ill effect of faults on TIM and the manufacturing industry (Sutar, 2017). Fault analysis and detection can be done by analyzing the current signature (Thomson, 2001). Real-time fault status analysis and mitigation can be easily done by using microcontrollers (Kunthong, 2017). Depending on the handling of these faults, one can predict the life cycle of the TIMs (IEEE S.A Standard board, 2016).

The IoT technology provides various advantages such as easy installation, low-cost solutions, and remotely upgradable software with alert notifications to operators (Goundar, 2015; Abid, 2020). ESP32 immerged as best solution for IoT application due to its swiftness, compactness and Low power consumption (Sharp, 2017).

BLOCK DIAGRAM AND FLOW CHARTS

The block diagram of the speed control and fault mitigation of the three-phase induction motor is shown in figure 1. It has a power unit, an intermediate unit, and a control unit.

Figure 1. The block diagram of proposed system

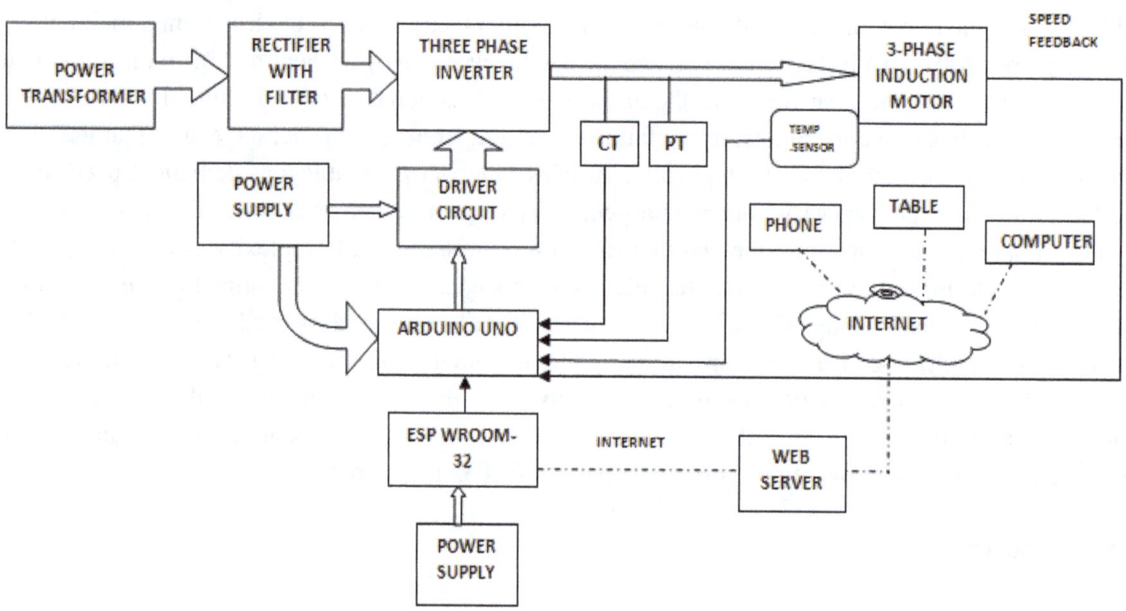

The power unit consists of a rectifier, DC-link filter, an inverter connected in the cascade. Microcontroller Arduino UNO control gate pulses of IGBTs with the help of the intermediate unit, thus controlling inverter output. The intermediate unit comprises optical Isolators and gate driver circuits for IGBTs. Esp32 microcontroller acts as a link between the web server and the proposed circuit for speed control and fault mitigation. Arduino UNO and ESP32 exchange information using serial communication (UART protocol). Any changes in measured values are directly uploaded to a web server. The mobile application fetches this data and makes it available for users over the internet.

Due to proteus8 software's limitations, the authors have used Arduino as a computational microcontroller. Otherwise, ESP32 is powerful enough for performing computational tasks quickly as well as web serving at the same time. Thus, it reduces complexity and further increases the reliability of the system.

Thus, ESP32 provides a low-cost solution for web-serving requirements and computational tasks.

Flow charts of the proposed system are shown in Figures 2, 3, and 4. Figure 2 shows the instructions flow for the Arduino module. Once the Arduino port pins are initialized, it reads data from C.T., P.T., and sensors and communicates with the ESP32 microcontroller accordingly. It receives commands from the web server and uploads updated data to the web server with the help of ESP32. This stored data can be analyzed and used to predict the preventive maintenance of an induction motor. In this way, the operating cost of certain industries can be minimized to a certain level. When the user sends a speed variation command to the Arduino via the mobile application, it corrects the speed by varying the voltage and frequency.

Figure 2. Flow chart of Arduino module working

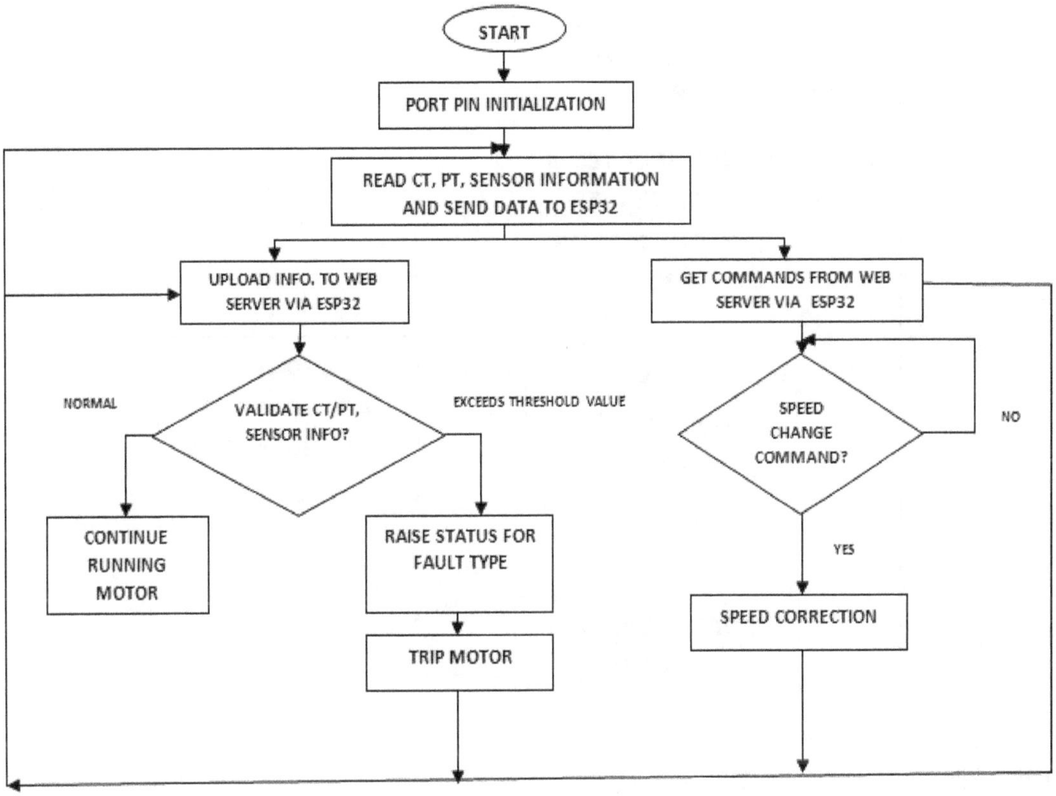

Figure 3 shows the instruction flow of the ESP32 module. It basically communicates with Arduino and the web server. If the user alters any value on the mobile application, altered data get updated on the server. Esp32 fetches this data and communicate with Arduino for further action. When any abnormal condition arises, Arduino communicates with ESP32. It updates the server with new conditions and thus user receives alert signals to the mobile application. For this purpose HTTP and TCP, protocols are used. Figure 4 shows an instruction flow chart for a mobile application. It was built using Native.io website services. It gets updated from the server and displays real-time values of three-phase voltages, current, frequency, speed, and temperature. When these values exceed their threshold value, it displays an alert and faults status command on the mobile screen.

When a user wants to change the speed of TIM, it updates data to the server. The server sends this command to Arduino via ESP32. Similarly, when the user wants to start or stop the TIM, it again updates the web server for further action.

Thus, these three flow charts completely explain the flow of instructions for the proposed system.

Figure 3. Flow Chart of ESP32 module working

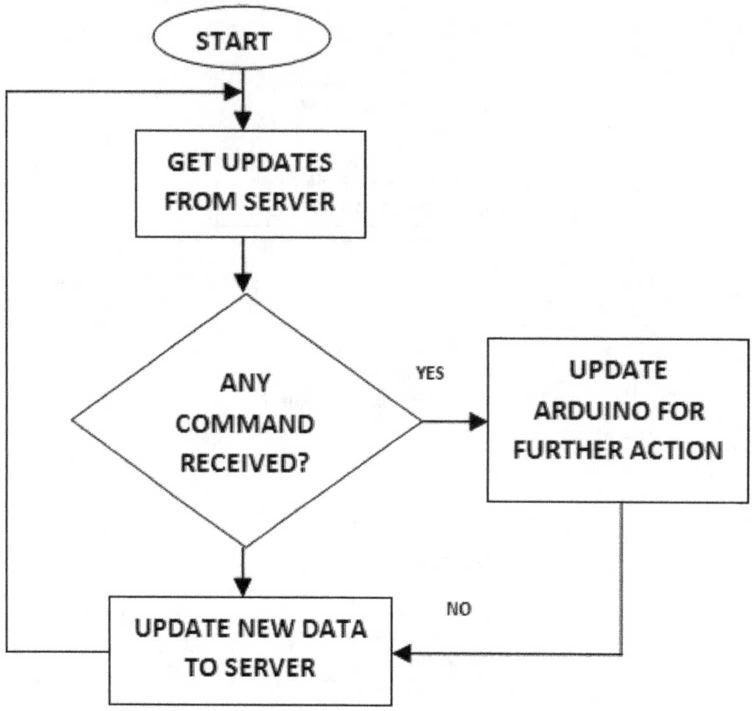

Figure 4. Flow chart of mobile application

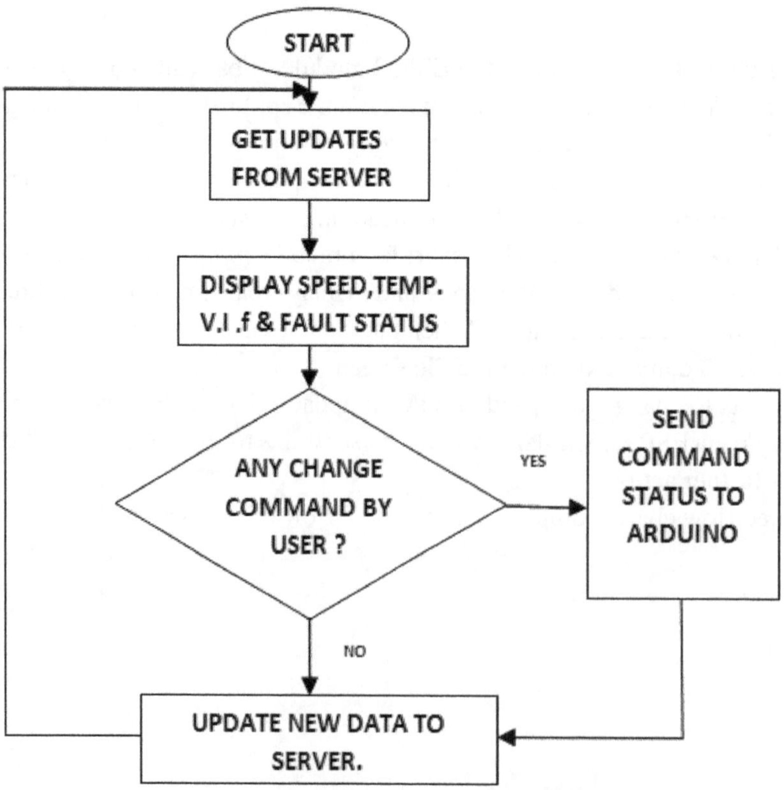

METHODOLOGY

Speed Control

The Traditional Scalar V/f control offers simplicity and practicality while applied along with low steady-state error (Garcia, 1998). In the V/f scalar method, the ratio of V/f kept constant to keep the flux constant. Since Flux is directly proportional to torque, it also remains constant.

Ø= V/f = constant

Therefore, to keep good torque performance at the time of speed alteration, it is essential to change the supply voltage in accordance with the change in frequency.

Rotor speed follows the Synchronous speed.
Thus, Nr = (1-s)*Ns,

Where s is the slip of the Asynchronous Motor.
There are two types of V/f controls, Viz; Open-loop and closed-loop controls.

Open-Loop V/F Control

The open-loop method is the simplest and most economical. But the speed of the motor in the Open-loop V/f method will never be exactly equal to the given speed command as induction motors can never run at synchronous speed (Peña, 2016). Due to the absence of any feedback signal, the system becomes more vulnerable to instability. Thus, in case of any disturbances in the system or speed of the motor, the System cannot rectify it by itself.

Closed-Loop V/F Control

Closed-Loop V/f control, also known as V/f feedback control resolves all the shortcomings of the Open-Loop V/f control system. In this method, the response or the actual speed is continuously compared with the desired speed command. The error signal generated by it is processed by the Controller, which generates the slip frequency command (Peña, 2016). The addition of this slip frequency and instantaneous speed provides the required Stator frequency. This frequency is then fed to the V/f controller, which computes the corresponding voltages. The remaining operation is similar to the Open-Loop V/f control method. The Speed Control loop restores the desiring speed whenever speed drops due to load torque or disturbances and maintains it at the desired speed. Consequently, Stability also Prevail.

Thus, this method is considered for speed control of TIMs and can be implemented with the help of VFD (variable frequency drive). VFD comprises a rectifier, a DC link filter, and a three-phase inverter connected in cascade.

A microcontroller is used to control the output of the inverter. The voltage waveform can be altered as per the requirement just by controlling the duration of the closing of the switches, i.e. by trimming the opening and closing time of switches. Similarly, by controlling the timing of the switches, the frequency can be controlled. Therefore, Speed can be controlled by maintaining the V/f constant (Peña, 2016).

Fault Mitigation

Three-phase induction motor predominantly suffers from the following electrical problems.

- Over-current /Overload
- Over-temperature
- Phase failure
- Under and Overvoltage

Thus, Protection from these problems is foremost important for the smooth operation of the induction motor. Besides this, it is good for its lifetime and its efficiency as due to these electrical faults motor gets overheated and further add to insulation failure.

Over Current / Overload

The excessive current flow, which can damage the electrical circuit or initiate fire, is termed as Over-current. Here "excessive current" refers to the flow of current whose value exceeds the nominal full load current value. This current is analogous to that of short circuit currents. Therefore, over current protection is necessary for obstructing serious damages and even saving people's lives from any hazardous situations. Due to prolonged overload on induction motor load torque increases, as a result, the value of current exceeds and becomes more than 120% of rated current (IEEE S.A Standard board, 2016). In such an instance, this over-current protection comes into action. This overload situation occurs when there is a proliferation in mechanical loads on the motor beyond the rated value which works as a stepping stone for other electrical failures too.

This over current protection works by first detecting and then limiting this exorbitant current flow as soon as possible. If the current value reaches more than the rated value it isolates the motor from the power circuit. But there must be exceptions for starting currents, permissible over-currents, and current surges. By providing time delay this objective can be achieved.

This protection scheme automatically includes short-circuiting protection as well. Generally, the short circuit includes Phase faults, Winding faults, or Earth faults and currents which are 5-20 times the full load current is known as short circuit current (IEEE S.A Standard board, 2016). So, fast clearances of these faults are always desirable.

Over Temperature

With Overloading excessive current flow takes place, if it allows for a long time temperature of the motor rises. This reduces the lifetime of IM up to 50% if it is allowed to reach and operate at a temperature above its maximum rating. Since the heating of the induction motor is directly proportional to the I2 (square of the current) which eventually results in its insulation failure. This also indicates the larger the magnitude of over-current, the lower the time required to cause severe damage to components due to overheating. Most of the induction motors fail or their components fail due to motor overheating. The major parts of induction motor like Stator, Rotor, Bearing, Shafts, and Frame fails potentially due to thermal stress. Stator winding deterioration and the melting of rotor conductors are two major risks of overheating. When the operating thermal limit of the induction motor exceeds by 10°C, the insulation lifetime reduces by half (IEEE S.A Standard board, 2016).

Thus, thermal protection is a must and it can be achieved by first detecting the temperature of stator winding on a real-time basis and then comparing it with the rated temperature. When the temperature goes far off its rated limit, it must trip the motor power circuit and the motor stops running. When the motor cools down and the temperature comes within a certain limit preferably near ambient temperature, it should start running again.

Phase Failure

In three-phase motors, if any phase fails or there is any unbalancing among these phases, other remaining phases either deliver more current or carry more phase voltage. This can damage the motor or can affect the operation of the motor (IEEE S.A Standard board, 2016). This unbalancing can burn the motor or other circuits with which it is attached. Thus, phase failure protection is used in this case.

It is achieved by first monitoring the current and the voltage levels of each phase. If their value crosses permissible limits of either phase voltage or current or both, it isolates the induction motor from the power circuit and trips the motor.

Under and Overvoltage

The under-voltage occurs when the supply voltage falls below 85% of the rated allowed value with a rated mechanical load on the motor (IEEE S.A Standard board, 2016). Some of its effects are excessive currents, exorbitant heating of machines, and an increase in stator and rotor losses. Therefore, it is needed to safeguard the machine from this condition, so under-voltage protection is used. It is achieved by monitoring voltage levels and comparing them in real-time.

Whenever it goes below the permissible rated value, the motor disconnects from the power circuit. The motor operates again when the voltage to normal value.

Whenever the line voltage is greater than 120% of the rated value, it is considered as overvoltage (IEEE S.A Standard board, 2016). This harms the machine insulation; therefore overvoltage protection is a must. This overvoltage protection is carried out by overlooking and comparing the voltage levels with the pre-set value. Whenever the voltage exceeds this threshold value control circuit shuts down the supply. It is more of a power supply feature and whenever voltage stabilizes within its normal range, the system starts again.

For Current and Voltage measurement CTs and PTs are used since they are commonly available in several industries. For temperature measurement DHT11 sensor is attached with TIM. Microcontroller further analyzes signals given by these measuring instruments and act as per the threshold conditions of above mention electrical faults. If any fault occurs, it trips the motor with the help of an inverter which is basically a switching device. Thus, the system doesn't need additional relays. This reduces the number of components and costs with an increase in reliability.

Computation Architecture

After comparing different microcontrollers on the basis of technical aspects authors came to the conclusion that ESP32 is the best solution for the proposed system (Sharp, 2017). Due to software limitations, the simulated system comprises two microcontrollers namely Arduino UNO and ESP32. Arduino UNO is integrated into the simulation circuit, which helps in controlling output voltage and frequency. Whereas ESP32 running a Web Server and acts as a communication channel between Mobile application and Arduino.

Table 1. Comparision of different microcontrollers

	Arduino UNO R3	Arduino Wifi R2	ESP32	ESP8266 V3	Raspberry PI 3 B+
Microprocessor	ATMEGA328P Single core	ATMEGA4809 Single core	Xtensa LX6 dual core(32 Bit)	Xtensa L106 single core(32 Bit)	**BCM2837 Quad core (64 bit)**
Operating Voltage	5V	5V	**3.3V**	**3.3V**	**3.3V**
Current consumption	45mA-80mA	50mA-150mA	**15µA-400mA**	**15µA-400mA**	16mA-80mA
Digital I/O pins with PWM	6	5	**36**	16	26
Analog input pins	6	6	**15**	1	0
SPI/I2C/I2S/UART	1/1/1/1	1/1/1/1	**4/2/2/2**	2/1/2/2	2/2/0/2
Flash memory	32KB	48KB	4MB	4MB	**16GB(SD CARD)**
RAM	2KB	6KB	520KB	64KB	**1GB**
Clock Speed	16MHz	16MHz	160/240MHz	80MHz	**1.2GHz**
Wifi*	No	Yes	**Yes**	**Yes**	**Yes**
Bluetooth	No	No	**Yes**	No	**Yes**
Ethernet	No	No	**Yes**	No	**Yes**
CAN	No	No	**Yes**	No	Yes
Temperature sensor	No	No	**Yes**	No	No
Hall effect sensor	No	No	**Yes**	No	No
Operating temperature	-45°C - +85°C	-45°C- +85°C	**-45°C-+125°C**	**-45°C-+125°C**	-45°C- +85°C
Price (in Rs.)	330-550	3000-4500	**380-600**	250-400	2500-3500

Table 1, shows a comparison of the different microcontrollers. As per the requirement of the proposed system ESP32 microcontroller emerged as the solution to the entire requirement. It has 15 analog pins, a Wi-fi module, a temperature sensor, a dual-core processor, a clock frequency of up to 240MHz, an operating temperature of -45°C to 125°C, etc. It is the lowest-cost solution covering all major requirements of the proposed system.

IoT Architecture and Mobile Application

As shown in figure 5, IoT architecture consists of four major components viz. sensed parameters, IoT gateway, Cloud server, and mobile application.

Parameters are sensed with the help of sensors, CT, and PT. The IoT gateway is networking hardware that communicates using more than one protocol and acts as a bridge between other networks. A cloud server is a virtual infrastructure that process and store data, thus made it available for mobile application and hardware circuit. The mobile application fetches data from the cloud server and displays updated information to the user. It also takes any command from the user and uploads it to the server. After it, the server conveys this information to the hardware unit via IoT gateway.

Figure 5. IoT architecture of proposed system

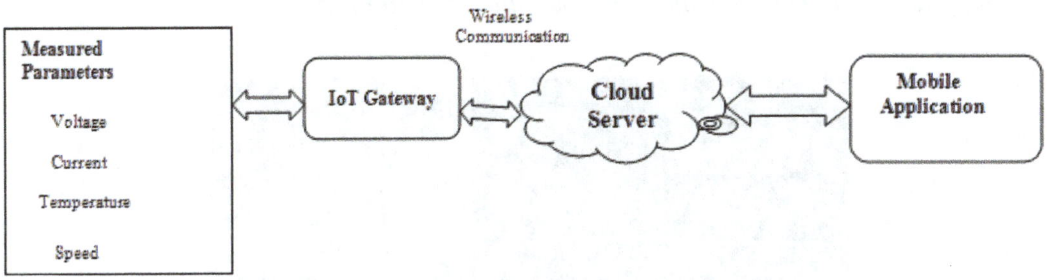

Simulation and Results

Figure 6 shows the simulation circuit of the proposed system using Proteus 8software. While simulation a three-phase, 5HP, 4 pole motor is considered. The initial slip is considered to be 1%. The circuit consists of an uncontrolled bridge rectifier, IGBTs driver circuits, three Phase inverter, Current, Voltage, temperature, and humidity measuring devices, Arduino UNO as a microcontroller, Optical Isolator to isolate the power unit from the control unit, resistors, and capacitors. Com-Port is used here to send computational instructions and sensed data to ESP32. ESP32 is acting as an IoT gateway in the proposed system. Arduino controls gate pulses of an inverter as per requirement and data feed by the user over the mobile application. It also continuously shares real-time Voltage, current, frequency, and speed data for monitoring. At the instance of an electrical fault, the system generates an alert signal for the user and trips off the induction motor.

Figure 6. Proteus simulation diagram of the proposed system

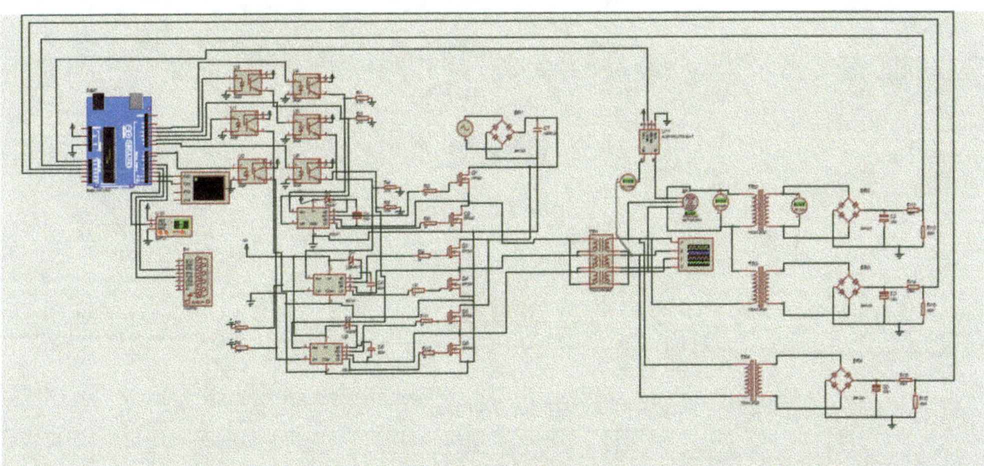

Thus, there is no need for relays or any other additional switching device in the circuit. Therefore, the proposed system is economical, fast, and reliable. Figure 7. Shows output voltage waveform of the inverter when the speed is set to 1050 RPM. An example of an electrical fault alert on a mobile application for the operator is shown in figure 8. It indicates low voltage between YB phases along with current, frequency, and speed of the motor at that particular instance.

Figure 7. Snapshot of VFD output voltage without filter at 1050 RPM

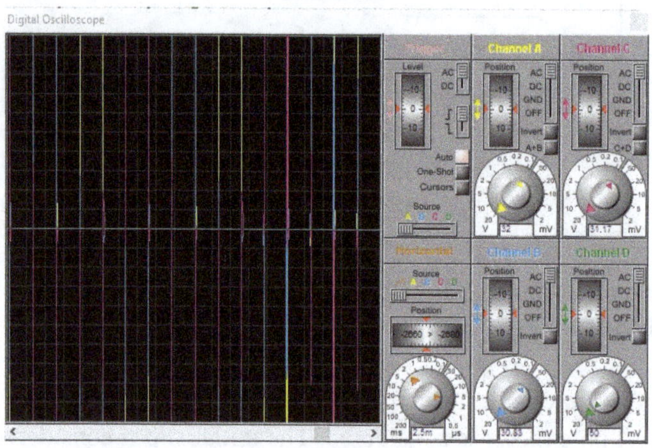

Figure 8. Mobile application Snapshot of Under-Voltage fault alert

Here they consider a 5hp motor along with a .866 power factor. Table 2 indicates a different electrical fault alerts and speed variations of the 3-phase induction motor with respect to frequency. The healthy motor condition is also indicated in the table along with line voltages, current, frequency, and speed. Here 497V, 350V, 7.2A are taken as the threshold values for Over voltage, under voltage, and over current respectively. Temperature sensor output is 28°C

Table 2: Speed variation and fault status indication of Induction motor

S.no	Voltage (in volts) V_L peak-peak			I_L	Speed (RPM)	Frequency(Hz)	Fault status
	RY	YB	B R				
1.	410	414	420	5.2	1475	49.8	Healthy
2.	411	410	408	7.5	312	15.5	Over current
3.	392	418	420	5.4	1015	44	Healthy
4.	320	419	317	7.6	1350	45	Phase Failure
5.	413	280	420	6.8	412	22	Under voltage
6.	498	362	390	4.1	1440	47.7	Over voltage
7.	420	408	411	5.13	1448	49.8	Healthy
8	418	315	407	6.9	380	17.8	Healthy
9.	422	345	413	6.1	720	28	Healthy

CONCLUSION

The Authors have developed a low-cost and user-friendly system design for speed control and fault mitigation of TIMs and the design is verified by the Proteus simulation software. This system can be implemented in any industry, even at places where human reach is strenuous or skilled people are not available. Researchers can control the speed of the induction motor in the range 380 to 1480 RPM and able to mitigate over current, over and under voltage, over temperature. Electrical Fault mitigation was fast, due to the use of a fast microcontroller i.e. ESP32. Traditionally, Relay is considered a must apparatus for fault clearance. But in the proposed system need there is no need for relays due to the use of a microcontroller with a power circuit. At the instance of fault, the microcontroller analyses the situation and generates a trip signal. In the simulation, they have used two microcontrollers while in actual implementation author explained how only one controller is needed to incorporate all major computations and they suggested using only ESP32 in the hardware for better performance. Since ESP32 is an economical yet powerful controller, it provides a low-cost solution for the proposed system. For speed control of the three-phase induction motor, the SVPWM technique can also be implemented for minimizing its THD.

REFERENCES

Abid, G., Shaikh, S. A., Rajput, S. H., & Mazid, U. A. (2020). IoT based smart industrial panel for controlling three phase induction motor. IEEE iCoMET, 1-8.

Garcia, A. M., Lipo, T. A., & Novotny, D. W. (1998). A new induction motor V/f control method capable of high-performance regulation at low speeds. *IEEE Transactions on Industry Applications, 34*(4), 813–821. doi:10.1109/28.703982

Goundar, S. S., Pillai, M. R., Mamun, K. A., Islam, F. R., & Deo, R. (2015). Real time condition monitoring system for industrial motors. *2nd Asia-Pacific World Congress on Computer Science and Engineering (APWC on CSE),* 1-9.

IEEE S. A Standard Board. (2016). *Recommended practice for motor protection in industrial power system.* https://mentor.ieee.org/3000-stds/dcn/18/stds-18-0003-00-PUBS-3004-8-2016.pdf

Kuthong, J., Sapaklom, T., Konghirun, M., Prapanavarat, C., Ayudhya, P. N., Mujjalinvimut, E., & Boonjeed, S. (2017). *IoT based traction motor drive condition monitoring in electric vehicles- Part 1.* IEEE PEDS.

Lüdtke, I., & Jayne, M. G. (1995). A comparative study of high performance speed control strategies for voltage source PWM inverter fed I.M drives. *Electrical Machines and Drives, Conference Publications No. 412.*

Negm, M. M., Bakhashwain, J. M., & Shwehdi, M. H. (2006). Speed control of three-phase induction motor based on optimal preview control theory. *IEEE Transactions on Energy Conversion, 21*(1), 77–84.

Peña, J. M., & Diaz, E. V. (2016). *Implementation of V/f Scalar Control for speed regulation of three phase induction motor.* IEEE ANDESCON Arequipa.

Sharp, M. A., & Vagapov, Y. (2017). *Comparative analysis and practical implementation of the ESP32 Microcontroller Module of the internet of things.* Internet Technologies and Applications.

Suetake, M., Da Silva, I. N., & Goedtel, A. (2011). Embedded DSP-Based Compact Fuzzy System and Its Application for Induction-Motor V/fV/f Speed Control. *IEEE Transactions on Industrial Electronics*, 750–760.

Sutar, P. P., & Panchade, V. M. (2017). Induction motor faults mitigation using microcontroller. *International Conference on Energy, Communication, Data Analytics and Soft Computing (ICECDS)*, 489-493.

Thomson, W. T., & Fenger, M. (2001). Current signature analysis to detect induction motor faults. *IEEE Industry Applications Magazine*, 7(4), 26–3.

Chapter 13
Reliability of Smart Grid Including Cyber Impact: A Case Study

Janavi Popat
Sardar Vallabhbhai National Institute of Technology, India

Harsh Kakadiya
Sardar Vallabhbhai National Institute of Technology, India

Lalit Tak
Sardar Vallabhbhai National Institute of Technology, India

Neeraj Kumar Singh
Sardar Vallabhbhai National Institute of Technology, India

Mahshooq Abdul Majeed
Sardar Vallabhbhai National Institute of Technology, India

Vasundhara Mahajan
https://orcid.org/0000-0002-2698-6096
Sardar Vallabhbhai National Institute of Technology, India

ABSTRACT

Smart grid has changed power systems and their reliability concerns. Along with that, cyber security issues are also introduced due to the use of intelligent electronic devices (IEDs), wireless sensory network (WSN), and internet of things (IoT) for two-way communication. This chapter presents a review of different methods used from 2010 to 2020 focusing on citation as the main criteria for reliability assessment of smart grids and proposals to improve reliability when it comes to assessing a practical transmission system. It shows that evolutionary techniques are the latest trend for smart grid security.

INTRODUCTION

As the world is aiming to lower carbon emissions, X. Zhang, Bie, and Li (2011) smart grid is becoming more important than conventional power grid due to its resources like renewable energy and distributed networks, as well as interference of information and communication (ICT) technology. Use of wireless sensor networks (WSN) (Singh, Gupta, & Mahajan, 2020), intelligent electronic devices (IEDs) and

DOI: 10.4018/978-1-7998-3327-7.ch013

modern communication systems has made it easier to operate distributed generations and transmission. It is also helpful to minimize the energy losses due to effective control and faster response time. But the great technical advances lead to security issues and it plays a crucial role in the reliability of the power system.

Enhancement of distributed generations and distributed networks will affect reliability of the system at huge extent. (X. Zhang et al., 2011) Improved reliability helps to make the power system cost effective and will be helpful to provide continuous supply to consumers. Cyber systems will be more vulnerable to malicious attacks and these affect the reliability of systems at huge scale. Availability, integrity and confidentiality of cyber systems are the main concerns. Due to the integration of cyber and physical systems, reliability concerns of the cyber system have brought many uncertainties in the modern power systems.

There are various methods to detect cyber-attacks in power system e.g. Particle Swarm Optimization (PSO), Grey-Wolf Optimization (GWO), (Singh & Mahajan, 2020) Genetic Algorithm (GA) and Fire-fly Algorithm (FA). Artificial neural networks (ANN) plays a vital role in load forecasting, generation forecasting for renewable energy, fault diagnosis, analysis of consumption behaviour, etc and thus enhances reliability and security smart grid. (Jiao, 2020)

APPROACHES TO ASSESS RELIABILITY

There are various methods to assess reliability which are classified as shown in the figure-1 (Al-Abdulwahab, Winter, & Winter, 2011; Chen, Lu, & Zhou, 2018; Refaat, Abu-Rub, Trabelsi, & Mohamed, 2018; Wadi, Baysal, Shobole, & Tur, 2018; Yuan, Li, Bie, & Arif, 2019). It is mainly classified as historical and probabilistic method. Probabilistic approach is further classified in two types: 1. Analytical assessment which includes event tree analysis (ETA), reliability block diagram (RBD) and Markov Modelling and 2. Through Simulation which includes Monte Carlo simulation method.

Figure 1. Classification of Approaches to Assess Reliability

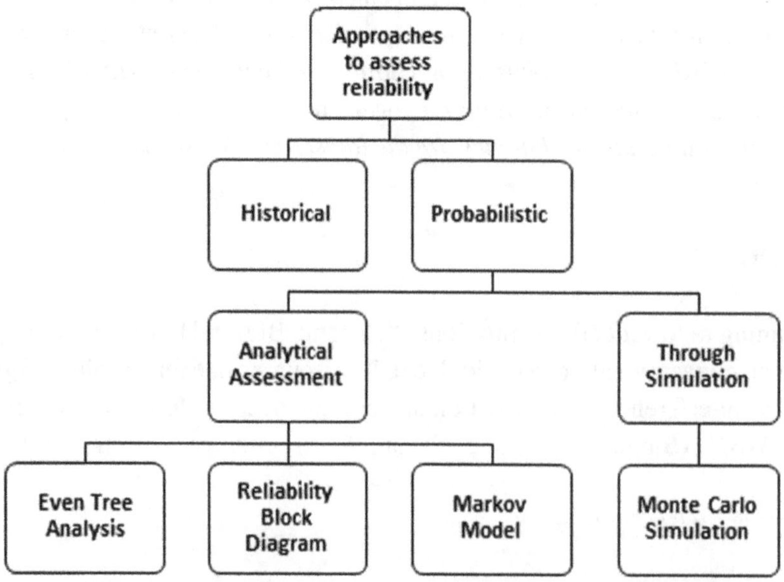

1. Historical

It is a method of assessing reliability from the experimental values recorded from a practical power system. Forecasting can be done using past data and time to repair (TTR), failure rate, Energy Not Supplied (ENS), Cost of Energy Not Supplied (CENS), etc can be estimated by this conventional method.

2. Probabilistic

Probabilistic methods includes assessment using mathematical models, graphs and trees, machine learning algorithms using python as well as simulation using softwares like MATLAB. It can be further classified in two parts:

a. Analytical Assessment

Analytical assessments are easy to be implemented as well as for understanding and teaching. Two analytical assessment methods are described below.

i. Event Tree Analysis

Event tree analysis, also known as ETA, is the most efficient method to teach reliability assessment. It is useful for both educators and trainees to analyze and compare different protection schemes and to understand the importance of smart grid (Alabdulwahab, 2017). The tree diagram defines potential accident sequences in the analyzed system which shows the failure and success of the system. Process of development of event tree can be divided into tasks like plant familiarization, defining event tree headings and safety functions, determining event tree and safety functions handlings, determining system success criteria, identifying initiating events, defining accident consequences, determining plant damage state and evaluating event tree. (Čepin, 2011a)

ii. Reliability Block Diagram

Reliability block diagram is also known as dependence diagram. It is easy to read. (Distefano, Puliafito, & Computing, 2009) It is implemented using representation of distinctive components of power system in form of blocks which are inter-connected depending upon their effect on the system. (Čepin, 2011b) Each block in RBD/DD is representation of a physical unit and it is arranged in such a way that failure of a particular unit will lead us to identify failure of other connected blocks and vice-versa. (Distefano et al., 2009)

iii. Markov Modeling

It is a famous method implemented to model randomly changing systems. It is widely used to assess reliability of distributed networks with distributed generations, variable weather conditions, hybrid AC/DC power system, etc. (Sarwat, Domijan, Amini, Damnjanovic, & Moghadasi, 2015)

b. Through Simulation:

Simulations are preferred when system is complex as well as parameters are varying depending upon each other. Two powerful and widely used simulation methods are discussed as follows.

i. Monte Carlo Simulation

A computerized mathematical technique widely used for quantitative analysis and decision making in various fields. In power systems it is used to determine reliability of closed ring topology(Wadi et al., 2018), impact of electrical energy storage and electrical vehicles on reliability of smart grid (Calderaro, Galdi, Graber, Graditi, & Lamberti, 2014) etc.

COMPARISON OF DIFFERENT METHODS AND ARTICLES

Table 1 shows comparison of various experiments and simulations carried out based on smart grid reliability as well as cyber securitty concerns. It describes the problem, method to obtain solution, result as well as future research scope. Articles and research papers from IEEE, ScienceDirect, IET and CIRED. The experiments were based on various reliability assessment methods discussed above.

Table 2 shows comparison of values of reliability indices obtained from various systems under various conditions and consideration. Reliability indices compared are, System Average Interruption Duration Index (SAIDI), Average Service Availability Index (ASAI), System Average Interruption Feeder Index (SAIFI), and Customer Average Interruption Duration Index (CAIDI). This indices are calculated using failure rate of system, number of customers and feeders, duration of interruption and time of usage. Reliability can also be assessed by availability, non-availability, Average Service Un-availability Index (ASUI), Energy Not Supplied (ENS) and Average Energy Not Supplied (AENS).

According to the citations, figure 2 shows popular reliability assessment methods and its citation values. In similar manner, figure 3 shows the latest trend in reliability assessment and its citation percentage.

FUTURE RESEARCH DIRECTIONS

Cyber security has great intrusion in reliability of smart grid, so advance monitoring methods can be adopted to monitor cyber-physical infrastructure of smart grid. . Figure 4, highlight research work done on topic related to reliability and cyber security of smart grid during the time period of 2010-2020 taking IEEExplorer as a reference. It can be achieved by graph theory (Singh, Gupta, Mahajan, Yadav, & Mudgal), weight reduction technique (Gupta, Singh, & Mahajan, 2020), using ensemble prediction algorithm based on time series (EPABT) (Li et al., 2019) and adopting real time monitoring and modeling methods (Facchinetti & Della Vedova, 2011). Internet of things can play a major role in development of smart grid and smart meter infrastructure as well as improving reliability and resilience of system. (Kimani, Oduol, & Langat, 2019) (Sharma, Saini, & Reviews, 2015) An optimization frame-work can be utilized with a view to assign the optimum points of reliability and cost (Honarmand et al., 2019).

Table 1. Comparison of Research Articles

Reference	Name of article	Citations	Problem description	Method used for solution	Result	Further scope
(Wang, Cai, Tang, Ni, & Systems)	Methods of cyber-attack identification for power system based on bilateral cyber-physical information	-	To accurately identify the fault causes by applying bi-lateral information from both cyber and physical side.	-	Quantization method and support threshold based sequence extraction method was developed.	To build up a strategy for preparing the model on restricted examples to upgrade its pertinence and consider more conceivable attack circumstances to protect the algorithm.
(Moslehi & Kumar, 2010)	A reliability perspective of the smart grid	943	To obtain combined reliability impacts of resources such as renewables, demand response and storage and electrical transportation.	By classification of resources and developing infrastructure and architecture accordingly.	The architecture which will be useful to make an ideal mix of the smart grid resources more reliable.	-
(Falahati, Fu, & Wu, 2012)	Reliability Assessment of Smart Grid Considering Direct Cyber-Power Interdependencies	153	To assess reliability considering direct element-element interdependencies and direct network-element interdependencies	Initializing probability table, merging equivalent states, state mapping, load shedding and then calculating the reliability indices.	Mapping procedure is proposed to assess the impact of failure of cyber network on power network.	Developing an optimization problem for cyber-network as well as simulating cyber-power network.
(Falahati & Fu, 2014)	Reliability Assessment of Smart Grids Considering Indirect Cyber-Power Interdependencies	96	To assess reliability considering indirect element-element interdependencies and indirect network-element interdependencies	Creating probability table, updating states and then calculating reliability indices.	The proposed reliability quality model updates the state probability for failure in a cyber-network. To expand the information network, an advancement model was designed.	-
(Albasrawi, Jarus, Joshi, & Sarvestani, 2014)	Analysis of Reliability and Resilience for Smart Grids	53	Effect of the reliability and resilience of smart grid on it's functionality	Demonstration of a reliability model to the IEEE 9-Bus test system (both, with and without an Intelligent control device).	Estimation of the reliability of constituent parts of the framework by the Markov model.	Examination of the impact of intelligent control on the resilience of a smart grid, investigation of the significance of assorted components of a smart grid, thought of non-determinism in remedial activities, and bouncing the time before recuperation activities have endeavored. Demonstrating of survivability, accessibility, and recuperation of framework will be likewise arranged.
(Xiang, Ding, Zhang, & Wang, 2016)	Power System Reliability Evaluation Considering Load Redistricution Attacks	53	Impact of load redistribution attacks on the long term power supply reliability	Semi-Markov model, IEEE RTS79 system, Monte Carlo simulation	Through regress attack and defense mechanism	-
(Sarwat et al., 2015)	Smart Grid Reliability Assessment Utilizing Boolean Driven Markov Process and Variable Weather Conditions	24	Demonstrate impact of variable weather conditions and distributed generation on reliability of smart grid.	Variable weather Boolean logic driven Markov process was used for smart grid reliability assessment	pertinent to evaluate the reliability of smart grid.	To obtain accurate weather forecast scheme and its impact on the operation of hybrid AC/DC power system while keeping a specific reliability level.

continues on following page

Table 1. Continued

Reference	Name of article	Citations	Problem description	Method used for solution	Result	Further scope
(Hashemi-Dezaki, Askarian-Abyaneh, Haeri-Khiavi, & Distribution, 2016)	Impacts of direct cyber-power interdependencies on smart grid reliability under various penetration levels of microturbine/wind/solar distributed generations	22	To evaluate the risk assessment of smart grid due to direct cyber power interdependencies under various distributed generation technologies.	Monte Carlo Simulation	network topology optimization will be very important in order to overcome the negative effective of DCPI,	Reliability assessment indirect cyber-power interdependencies
(R. Zhang, Zhao, & Chen, 2010)	An Overall Reliability and Security Assessment Architecture for Electric Power Communication Network in Smart Grid	17	To assess reliability and security of information and communication technology in reliability of grid.	Three level assessment hierarchical architecture: 1. Device level, 2. Network level, (Reliability assessment - RBD based path tracing) 3. Service level (Security assessment - service severity)	Application framework configuration created to assess the reliability and security of the smart grid.	Need improvement at service level for risk analysis.
(Calderaro et al., 2014)	Impact Assessment of Energy Storage and Electric Vehicles on Smart Grids	17	Non-scheduled renewable energy sources and deployment of electrical vehicles can lead to peak demand and thus, there will be a significant impact on reliability of distribution system	Monte Carlo simulation method was used to analyse energy storage system.	Optimum value of size & cost of ESS was obtained.	-
(Al-Abdulwahab et al., 2011)	Reliability assessment of distribution system with innovative Smart Grid technology implementation	8	To study the imapct of the protection schemes in the reliability of distribution network	Ground Fault Neutraliser (GFN) scheme. In GFN, ASC and RCC is used to compensate fault current. RBTS 6 bus system is used as test system	as SAIFI, SAIDI and ENS.	The method is very useful in reliability of underground cables.
(Chen et al., 2018)	Reliability Assessment of Distribution Network Considering Cyber Attacks	6	Integration of cyber system with physical distribution system and increased use of intelligent electronic devices, makes the system more vulnerable to malicious attacks.	IEEE-33 bus test system	Reliability evaluation method of distribution network CPS	-
(Refaat et al., 2018)	Reliability Evaluation of Smart Grid System with Large Penetration of Distributed Energy Resources	6	To assess reliability indices of smart grid with huge penetration of distributed energy resources	IEEE-30 bus test system with large scale photo voltaic and turbines	Integration of distributed generation resources can improve reliability indices of smart grid.	-
(Liu, Deng, Gao, & Sun, 2019)	A reliability assessment method of cyber physical distribution system	5	To analyze the effect of failure in cyber system on fault processing of distribution network..	Event tree model, Sequential monte carlo method	Cyber system will be helpful to make fault processing in power system more reliable.	Consideration of demand response and distributed generation.

continues on following page

Table 1. Continued

Reference	Name of article	Citations	Problem description	Method used for solution	Result	Further scope
(Wadi et al., 2018)	Reliability Evaluation in Smart Grids via Modified Monte Carlo simulation method	4	To determine the reliability of closed-ring topology using Monte Carlo simulation method.	Modified Monte Carlo simulation, RBTS 2 & 4 bus system is used as test system	The method is modified by integration of total loss of continuity (TLOC) which is suitable for both complex closed ring networks as well as smart grids. Reliability indices such as SAIDI, SAIFI, CAIDI, ASAI, ENS, AENS, TTF, TTR are calculated	Modify method using both Total loss of continuity (TLOC) and partial loss of continuity (PLOC)
(Al-Wafi, Al-Subhi, & Al-Muhaini, 2015)	Reliability Evaluation of a practical power system using Monte Carlo simulation method	4	To study the reliability aspects of a practical transmission system in kingdom of Saudi Arabia	Monte Carlo simulation, Next event method to generate random number for different iterations	Effectiveness of upgrades in transmission line as per the deamand by indices such as system availability, non-availability and ENS	we can include more simulation on the power flow and then calculate probability of failure of more than one components and to calculate different system indices
(Honarmand, Ghazizadeh, Hosseinnezhad, Siano, & Systems, 2019)	Reliability Modeling of Process-oriented smart monitoring in the distribution systems	4	Assessing improvement in various reliability indices with use of smart monitoring system.	Markov method for 1. Minimal monitoring and 2. Smart monitoring	Reliability indices will drastically increase by the use of smart-monitoring system.	An optimization frame-work can be utilized with a view to assign the optimum points of reliability and cost.
(Yuan et al., 2019)	Distribution System Reliability Assessment Considering Cyber-Physical Integration	3	To decrease the impact of Cyber-induced dependent failures.	Monte-Carlo Simulation, Test system- IEC 61850, including the modified Roy Billinton Test System (RBTS) with upstream substation.	Values of reliability indices were obtained considering cyber physical integration.	-
(Alabdulwahab, 2017)	Teaching reliability assessment of smart grid protection systems	2	Explaining and understanding smart grid is difficult for both trainer and trainee	Event Tree Analysis (ETA)	Improvement in reliability of breaker-oriented systems can be carried out using intelligent electronic devices in advance protection scheme instead of conventional.	The method helps trainees to work with researchers and to understand and design own event tree for different purposes.

Table 2: Comparison of reliability indices obtained by various methods

Article	Conditions and considerations	System used for calculation	Value of various system indices			
			SAIDI	SAIFI	ASAI	CAIDI
(Al-Abdulwahab et al., 2011)	Conventional+ 100% reliable GFN protection scheme	RBTS bus 2 system	0.55	0.025	-	-
(Chen et al., 2018)	Cyber attack	IEEE 33 bus distribution system	3.3917	-	-	-
(Wadi et al., 2018)	MMCS method for closed ring	RBTS bus 4	4.296624	0.190451	0.9999367	22.56026
(Yuan et al., 2019)	3/2 circuit breakers configuration	RBTS bus 2 system	3.69	0.295	0.999578	12.52
(Refaat et al., 2018)	Large scale PV and gas turbine	Modified IEEE 13 bus radial distribution system				
	1. Grid connected		6.5555	0.1388	0.9993	47.225
	2. Islanded		4.2235	0.0747	0.9995	56.504

Figure 2. Year wise methods citation %

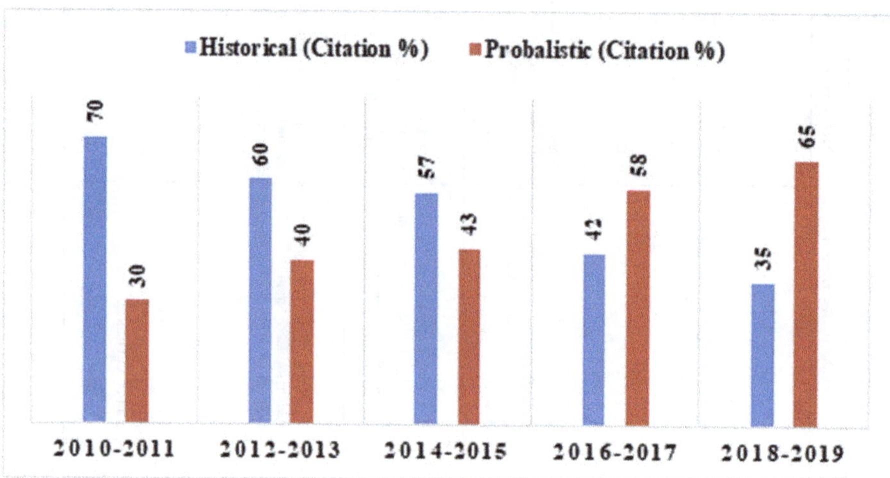

Figure 3. New methods with citation history

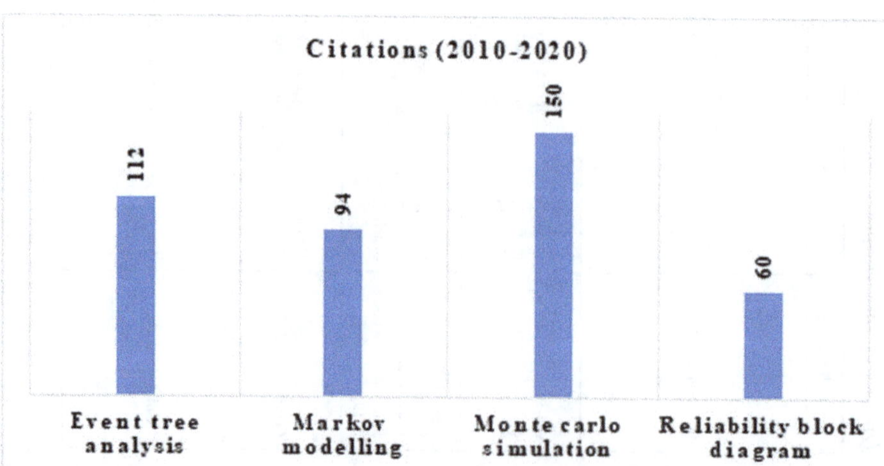

Figure 4. Research work done on topic related to reliability and cyber security of smart grid

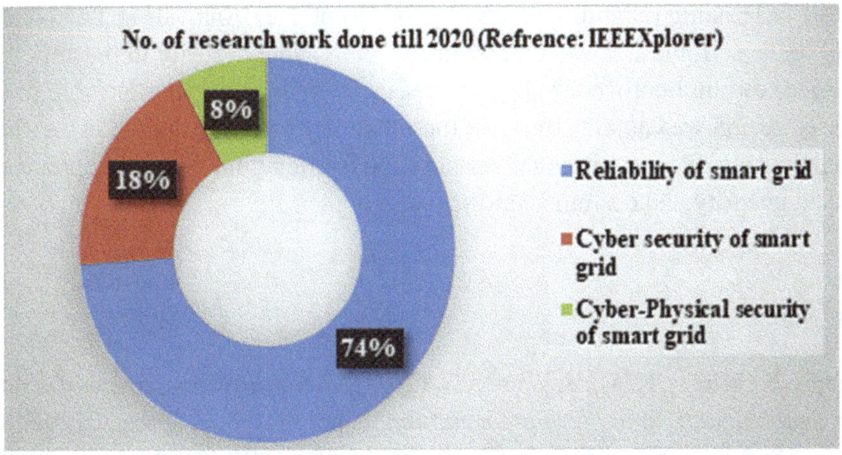

Figure 5 shows the evolution of new techniques used for reliability assessment. In the first stage fuzzy logic plays a vital role for smart grid assessment. Due to some limitations like dependency on human knowledge make it's sluggish. Artificial neural network take the place of fuzzy and remove the regular dependency on human expertise. As smart grid is a complex network which required fast and reliable controllers, which is possible through the new and advance evolutionary theories.

Figure 5. Evolution of new techniques for reliability assessment

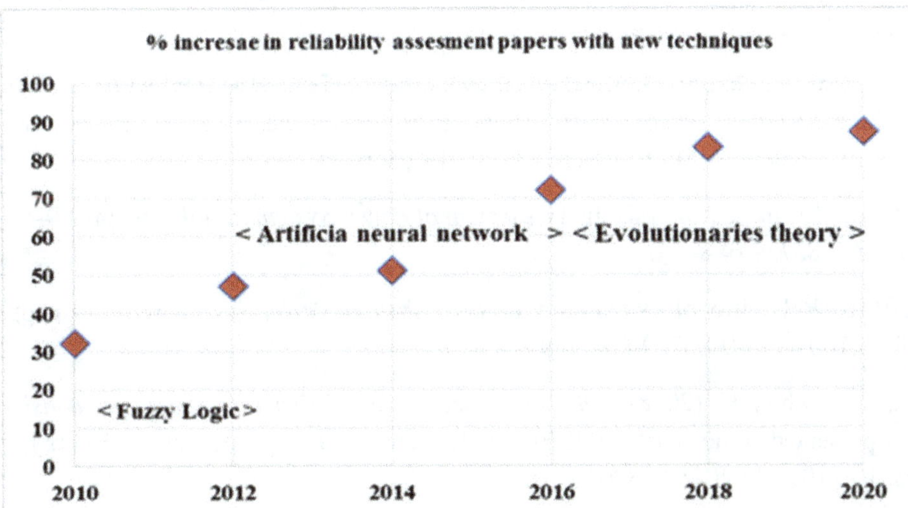

CONCLUSION

We need interference of a two-way communication system in the power grid due to the increased numbers of renewable resources which leads to distributed generations and distributed networks. This system is known as smart grid. Table 1 shows that among various reliability assessment methods reviewed in this paper, Monte Carlo simulation method has many advantages compared to other methods but when it comes to teaching or learning reliability of smart grid, Event Tree Analysis (ETA) is quite simpler for both trainee and trainers. Table 2 shows that by introducing cyber security in the smart grid, the value of reliability indices can be improved. Thus, cyber security plays an important role in the smart grid. Including cyber extrusion we can also improve the reliability by detecting attacks earlier and making lower time to repair using different machine learning and artificial intelligence approaches. More focus is required on data integrity and customer safety.

REFERENCES

Al-Abdulwahab, A. S., Winter, K. M., & Winter, N. (2011). *Reliability assessment of distribution system with innovative Smart Grid technology implementation.* Paper presented at the 2011 IEEE PES Conference on Innovative Smart Grid Technologies-Middle East. 10.1109/ISGT-MidEast.2011.6220780

Al-Wafi, E., Al-Subhi, A., & Al-Muhaini, M. (2015). *Reliability assessment of a practical power system using Monte Carlo simulation.* Paper presented at the 2015 Saudi Arabia Smart Grid (SASG). 10.1109/SASG.2015.7449276

Alabdulwahab, A. S. (2017). *Teaching reliability assessment of smart grid protection systems.* Paper presented at the 2017 IEEE Global Engineering Education Conference (EDUCON). 10.1109/EDUCON.2017.7943036

Albasrawi, M. N., Jarus, N., Joshi, K. A., & Sarvestani, S. S. (2014). *Analysis of reliability and resilience for smart grids.* Paper presented at the 2014 IEEE 38th Annual Computer Software and Applications Conference. 10.1109/COMPSAC.2014.75

Calderaro, V., Galdi, V., Graber, G., Graditi, G., & Lamberti, F. (2014). *Impact assessment of energy storage and electric vehicles on smart grids.* Paper presented at the 2014 Electric Power Quality and Supply Reliability Conference (PQ). 10.1109/PQ.2014.6866775

Čepin, M. (2011a). Event tree analysis. In *Assessment of Power System Reliability* (pp. 89–99). Springer. doi:10.1007/978-0-85729-688-7_6

Čepin, M. (2011b). Reliability block diagram. In *Assessment of Power System Reliability* (pp. 119–123). Springer. doi:10.1007/978-0-85729-688-7_9

Chen, B., Lu, Z., & Zhou, H. (2018). *Reliability assessment of distribution network considering cyber attacks.* Paper presented at the 2018 2nd IEEE Conference on Energy Internet and Energy System Integration (EI2). 10.1109/EI2.2018.8582610

Distefano, S., Puliafito, A., & Computing, S. (2009). *Dependability evaluation with dynamic reliability block diagrams and dynamic fault trees.* Academic Press.

Facchinetti, T., & Della Vedova, M. (2011). *Real-time modeling for direct load control in cyber-physical power systems.* Academic Press.

Falahati, B., & Fu, Y. (2014). *Reliability assessment of smart grids considering indirect cyber-power interdependencies.* Academic Press.

Falahati, B., Fu, Y., & Wu, L. (2012). *Reliability assessment of smart grid considering direct cyber-power interdependencies*. Academic Press.

Gupta, P. K., Singh, N. K., & Mahajan, V. (2020). *Monitoring of Cyber Intrusion in Wireless Smart Grid Network Using Weight Reduction Technique*. Paper presented at the 2020 International Conference on Electrical and Electronics Engineering (ICE3). 10.1109/ICE348803.2020.9122981

Hashemi-Dezaki, H., Askarian-Abyaneh, H., Haeri-Khiavi, H. (2016). *Impacts of direct cyber-power interdependencies on smart grid reliability under various penetration levels of microturbine/wind/solar distributed generations*. Academic Press.

Honarmand, M. E., Ghazizadeh, M. S., Hosseinnezhad, V., Siano, P., & Systems, E. (2019). *Reliability modeling of process-oriented smart monitoring in the distribution systems*. Academic Press.

Jiao, J. (2020). *Application and prospect of artificial intelligence in smart grid*. Paper presented at the IOP Conference Series: Earth and Environmental Science.

Kimani, K., Oduol, V., & Langat, K. J. (2019). *Cyber security challenges for IoT-based smart grid networks*. Academic Press.

Li, Q., Meng, S., Zhang, S., Wu, M., Zhang, J., Ahvanooey, M. T., & Aslam, M. S. (2019). *Safety risk monitoring of cyber-physical power systems based on ensemble learning algorithm*. Academic Press.

Liu, Y., Deng, L., Gao, N., & Sun, X. (2019). *A reliability assessment method of cyber physical distribution system*. Academic Press.

Moslehi, K., & Kumar, R. (2010). *A reliability perspective of the smart grid*. Academic Press.

Refaat, S. S., Abu-Rub, H., Trabelsi, M., & Mohamed, A. (2018). *Reliability evaluation of smart grid system with large penetration of distributed energy resources*. Paper presented at the 2018 IEEE International Conference on Industrial Technology (ICIT). 10.1109/ICIT.2018.8352362

Sarwat, A. I., Domijan, A., Amini, M. H., Damnjanovic, A., & Moghadasi, A. (2015). *Smart Grid reliability assessment utilizing Boolean Driven Markov Process and variable weather conditions*. Paper presented at the 2015 North American Power Symposium (NAPS). 10.1109/NAPS.2015.7335101

Sharma, K., Saini, L., & Reviews, S. E. (2015). *Performance analysis of smart metering for smart grid: An overview*. Academic Press.

Singh, N. K., Gupta, P. K., & Mahajan, V. (2020). *Intrusion Detection in Wireless Network of Smart Grid Using Intelligent Trust-Weight Method*. Academic Press.

Singh, N. K., Gupta, P. K., Mahajan, V., Yadav, A. K., & Mudgal, S. Monitoring Cyber-Physical Layer of Smart Grid Using Graph Theory Approach. In *Control Applications in Modern Power System* (pp. 519–525). Springer.

Singh, N. K., & Mahajan, V. (2020). *Detection of cyber cascade failure in smart grid substation using advance grey wolf optimization*. Academic Press.

Wadi, M., Baysal, M., Shobole, A., & Tur, M. R. (2018). *Reliability Evaluation in Smart Grids via Modified Monte Carlo Simulation Method.* Paper presented at the 2018 7th International Conference on Renewable Energy Research and Applications (ICRERA). 10.1109/ICRERA.2018.8566982

Wang, Q., Cai, X., Tang, Y., Ni, M., & Systems, E. (n.d.). *Methods of cyber-attack identification for power systems based on bilateral cyber-physical information*. Academic Press.

Xiang, Y., Ding, Z., Zhang, Y., & Wang, L. (2016). *Power system reliability evaluation considering load redistribution attacks*. Academic Press.

Yuan, H., Li, G., Bie, Z., & Arif, M. (2019). *Distribution system reliability assessment considering cyber-physical integration*. Academic Press.

Zhang, R., Zhao, Z., & Chen, X. (2010). *An overall reliability and security assessment architecture for electric power communication network in smart grid.* Paper presented at the 2010 International Conference on Power System Technology.

Zhang, X., Bie, Z., & Li, G. (2011). *Reliability assessment of distribution networks with distributed generations using Monte Carlo method*. Academic Press.

Chapter 14

Customer-Operated Solar Photovoltaic System to Improve the System and Customer Reliability:

Solar Photovoltaic System Incorporation for Reliability Analysis of Composite System

Atul Kumar Yadav
Sardar Vallabhbhai National Institute of Technology, India

Lalit Tak
Sardar Vallabhbhai National Institute of Technology, India

Vasundhara Mahajan
https://orcid.org/0000-0002-2698-6096
Sardar Vallabhbhai National Institute of Technology, India

ABSTRACT

In this chapter, the advantage of distributed generation can be seen in terms of system reliability and reliability of customer load. The solar photovoltaic (SPV) system is one of the distributed generations that may lead to the supply of electrical energy. The customer at the site of load demand mainly uses the SPV system. The installation of the SPV system is advantageous for the electrical load demand. Solar systems have greater efficiency for supplying both types of load (i.e., thermal and electrical) simultaneously. The modeling of two power system components (i.e., generation and distribution) can be performed using the Monte Carlo simulation (MCS) technique. The data used for generation modeling is taken from IEEE-RTS (reliability test system) and data for the distribution system is obtained from IEEE-RBTS (reliability busbar test system). The reliability parameters such as average energy not supplied (AENS) and loss of energy expectation (LOEE) are evaluated for the analysis of individual customer reliability and overall system reliability simultaneously.

DOI: 10.4018/978-1-7998-3327-7.ch014

INTRODUCTION

The customer load demand should approaches to meet with the utility generation of available capacity. However, there is a shortage of available power in the area for a particular time zone. Distributed generation (DG) plays a significant role to fulfil the requirements in such an unavailability situation. SPV system is advantageous among the existing distributed generations because solar system can generate the thermal and electrical energy simultaneously (Chamandoust, Bahramara, & Derakhshan, 2020; Xu & Singh, 2012). However, other DG's systems are involve with generation of electrical energy only. As of the customer concern, many of them have installation of DG units so maximum time they used it to serve their load requirements (Manohar, 2009; Selinger-Lutz, GroÃŸ, Wille-Haussmann, & Wittwer, 2020). This will results that the substation is unaware of such load or the load is undetectable to the grid. However, under the failure situation of DG's, the customer used the utility power to meet the load requirements (Obi, Slay, & Bass, 2020; Ridzuan, Roslan, Fauzi, & Rusli, 2020). The customers have different types of load, the solar system is advantageous for the customer those have electrical and thermal load simultaneously. The SPV system contains the heat engine that are used to produce the thermal energy from sun light. Thermal energy is used by prime mover to rotate it and mechanical energy obtain by rotation is converted into electrical energy through turbine (Haghifam & Manbachi, 2011; Naderipour et al., 2020). The unused heat energy of prime mover or the remaining thermal energy can directly used to fulfil the thermal energy requirements of the customer. As of the increment of load demand, DG are more useful to serve the load at the local point of demand. The modelling of different section of power system can be elaborated below with inclusion of SPV generation in the system (Anastasiadis, Poulimenos, Polyzakis, Manousakis, & Vokas, 2019). Transmission system reliability is also a major concern to analysis the overall performance of the system (Yadav & Mahajan, 2019a, 2019b; Yadav, Mudgal, & Mahajan, 2020b). Reliability examination of power system network is a multifaceted process. The whole system analysis will take more time and a lot of processing memory for a computer. Therefore, the power system is divided into sub-systems for reliability analysis. The subsystem of the power system network is separated into generation part, composite part (Generation and transmission), and distribution part for reliability evaluation and planning. Most of the failures in the distribution system are not that much effect on the reliability operation because of the local balancing of load, cheaper structure in comparison with the composite system, and components can be controlled over there. In the last, the overall performance index can be evaluated by combining all three individual subsystem reliability performances (Rai, 2020; Yadav, Mudgal, & Mahajan, 2020a). Here, the SPV system has an advantage to improve the distribution system reliability, as the extra load is compensated through the local customer at the load location or bus by solar activities. As the DG's have very low generating capacity in comparison with the conventional generator. Therefore, this can be served under small changes in load demand. For the drastic load change, the available conventional generator will feed all the load requirements because SPV is used for the small load variations. Therefore, the main aim of the DG is used in the reliable operation of the system. The expected demand that is not supplied by the conventional generator is mainly maintained through the SPV system. Renewable energy sources can play a major role in the future to fulfil the incremental load demand. Solar energy is the one of generating source that can be available at more of the customer. In rural areas, the electricity availability is not sufficient as per the customer requirements. Therefore, the reliability of system and customer in rural areas is very poor. The SPV installation at customer location gives the flexibility to the individual customer (Quiles, RoldÃ¡n-Blay, EscrivÃ¡-EscrivÃ¡, & RoldÃ¡n-Porta, 2020). The distribution system with SPV has various objectives to study such as the system's overall reliability assessment and the customer individual operation and planning for the placement of these SPV units at various load points. The reliability parameters are evaluated in MATLAB by obtaining the probability of individual available and unavailable units in the system with selective load points where the SPV units are connected. The main objective of this paper is classified as follows:

- The reliability analysis of distribution system with SPV units taken at various load points.
- Effect on complete system reliability and distinct customer reliability with incorporation of SPV units in the system at various load points.
- Optimal location of customer operated SPV units.

The paper is summarized as follows: Section I gives the basic idea of the introduction part and case studies regarding the several publications. Section II elaborates the backgrounds of different state modeling of all the subsystems including the SPV generation modeling. Section III gives the system description, which is taken for the analysis of reliability. Section IV discusses the analysis of the results with different reliability parameters. Section V presents the overall concluding remarks of the analysis.

Background

Distribution grids are always a path for electricity distribution at different customer load points. The current situation in the power system network slightly changes due to the incorporation of DG's at different customer sites. The customer who is using the DG's can manage some loads as per requirements. Therefore, some load is invisible to the electric grid when the DG's are under the operating condition. Whenever the DG's failed to operate, a load of such customer can be served through grid utility. Therefore, such a situation in the current scenarios leads to make system in uncontrolled operation. The DG's used here for modelling is taken as the SPV system. The operation of generation and distribution system reliability can be evaluated through reliability indices at different customer location. Further, the reliability analysis of the system can be evaluated by incorporating the SPV system at different customer load points (Yadav et al., 2020a). The paper describes the technique to model generation and distribution system that is further used to estimate the effect of SPV units on the utility controlled system and also on the reliability of power supply to the different customers. In this section, all the subsystems modelling are elaborated including with the SPV generation modelling.

- **System modelling:** The power system components including SPV system modeling can be elaborated in this section which are as follows.
    - **Generation Modelling:** The generator model have two stages for its operation i.e. UP state and DOWN state. UP state indicate the healthy or operating condition for the generating units and DOWN state indicate the unhealthy or shut down condition for the units. The transition from one state to another state can be obtained through factors Mean Time to Failure (MTTF) i.e. from UP to DOWN and Mean Time to Repair (MTTR) i.e. from DOWN to UP. Figure 1 shows the two state modelling of generating units with given transition (S. Mudgal & Mahajan, 2019; S. M. Mudgal, Yadav, & Mahajan, 2019).

$$\text{Mean Time to Repair (MTTR)} = \frac{1}{\mu} \quad (1)$$

$$\text{Mean Time to Failure (MTTF)} = \frac{1}{\lambda} \quad (2)$$

Figure 1. Two state modelling of generating unit

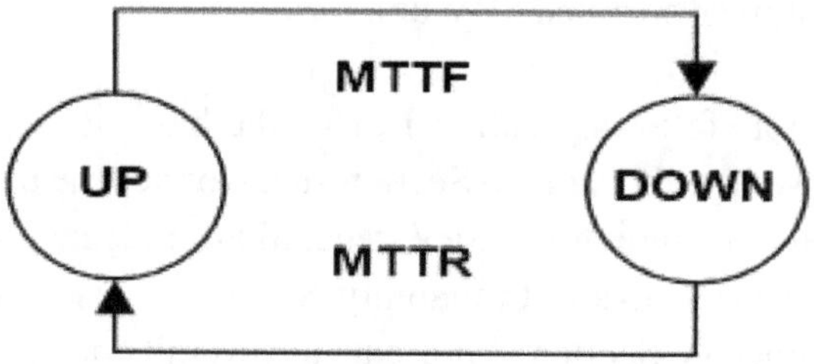

Where, μ is expected repair rate and λ is expected failure rate.

- **Lode Modelling:** The reliability test can be perform by collecting the one-year data of total system load and distinct customer load. As the hourly load details for every customer is also available which can be served by utility or customer themselves. The total distribution load is obtain by summing the individual customer load, which can be served by using the utility not by the customer themselves.
- **Distribution system modelling:** The distribution system modelling has two stages same as the generation model. The UP state of distribution system indicates the operational condition of the distribution line, which indicates the load point is associated through the source or supply point. The DOWN state of distribution system indicate the failure of distribution line connected between load point and the source or supply point (Kumar & Mahajan, 2018).
- **SPV generation modelling:** The SPV model is operated in four states i.e. UP, DOWN, DERATED and FAILED states. The UP state specifies the customer is operating the SPV unit at full generating capacity. The SPV unit can transit from UP state to all other three states. The DOWN state indicates that the customer shut down the SPV units. The DERATED states indicate that the SPV units are operating at derated capacity that is the some percentage of full generating capacity. The FAILED sate occurs in due to unscheduled outages of SPV units. Figure 2 shows the four state modelling of SPV generating units (Pazouki, Mohsenzadeh, Ardalan, & Haghifam, 2016).

For the determination of this paper, the generation and distribution system under consideration is modeled from IEEE - RTS and IEEE - RBTS. Figure 3 shows Block diagram for modelling of different section of power system including SPV units modelling (Kim, James, & Crittenden, 2017). Table 1 shows the customer margin without the SPV units. In this scenario, two conditions are compared that results in the existence of the customer load curtailments. Here, only one combination of the four state results the no load curtailments otherwise, in remaining three states load curtailments will presents. Table 2 shows the customer margin with the SPV units. In this scenario, three conditions are compared. All three conditions including SPV units in the system have seven customer margin states. In which, four combinations of the seven state results into the no load curtailments otherwise, in the remaining three states load curtailments will presents. The states of several combinations will be useful in the analysis of reliability because if there are no load curtailments in the system then the system will be more reliable. Otherwise, if load curtailments have existed in the system, it will system unreliable or less reliable. The reliability level will be identified through the reliability parameters. In general, if load curtailment will be more then the system will be less reliable and if load curtailment is less then the system will be more reliable. The different states of the system and the SPV system combined effect can be seen in terms of reliability indices in the results section. The reliability data related to the SPV units are taken from the paper (Manohar, 2009; Paliwal, 2020).

Reliability Indices: The reliability parameters can be determined through customer margin. The reliability index Loss of Energy Expectation (LOEE) gives the system reliability and index Average Energy Not Supplied (AENS) gives the reliability of the customer. The equation (3-4) shows the formula to evaluate the reliability indices.

$$AENS_k = \frac{\sum_{m=1}^{I_n} ENS_{(customer_k)m}}{I_n} \quad (3)$$

$$LOEE = \frac{\sum_{m=1}^{I_n} ENS_{(system)m}}{I_n} \quad (4)$$

Figure 2. Four state modelling of SPV generating units

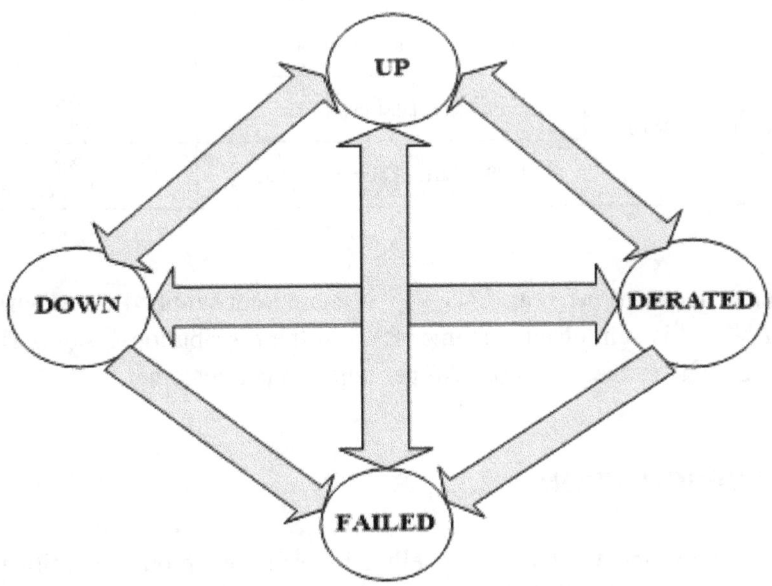

Figure 3. Block diagram for modelling of different section of power system including SPV

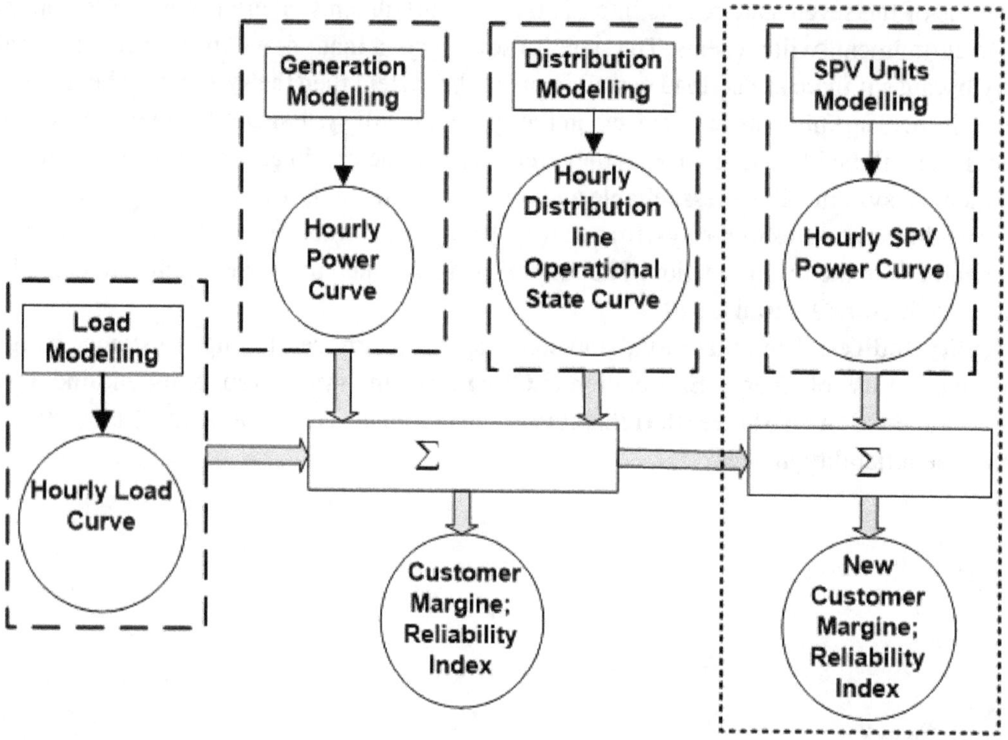

Table 1. Customer margin without SPV units

Scenarios 1	Scenarios 2	Customer Margin
Aggregate generated power by available units > Net load demand	UP State of distribution line	Zero
	DOWN State of distribution line	Portion of customer load curtailment occur
Aggregate generated power by available units < Net load demand	UP State of distribution line	Portion of customer load curtailment occur
	DOWN State of distribution line	Portion of customer load curtailment occur

I_n is the number of iterations per year, $ENS_{customer}$ is energy not supplied (in MWh) to customer k in a specified year, ENS_{system} is sum of all customer ENS in the distribution system. The sum of hourly Customer Margin (CM) gives the values of *ENS* per iteration per customer.

TEST SYSTEM DESCRIPTION

In this paper, the generation and distribution modelling is performed through additional DG's (i.e. SPV) which is used to evaluate the reliability of power supply to the customer and the influence on the utility of the controlled system. The reliability parameters can be obtained and compared with and without customer-controlled SPV units. The data used for generation modelling is taken from IEEE- RTS (Reliability test system) and data for distribution system is obtained from IEEE- RBTS (Reliability Bus-bar test system) (Manohar, 2009; Subcommittee, 1979). Case study involves with installation of DG's (SPV) at different customer load points. The location of customer in distribution system are located at bus bar 2. The distribution test system network and load profile at various customer location is taken from RBTS system (Manohar, 2009). The data of reliability parameters for distribution system are taken from Table 3 that can help in modelling stage of distribution system. Table 4 shows the data of reliability parameter for SPV system units of different capacity. The modelling of SPV system to find the new reliability index have uses the following reliability data of different capacity of DG's. The reliability data of different capacity of SPV system is used to find the availability of the units.

Table 2. Customer margin with SPV units

Scenarios 1	Scenarios 2	Scenarios 3	Customer Margin
Aggregate generated power by available units > Net load demand	UP State of distribution line	For any values of SPV power	Zero
	DOWN State of distribution line	SPV power ≥ Customer load	Zero
		SPV power is either zero or less than the customer load demand	Load curtailment occur
Aggregate generated power by available units < Net load demand	UP State of distribution line	(Power generated by utility units + SPV power) ≥ Customer load demand	Zero
		(Power generated by utility units + SPV power) < Customer load	Load curtailment occur
	DOWN State of distribution line	SPV power ≥ Customer load	Zero
		SPV power is either zero or less than the customer load demand	Load curtailment occur

Table 3. Data of reliability for different components of distribution system network

System components	Failure Rate (failures/yr.)	Repair Time (in hours)
Transmission lines	0.056	4.5
Circuit Breaker (11 kV)	0.008	3
Circuit Breaker (33 kV)	0.001	3
Circuit Breaker (138 kV)	0.0049	6
Transformer (11/0.415)	0.018	180
Transformer (33/11)	0.016	100
Transformer (138/33)	0.02	150
11 kV Busbar	0.002	1
33 kV Busbar	0.002	3

Table 4. Reliability parameter data for SPV System units

Unit size (MW)	MTTF (Hour)	MTTR (Hour)
0.375	1146.35	26.40
0.250	948.04	26.11
0.300	1028.24	26.2277
0.325	1067.91	26.2843
0.550	1411.57	26.7542
0.625	1520.81	26.8947

RESULTS AND ANALYSIS

The renewable integration of power availability to the local customer site lead to secure the burden less power transmission for the existing system. The SPV unit characteristics can be observed through various parameter according to the customer load location and requirements. The reliability indices AENS and LOEE are estimated with no SPV units operating in the system by the customer and further with the inclusion of SPV units in the system. The system reliability parameter LOEE can be evaluated at different percentage of availability of SPV units as shown in Table 5. The customer reliability parameter AENS can be evaluated with and without installation of SPV units in the distribution system at different customer load point. Figure 4 shows the variation in AENS at different load point of distribution system. The minimum value of the AENS index can be found at customer load point 16, which gets improved when the SPV system can be included in the system. The maximum value of the AENS index can be obtained at load point 9, which gets reduce (i.e. reliability is improved) with the incorporation of SPV units in the system by the customer. From the Table 5 we conclude that when the percentage of SPV unit capacity will increase, it will result in to decrement in the LOEE (i.e. Reliability of the system is improved). If SPV generation changes from 5% to 50% of the total distribution load, the improvement in reliability are observed from 4.58% to 49.43%. The percentage change in reliability with SPV generation is almost linear in the graphical characteristics. The smallest decrement in percentage reliability is due to the power loss in the system due to the generation of SPV units. The power loss is negative slope characteristics for the linear operation of SPV generation and percentage change in reliability.

Table 5. Variation in LOEE with different percentage SPV generation w.r.t total distribution system load

SPV Generation in percentage of net System load demand	LOEE (MWh/year)	Percentage Improvement in Reliability (%)
5%	48.53	4.58
15%	43.49	14.49
25%	38.49	24.32
50%	25.72	49.43

Figure 4. Variation in AENS with and without SPV at different customer location of load point

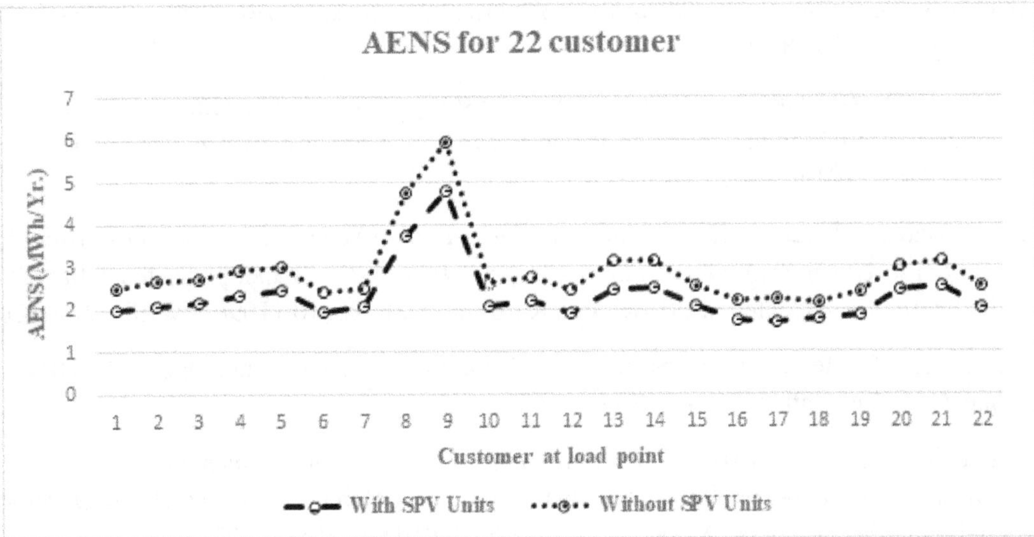

CONCLUSION

This paper discusses a method i.e. SPV generation (one of the DG) by the customer, which may result in the improvement in system reliability as well as customer reliability simultaneously. Rural areas customer have not sufficient electricity fulfilment as per the load and time. Therefore, Solar system at the rural areas in individual customer makes the system reliable and availability of power for the load demand will be more, which can served the load in the time of unavailability of electricity. The capacity of the SPV system can be taken as a different percentage of system load. The results conclude that the increment in the capacity of SPV units will result in the decrement of LOEE index, which may lead to better reliability for the system. The optimal location of SPV units at the customer side can be obtained as the function of the location of the customer from the supply or source point. This analysis is performed by using the data of IEEE-RTS and IEEE-RBTS for the evaluation of reliability indices at different distribution load points. The modelling of the generation and distribution system is combined here to perform the reliability test. In future, including a transmission system for the overall analysis of the system may play an important sign in the field of system modeling.

REFERENCES

Anastasiadis, A. G., Poulimenos, G., Polyzakis, A., Manousakis, N., & Vokas, G. (2019). Algorithms development for the energy management of a micro combined heat and power unit in an AC-DC microgrid. *AIP Conference Proceedings*. 10.1063/1.5117020

Chamandoust, H., Bahramara, S., & Derakhshan, G. (2020). Multi-objective operation of smart stand-alone microgrid with the optimal performance of customers to improve economic and technical indices. *Journal of Energy Storage*, *31*, 101738. doi:10.1016/j.est.2020.101738

Haghifam, M. R., & Manbachi, M. (2011). Reliability and availability modelling of combined heat and power (CHP) systems. *International Journal of Electrical Power & Energy Systems*, *33*(3), 385–393. doi:10.1016/j.ijepes.2010.08.035

Kim, I., James, J.-A., & Crittenden, J. (2017). The case study of combined cooling heat and power and photovoltaic systems for building customers using HOMER software. *Electric Power Systems Research*, *143*, 490–502. doi:10.1016/j.epsr.2016.10.061

Kumar, N., & Mahajan, V. (2018). *Reconfiguration of Distribution Network For Power Loss Minimization & Reliability Improvement using Binary Particle Swarm Optimization.* Paper presented at the IEEE 8th Power India International Conference (PIICON), Kurukshetra, India. 10.1109/POWERI.2018.8704466

Manohar, L. P. (2009). Reliability Assessment of a Power Grid with Customer Operated CHP Systems Using Monte Carlo Simulation. *Masters Theses*, 348.

Mudgal, S., & Mahajan, V. (2019). *Reliability and Active Power Loss Assessment Of Power System Network With Wind Energy.* Paper presented at the IEEE Student Conference on Research and Development (SCOReD), Bandar Seri Iskandar, Malaysia. 10.1109/SCORED.2019.8896327

Mudgal, S. M., Yadav, A. K., & Mahajan, V. (2019). *Reliability Evaluation Of Power System Network With Solar Energy.* Paper presented at the 8th International Conference on Power Systems (ICPS), Jaipur. 10.1109/ICPS48983.2019.9067364

Naderipour, A., Abdul-Malek, Z., Nowdeh, S. A., Ramachandaramurthy, V. K., Kalam, A., & Guerrero, J. M. (2020). Optimal allocation for combined heat and power system with respect to maximum allowable capacity for reduced losses and improved voltage profile and reliability of microgrids considering loading condition. *Energy*, *196*, 117124. doi:10.1016/j.energy.2020.117124

Obi, M., Slay, T., & Bass, R. (2020). Distributed energy resource aggregation using customer-owned equipment: A review of literature and standards. *Energy Reports*, *6*, 2358–2369. doi:10.1016/j.egyr.2020.08.035

Paliwal, P. (2020). Reliability constrained planning and sensitivity analysis for Solar-Wind-Battery based Isolated Power System. *International Journal of Sustainable Energy Planning and Management*, *29*, 109–126.

Pazouki, S., Mohsenzadeh, A., Ardalan, S., & Haghifam, M.-R. (2016). Optimal place, size, and operation of combined heat and power in multi carrier energy networks considering network reliability, power loss, and voltage profile. *IET Generation, Transmission & Distribution*, *10*(7), 1615–1621. doi:10.1049/iet-gtd.2015.0888

Quiles, E., & Rold, Ã. (2020). Accurate sizing of residential stand-alone photovoltaic systems considering system reliability. *Sustainability*, *12*(3), 1274. doi:10.3390u12031274

Rai, A. K. (2020). Advances in Reliability of Solar PV Systems. In *Advances in Energy and Built Environment* (pp. 13–21). Springer. doi:10.1007/978-981-13-7557-6_2

Ridzuan, M. I. M., Roslan, N. N. R., Fauzi, N. F. M., & Rusli, M. A. Z. (2020). Reliability-based DG location using Monte-Carlo simulation technique. *SN Applied Sciences*, *2*(2), 1–11.

Selinger-Lutz, O., & Gro, Ã. Ÿ. (2020). Dynamic feed-in tariffs with reduced complexity and their impact on the optimal operation of a combined heat and power plant. *International Journal of Electrical Power & Energy Systems*, *118*, 105770. doi:10.1016/j.ijepes.2019.105770

Subcommittee, P. M. (1979). IEEE reliability test system. *IEEE Transactions on Power Apparatus and Systems*, *PAS-98*(6), 2047–2054. doi:10.1109/TPAS.1979.319398

Xu, Y., & Singh, C. (2012). Adequacy and economy analysis of distribution systems integrated with electric energy storage and renewable energy resources. *IEEE Transactions on Power Systems*, *27*(4), 2332–2341. doi:10.1109/TPWRS.2012.2186830

Yadav, A. K., & Mahajan, V. (2019a). *Reliability Improvement of Power System Network With Optimal Transmission Switching.* Paper presented at the IEEE 1st International Conference on Energy, Systems and Information Processing (ICESIP), Kancheepuram. 10.1109/ICESIP46348.2019.8938283

Yadav, A. K., & Mahajan, V. (2019b). *Transmission System Reliability Evaluation by Incorporating STATCOM in the System Network.* Paper presented at the IEEE Student Conference on Research and Development (SCOReD), Perak, Malaysia. 10.1109/SCORED.2019.8896263

Yadav, A. K., Mudgal, S., & Mahajan, V. (2020a). Monte Carlo Simulation Application in Composite Power System Reliability Analysis. In *Control Applications in Modern Power System* (pp. 379–385). Springer.

Yadav, A. K., Mudgal, S., & Mahajan, V. (2020b). *Transmission Switching Based Available Transfer Capability Assessment to Make System Network Reliable.* Paper presented at the International Conference on Electrical and Electronics Engineering (ICE3), Gorakhpur.

Chapter 15
The Growing Need of Renewable Energy in India

Anurag Kumar
UNSIET, VBS Purvanchal University, India

Anurag Singh
UNSIET, VBS Purvanchal University, India

ABSTRACT

This chapter discusses the current situation of renewable energy and the growing need for renewable energy. The present and past research revealed that the Ministry of New and Renewable Energy has done a great job and heading India towards renewable energy, but this is not done yet. India has not only sufficient climate condition, but also a large surface area which set a good chance for India to lead in the renewable sector in the world. An effort has been made to summarize the current scenario, benefits, and recent development of renewable energy.

INTRODUCTION

India is one of the fast-developing nations in the world, and the electrical energy consumption of India is continuously increasing from the last few years. As in 2016 per-capita, energy consumption is 1075 kWh which is increased to 1181 kWh in the year 2019, and it is continuously increasing. The growth of power sector plays an important role for the country to become self-reliant and economically stronger in energy outlook. The shares of different sources as of 2019 are, Coal 54.2%, Hydro12.6%, Renewable Energy Sources (RES) 23.1%, Gas6.9%, Nuclear-1.9%, Diesel-.1%. To full its energy demand India heavily depends upon coal (Paul J S, Sivan A P, 2013). . Coal causes pollution and the emission of greenhouse gas. This affects the environment and ecology to achieve sustainable energy platforms. The future consumption of non-convenient energy should be decrease and dependence on fossil fuel is also replaced by renewable energy. Growing oil prices and oil import dependency is giving an additional economic burden on the country. The country has to focus on RES and reduce coal and oil import to get a safe and secure energy path. It was estimated that India required 174521 MU of electrical energy

DOI: 10.4018/978-1-7998-3327-7.ch015

for year 2021-2022 and 2335987 MU for year 2026-2027 (Paul J S, Sivan A P, 2013). By focusing on renewable energy, India gets an efficient, sustainable and clean energy sources.

The importance of renewable energy sources cannot be overemphasized, for sustainable development and to protect the environment. Though conventional sources might last a little longer, it is not only a question of availability but also affordability. Renewable energy is more important than Non-renewable energy because they are cleaner and easy to use.

CURRENT SCENARIO OF ENERGY SECTOR IN INDIA

In the upcoming future, India will be suffering from an energy crisis if it will not fulfill the growing energy demand. For a developing country like India, this is not only a matter of concern but also a major challenge. It is quite surprising to know that India is the biggest consumer of energy resources in the world after The People's Republic of China and the United States of America yet per capita energy consumption is low in comparison with most of the countries.

Figure 1. Major Energy Sources in India

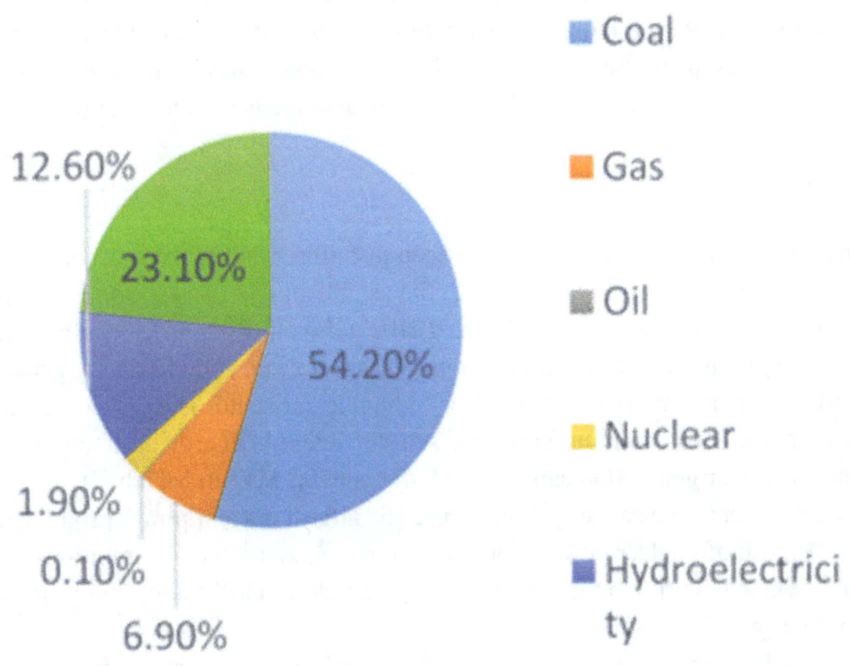

Research and statistics show that India can eliminate poverty and fulfill the energy demand of all the citizens just to uphold 8% to 10% economic growth until 2031-2032. India needs to increase 2-3 times its primary energy supply and 3-4 times its electricity generation capacity (Shyam, Kanakasabapathy p, 2017). India ranked 74th out of 115th in Global Energy Transition Index 2019.

Power Sector

The power sector of India is one of the most diversified in the world. In the Asia Pacific region on an index of measure their overall power, India ranked 4th out of 25 nations in March 2018. In the fiscal year, 2018-19 the gross electricity consumption of India is 1,181 kWh per capita (Digambar S, Saini K J, Sood Y R, 2016) Though India having a low electricity tariff yet per capita electricity consumption was low in comparison with most of the countries. According to IEA (International Energy Agency) if India wants to provide electricity to all its countrymen, it ought to invest at least 135 billion dollars, and it also estimates that country will increase its power generation capacity between 600 GW to 1200 GW before 2050 by installing additional generation plant.

Coal Sector

Coal is the main energy source of India, the total percentage of coal used in India's energy mix is 54.20%, installed power capacity is 56% and the electricity generated is of 70% (Ramana, 2007). India is the 2nd largest coal producer after the People's Republic of China. Fossil fuels dominate India's electricity sector, especially coal. In the fiscal year, 2018-2019, India produces three-quarters of electricity by coal.

In India lignite coal are present in a large amount, the carbon content of lignite coal is low only 40% to 55% and high ash content, and that is why Indian coal sector is one of the most resource-wasteful and polluting sectors in the world. India's Ministry of Environment and Forest has issued a guideline to use high carbon content coals or use those coals whose ash content has been reduced to at most 34% in power plants, especially in polluted and ecologically sensitive areas (Ramana, 2007).

Nuclear Sector

India's excitement towards nuclear energy can be seen just after one year of independence with the development of atomic energy commission and with the establishment of the Atomic energy department in 1954 (National Institute of Bio Energy). In the 1970s only a few countries can produce a complete fuel cycle from the minimum fuel fabrication, Uranium exploration, and electricity generation to waste management and India is among one of them. For India nuclear sector is an important tool to achieve fossil-fuel-free energy in the upcoming future and self-sufficient in energy production

The total nuclear power generation capacity of India is 6780 MW in March 2019, which is close to 2% of the total power generation capacity. India is the 15th largest nuclear power producer country in the world with a 1.2% share of nuclear power plant generation capacity. India's ambitious plan is to fulfill 9% of electricity demands with atomic power by the year 2032 and 25% by the year 2025 (National Institute of Bio Energy).

In India to establish a thermal power generation plant is quite a difficult process because it has to cross the different levels of safeguards and protection standards. Sometimes this process becomes challenging too because the owner has to overcome the public and pressure group perception about its safety after the nuclear disaster at Fukushima Daiichi in Japan.

Oil and Natural Gas

India is one of the biggest oil importers in the world after the United States and China, and also the third-largest oil consumer. The increasing oil demand will lead India towards the largest oil consumer in the world in the upcoming future. It imports about 82.8% of crude oil, accounts for 0.1% of oil, and 6.9% of gas of the total energy consumption in India (Ministry of New and Renewable Energy). Though India suffers from the decreasing global economy, yet demand for energy in India is continually rising.

India has a lack of adequate petroleum reserves, and petroleum demand increases day by day due to the rapid expansion of vehicle ownership, and that's why it depends mostly on crude oil import.

Renewable Energy Sector

The most effective and efficient solution for India to meets its growing power need, and to boost up the economy is to switch towards sustainable energy. Renewable energy contributes about 69,022.39 MW in January 2020, which is 26% of the total installed capacity (Singh D, Sood Y R, 2016) Though the country is blessed with a different type of renewable energy resources (i.e. hydro, wind, and sun), a large portion of it still unbuckle. The main focus of Indian companies is on traditional forms of energy rather than renewable energy.

The Government of India shows great interest in renewable energy and does a lot of work in this sector. There are so many government initiatives for sustainable renewable energy for the future i.e. UDAY Scheme, Solar Park Scheme, Rooftop Scheme, Government Yojana Solar Energy Subsidy Scheme, etc.

RAPIDLY RISING ENERGY DEMAND

India is one of the fastest developing countries in the world, with an increase in developments; demand will also increase i.e. electricity demand and other energy demands. Several scenarios reveal that India has only two paths to fulfill its growing energy demand, first is to save more and more energy in all sectors. In this path, there is more use of fossil fuel and locking of the energy system into today's pattern – with an increased level of air pollution and uncertainties around meeting its sustainable target, and in the second path, India has to increase industrial growth with high utilization of renewable energy.

In 2030, the demand for energy in India will more than double while the electricity demand will be triple (Bhattacharya S C, Jana C, 2009). India's population is the major problem for the country's development, and also for the country's economic growth.

India's large land area and different monsoon season enriched it with a huge potential of renewable energy generation. Renewable energy capacity of India is more than Germany, Japan and some other developed countries, but the overall install power-generation is quite low. Over the past decade, India's electricity demand has grown by 10% (International Renewable Energy Agency). Almost every village in India now has electricity but still millions are living in darkness. According to the World Bank report, around 200 million people are still struggling for electricity in India less than 50% of households have access to electricity for more than 12 hours. If in a village 10% of all public buildings and homes are connected to the grid than that village is considered electrified. To satisfy its electricity demand India imported some electricity from Bhutan in 2015.

Figure 2. Growth in the key economic and energy indicator of India from 2014 to 2030

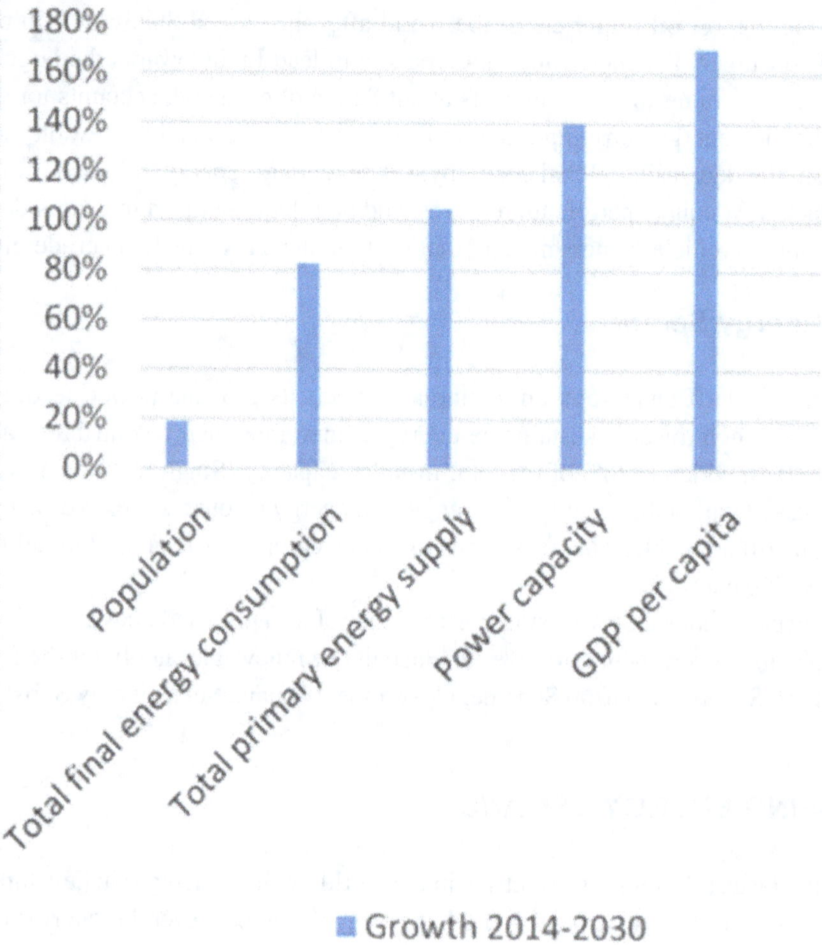

India still facing the growing demands for electricity because of some reasons which are as follow –

1. Low tariff and the problem of allocation of coal have prevented private companies from entering the market.
2. Banks are not lending working capital to the power sector that's why all the under-recovery of fixed and variable costs are pending and delayed.
3. Uttar Pradesh, Andra Pradesh, Tamil Nadu, Maharashtra and Karnataka, are unable to settle all the electric bills.
4. State governments imposed a charge on a consumer who buys electricity from other sources instead of government utilities, and the charge is almost double than the electricity bills.
5. Payments to coal companies and railways are done in advance that's why power plants are unable to purchase it.

In spite of rapid growth in adding power capacity, India still facing extreme energy poverty. If the business continues goes on with the present energy resources and existing environmental policies fossil fuel still dominates over renewable energy. And an industry may also increase the use of coal faster than expected.

POWER SYSTEM CAPACITY AND GENERATION

India's demand for power is rapidly increasing, studies and research reveals that the total power capacity of India will increase from 284 GW in 2015 to 670 GW by 2030 which is more than double while electricity generation increase from 1100 TWh per year to 3450 TWh per year which is more than triple is shown in *Figure 3*. If the current plan and policies persist, the planned growth of renewable power capacity is significantly low (Press Information Bureau, G. O. I). If the development in the renewable sector did not boost up, more than three-quarters of this new production is fulfilled by new coal-based production.

Figure 3. Power system capacity and generation, 2015-2030

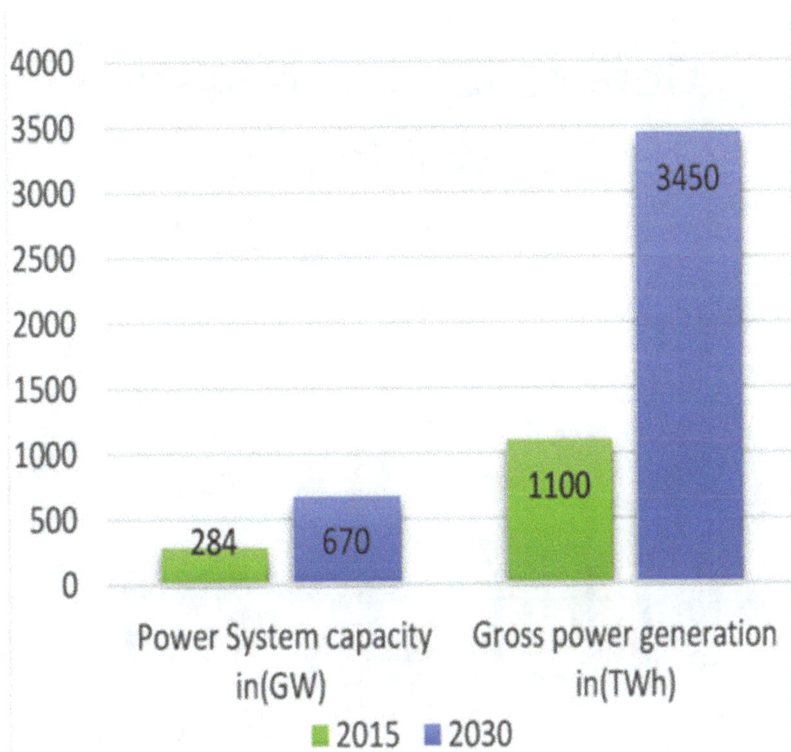

India is blessed with several types of renewable energy, i.e. hydropower, ocean power, wind energy, bio-energy, and mainly with solar energy.

RECENT DEVELOPMENT IN RENEWABLE SECTOR

India has installed a total capacity of 84GW of Renewable Energy by the end of November 2019. This includes 37.28 GW from wind energy, 32.52 GW from solar energy, 9.94 GW from bio-energy, and 4.65 GW of small hydro plants (Ministry of New and Renewable Energy). In the last six years, i.e. March 2014 to October 2019, India had installed a total capacity of 47.86 GW of renewable energy.

The growth rate of electricity generation from renewable energy source falls to 5.22% in the year 2019 Between (April-November) months. The country did not see this growth in the past four years. (National Institute of Wind Energy). The generation of electricity from different renewable energy sources in the last six-year is shown in Fig. 4. As shown in Fig.4. The generation from renewable energy is lowest in the year 2019-2020 in comparison with the last six year data. It is also shown from *Figure 4*. that the generation from bio mass and bio gas is declining, and their contribution in renewable sector is almost negligible in comparison with other renewable energy resource. The latest report shows that procurement for around 30-40 GW is in different stages.

Figure 4. Electricity generation from various energy sources

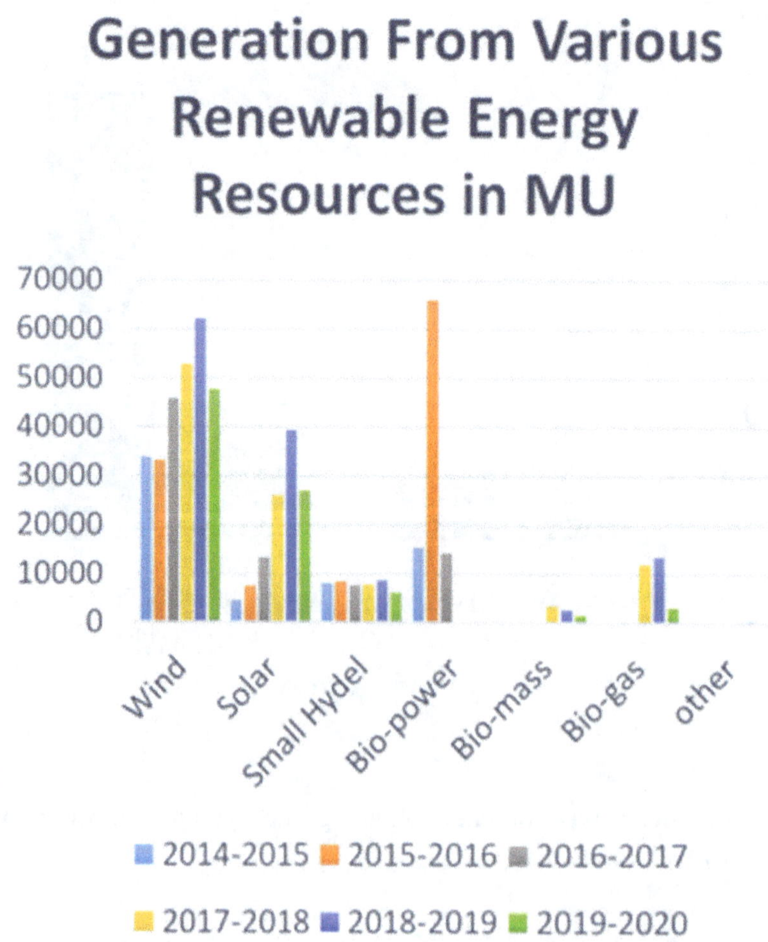

The Government of India has set an ambitious target of installing 175 GW additional renewable energy till the year 2022 which includes 100GW, 60GW, 10GW and 5 GW from solar, wind, bio power and small hydro (Akella A K, Saini R P, Sharma M P, 2009).

DRAWBACKS OF NON-RENEWABLE ENERGY

Causes Air Pollution

The total health impact of outdoor air pollution on India's budget is about 3% of total GDP per year, and indoor air pollution adds significantly to this total. Both urban and rural air pollution can be suppressed if non-renewable energy is replaced by green and sustainable energy forms of energy (Global Alliance on Heath and Pollution, 2020).

The basic cause of premature death in developing and undeveloped countries (i.e. India, Pakistan, etc.) is pollution. Eventually, the death rates are much higher in densely populated and lower-middle-income countries. After China, India is the most densely populated country in the world, it not only appears with the highest number of deaths but also the 10th highest death rate (Ul-Haq A, Jalal M, Sindi H F, Ahmed S, 2020)

In India, total annual premature air pollution-related deaths are1,240,529. India stands second in this list after China. The total number of deaths in India, due to air pollution is very high in comparison with other sources of pollution as shown in *Figure 5*. India ranked 168th out of 180 countries in Environment Performance Index 2020, whereas in air quality, it's ranked 179 which is worst (Kothari D P, 2000). The data on air pollution-related death is quite alarming for a country like India.

Figure 5. Deaths in India due to different types of pollution

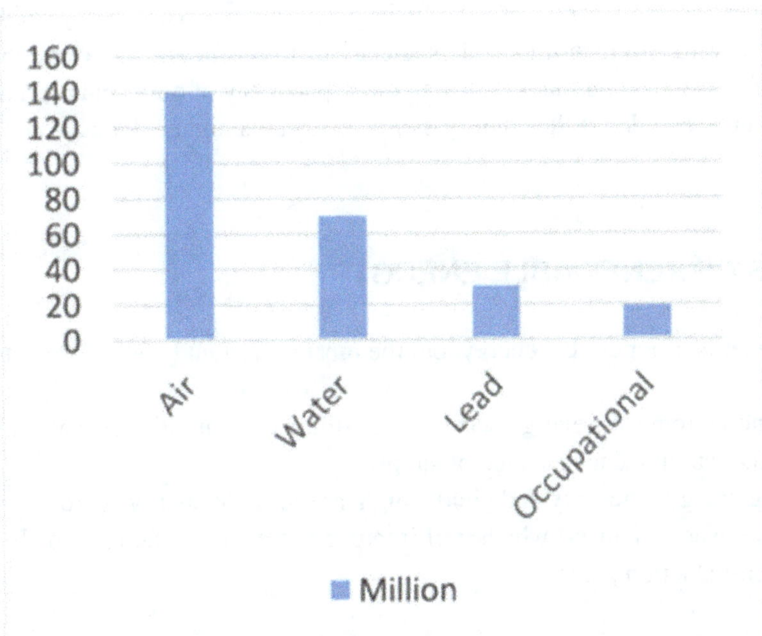

Lowers the Economy

In the last few decades, it is noticed that India's energy market is one of the fastest-growing energy markets in the world, and it is also expected that soon India's global energy consumption will increase to 18% and become the second-largest contributor by the year 2035. India both produces, and imports fossil fuels. The import percentage of fossil fuel is high in India, which affects India's economy (Elavarasan R M, Shafiullah G, Padmanaban S, Kumar N M, 2020). According to the World Bank, the Indian economy suffers about 86 billion dollar a year equivalent to 4% of the annual gross domestic product in the year 2016 due to the efficiency gap in the power sector.

The production rate of fossil fuel is quite low in comparison with the import rate which adversely affects the country's economy. India has one of the world's largest coal reserves with 10% of global coal reserves. In the future coal, demand will be expected to reach 1300 Mt/yr, with 15% of it (200 Mt) being met from other countries. Coal demand is expected to become twice in the next 10 years if the structure of the coal market remains the same. Though the country produces 1.6 EJ of crude oil, it imports an additional 7.9 EJ of crude oil . In the last few years, the annual domestic production of crude oil has been stagnant to 0.83 million barrels per day. The country's natural gas production has reached 1.3 EJ and it imports an additional 0.3 EJ. The consumption of natural gas, including LNG, its share in the total energy system of India is expected to increase 20% by the year 2025. For natural gas import India has to construct cross-border pipelines (e.g. from Turkmenistan and Trans-Afghanistan) which affects the Indian economy. The demand for natural gas is growing worldwide because it is an efficient and clean form of energy, natural gas consumption increase by 12% per year in India. An enormous populated country like India whose energy system mainly depends on non-renewable energy (i.e. coal) will be subjected to many risk and bring a larger energy bill. India's less carbon content coal poses multiple problems such as low calorific value, produces less energy, and a high amount of ash (Jaswinder S, 2016)

In India, demands for all types of fossil fuel are increasing, whereas domestic production is reduced which causes the country to face import dependency. The shares of crude oil are the highest of 82.8% followed by natural gas of 45% and coal of 13%. In the recent study of The Energy and Resource Institute (TERI), concludes that the average import dependency of fossil fuel in India will increase to 74% by year, 2030-2031. If a business continues, as usual, the dependency of both coal and natural gas increase enormously, this can be avoided if the country switches towards higher shares of renewable (Sarkar A N, 2010)

THE BENEFITS OF RENEWABLE ENERGY

There are many benefits of renewable energy, but the most important benefits are listed below:-

- Renewable energy improves energy access for destitute communities as well as reinforces energy security through an indigenous source of supply.
- It accelerating energy efficiency and yield a higher renewable energy share.
- It restrains the overall demand which will improve the energy efficiency of India to achieve its GHG emission-reduction goals.

- Studies and research indicate that renewable energy creates millions of jobs in India, over 180000 jobs in wind energy (National Institute of Solar Energy), and millions of jobs in solar energy by the year 2022. Renewable transition benefits India at the macroeconomics level.
- India's contribution towards the renewable sector can increase India's future economy. Higher investments in the renewable sector would increase twelve times higher saving than the costs. As a result, the cost of fossil fuel-related air pollution will reduce, and the increase of renewable energy will able to save climate change-related issues.

CONCLUSION

Energy is one of the most important sectors for every nation to boost up its economy and for human welfare development. Some country has fulfilled its energy need by investing a large amount of money in the energy sector, whereas some country like India still faces the problem of an energy crisis. The growing demand for energy in India could not be fulfilled by fossil fuels alone, India needs to focus on a clean and efficient form of energy, i.e. renewable energy and natural gas.

India has a huge potential for sustainable energy resources and from the last few years The Government of India do a lot of development in this sector, but the overall development in the renewable sector is significantly low because the county is still not able to reach its estimated potential. Renewable energy plays a crucial role in India's present and future energy policies. If India heads towards its development plans then it will come one step forward towards its goal of economic development and green energy production which certainly increases the sphere of a possibility for India's growth and development.

REFERENCES

Akella, Saini, & Sharma. (2009). Social, economical and environmental impacts of renewable energy systems. *Renewable Energy*, *34*(2), 390–396.

Bhattacharya & Jana. (2009). Renewable energy in India: Historical developments and prospects. *Energy*, *34*(8), 981–991.

Central Electricity Authority of India. (2019). *Executive Summary on Power Sector Jan-2019*. https://cea.nic.in/reports/monthly/executivesummary/2019/exe_summary-01.pdf

Elavarasan, Shafiullah, Padmanaban, & Kumar. (2020). A Comprehensive Review on Renewable Energy Development, Challenges, and Policies of Leading Indian States With an International Perspective. *IEEE Access : Practical Innovations, Open Solutions*.

Global Alliance on Health and Pollution. (2019). *Pollution and Metrics*. Available: http://gahp.net/wp-content/uploads/2019/12/PollutionandHealthMetrics-final-12_18_2019.pdf

International Renewable Energy Agency. (2017). *Renewable Energy Prospects for India*. Available: http://www.irena.org/-/media/Files/IRENA/Agency/Publication/2017/May/IRENA_REmap_India_paper_2017.pdf

Kothari. (2000). Renewable energy scenario in India. *2000 IEEE Power Engineering Society Winter Meeting. Conference Proceedings.*

Ministry of New and Renewable Energy. (2020). *Annual Reports 2019-2020.* Available: http:// mnre.gov.in/img/documents/uploads/file_f-1585710569965.pdf

Ministry of New and Renewable Energy Government of India. (n.d.). *Vision.and.Mission.* Available: http://mnre.gov.in/mission-andvision-2/mission-and-vision/

National Institute of Bio-Energy. (n.d.). Available: http://nibe.res.in/

National Institute of Solar Energy. (n.d.). Available: https://nise.res.in/

National Institute of Wind Energy. (n.d.). Available: http://niwe.res.in/

Paul, Sivan, & Balachandran. (2013). Energy sector in India: Challenges and solutions. *2013 International Conference on Green Computing,Communication and Conservation of Energy (ICGCE).*

Press Information Bureau, Government of India, Ministry of Power. (2019). *47.86 GW ofRenewable Energy Capacity Installed in Last Six Years.* Available: https://pib.gov.in/Pressreleaseshare.aspx?PRID=1596217

Ramana, M. V. (2007). *Nuclear power in India: Failed past, dubious future.* Available at www. npec-web. org/Frameset. asp

Sarkar. (2010). *Global Climate Change and Sustainable Energy Development: Focus on Emerging Issues and Strategies for the Asia- Pacific Region.* Strategic Planning for Energy and the Environment.

Shyam, B., & Kanakasabapathy, P. (2017). Renewable energy utilization in India—policies, opportunities and challenges. In *2017 International Conference on Technological Advancements in Power and Energy (TAP Energy).* IEEE. 10.1109/TAPENERGY.2017.8397311

Singh, D., Saini, J. K., & Sood, Y. R. (2016). The development and potential of wind power sector in India. In *2016 International Conference on Electrical Power and Energy Systems (ICEPES).* IEEE.

Singh, J. (2016). Management of the agricultural biomass on decentralized basis for producing sustainable power in India. *Journal of Cleaner Production.*

Singh & Sood. (2016). Technoeconomic development of renewable energy sector in India. *2016 IEEE Region 10 Humanitarian Technology Conference (R10-HTC).*

Ul-Haq, Jalal, Sindi, & Ahmad. (2020). Energy Scenario in South Asia: Analytical Assessment and Policy Implications. *IEEE Access : Practical Innovations, Open Solutions.*

Chapter 16
Solar PV Installation for Conventional Shutdown Units of Delhi

Pavan Gangwar
Indian Institute of Information Technology, Allahabad, India

Sandhya Prajapati
Indian Institute of Technology, Roorkee, India

E. Fernandez
Indian Institute of Technology, Roorkee, India

Ashutosh Kumar Singh
Indian Institute of Information Technology, Allahabad, India

ABSTRACT

Conventional generation is the most reliable option to meet the increased energy consumption in terms of operating performance. However, the increased greenhouse gas emission is a major threat from the conventional generating units due to fuel pollution. Although to meet the increased energy consumption, reliably conventional generating units are inevitable. So the government has taken the initiative to shut down the conventional generating units with higher pollution levels than the defined norms. This imposes the overall load burden to the other state generating units. As Delhi is sufficiently rich with solar radiation, the chapter proposes the solar PV installation to meet the generation gap of shutdown units.

1. INTRODUCTION

With the advancement of technology, the electricity sector has incorporated many renewable energy-based installations to reduce the share of the conventional sources, as the conventional sources are a cause of carbon emission and air pollution. As the electricity sector plays an essential role in greenhouse gas emission, in the light of U.K. governmental targets of securing 10% of electricity generation from renewable resources by 2010 and 20% by 2020 with widespread public support for renewable energy, distributed generators (DGs) are seen to be rapidly increasing in electrical power systems (Brinkman G. et al., 2016, Marcus W. et al., 2015, Miller N. et al., 2015). With the continuous integration of distributed

DOI: 10.4018/978-1-7998-3327-7.ch016

generation systems as the most distributed generation is based on renewable energy sources, it makes the power output of the DG highly uncertain (Ackerman T. et al., 2000). As the availability of renewable energy varies widely with environmental factors, which are difficult to predict. Therefore, the increased renewable penetration becomes challenging to the system reliability. The capacity contribution of renewable generation is not precise (Hoff T. et al., 2008, Madaeni. et al., 2012). It needs to be calculated to achieve a sufficient generation adequacy assessment (whether the installed capacity is sufficient to meet the load) for the renewable generation. To do such a capacity value needs to be calculated, defined as the amount of additional load that can be served with the new generation. The sufficient load-carrying capacity (ELCC) is used as the paper's capacity credit (CC). This capacity is essentially interpreted as the capacity credit (CC). Many predefined indices exist in the literature for capacity credit calculations. Methodology investigated for capacity credit calculation [7] Chronological methods are suited for system operation, and the probabilistic method is helpful in system planning (Castro & Ferreira, 2001). A detailed literature review is reported with the different approach used for the capacity credit broadly classified as reliability-based methods, based on the reliability indices such as loss of load probability (LOLP), loss of load expectation (LOLE) and expected energy does not serve (EENS) (sandhya & E. Fernandez, 2019, Haslett & M.Diesendorf, 1981, Chowdhury A. et al., 2003) and approximation methods like- Garver's approximation (GARVER L., 1986), ELCC approximation for multistate generator and Z method (Dragoon &. Dvortsov, 2006). It has been found out that the ELCC method is accurate and generally regarded as the benchmark for other capacity credit estimation methods (Pelland and Abboud, 2014). Many factors influence the capacity credit of PV systems. Some of them can be broadly classified as -PV penetration (Mills A. et al., 2011, Ding M et al., 2015), the variability of PV load profile time interval and the correlation between PV production and load demand (Richardson &. Harvey, 2015). However, renewables' time interval factor can be solved with the storage batteries (Ungjin Oh et al., 2016, Ungjin Oh et al., 2015, Hu P., 2009, Hu P. et al., 2009) and with an active load interaction between user and utility also known as the demand response (Zhou Y.,2016, Aghaei & Alizadeh,2013, Zheng X. et al.,2015). The reliability indices that define the level of CC need to be chosen carefully for effective system design, as the system reliability can differ with the reliability indices chosen it could be a probability of the period during which supply is not sufficient for the load in hours/year or it could be the measure of not supplied energy in MW/year during the same period. As the reliability include a wide range of definitions, to make it more concise, the reliability assessment for the proposed work includes the system adequacy which relates to the existence of sufficient facilities within the system to satisfy the consumer demand and therefore associated with the static conditions which do not include system dynamic and transient disturbances. Different zones are created for the system to study its main functional zones (Billinton & Allan, 1994). These are- generation systems, composite generation and transmission system and distribution system, as shown in figure 1. Each of these zones can be subdivided to study a subset of the problem.

For the proposed problem, the capacity contribution of the shutdown units can be replaced with the solar PV installation for a sustainable solution. To do so, the capacity contribution of the solar PV needs to be evaluated with the effective load carrying capability (ELCC). Effective load-carrying capability method is adopted for capacity credit calculated based on the hourly sampling; instead of the increased load for the renewable generation, the equivalent conventional generation is calculated while preserving the same reliability level. The reliability indices considered for the study is the loss of load probability (LOLP).

Figure 1. Different hierarchical levels in power system

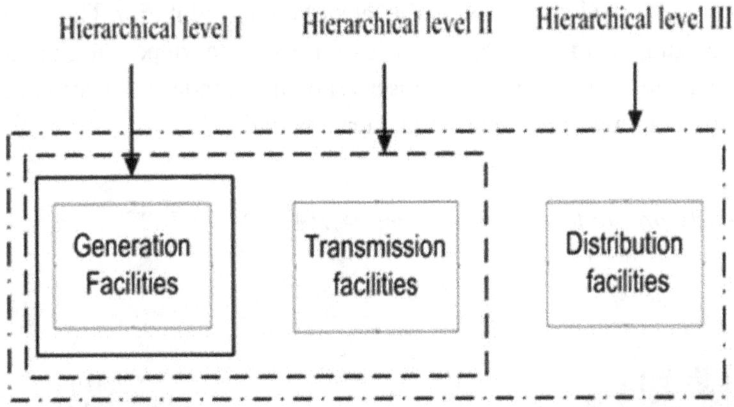

2. PROBLEM STATEMENT

For the proposed work, a case study is considered for New Delhi. Many generating units contribute to the overall generation in New Delhi. The total generation constitutes an extensive network involving units from the Delhi state generating station, other Central government generating units and also some units of the Eastern regions of the country. Of late environmental policies in force and pollution control schemes, there are many generating units that have been shut down due to excessive pollution generation exceeding the defined norms. The paper proposes the solar PV installation due to the rich solar radiation availability of Delhi to fill the generation gap of shutdown units and avoid generation burden on other generating units for a sustainable solution. This is proposed using rooftop solar PV panels to utilize the existing roof space of the households which further supports the government schemes of rooftop solar PV installation. The aggregate output can be used to support the existing generation in Delhi. (Delhi Solar Energy Policy, 2015). The abundant availability of the solar radiations for the Delhi location makes solar energy an appropriate choice for solar PV installation. The power output of the solar panel depends upon several environmental factors. The power available can be formulated as in equation (1).

$$P_{pv}(t) = \eta_{pv} \times A_{eff} \times G(t) \times \left(1 - 0.005(T_a(t) - 25)\right) \tag{1}$$

Here the A_{eff} is the effective area of the panel in m2, η_{pv} is the efficiency of the solar panel, $T_a(t)$ and G(t) are the ambient and solar radiation temperature at time t in °C and W/m^2 respectively, η_{pv} is the number of solar PV modules. The available power can be calculated using the annual radiation and ambient temperature load profiles shown in Figure 2.

3. LOAD MODELLING

For load modelling, the historical data of the annual load profile of Delhi has been considered (http://www.delhisldc.org/) with the peak load of 6570.6MW, while the minimum load 1416MW. The annual load demand data has been divided into different steps, considering a constant load for each step. The step size is taken as one hour; thus, the total number of steps for the overall load profile is 8760. The annual data is considered for the load profile in the lack of data availability. The same data is considered for the simulation procedure as the generation varies for each year depending upon the state sampling approach. With the increased number of simulation years, the standard deviation can be maintained in the desired range corresponding to the generation curve variation.

Figure 2. Annual hourly profile for Delhi location 28.7041° N, 77.1025° E (a) Ambient Temperature (b) Solar radiation

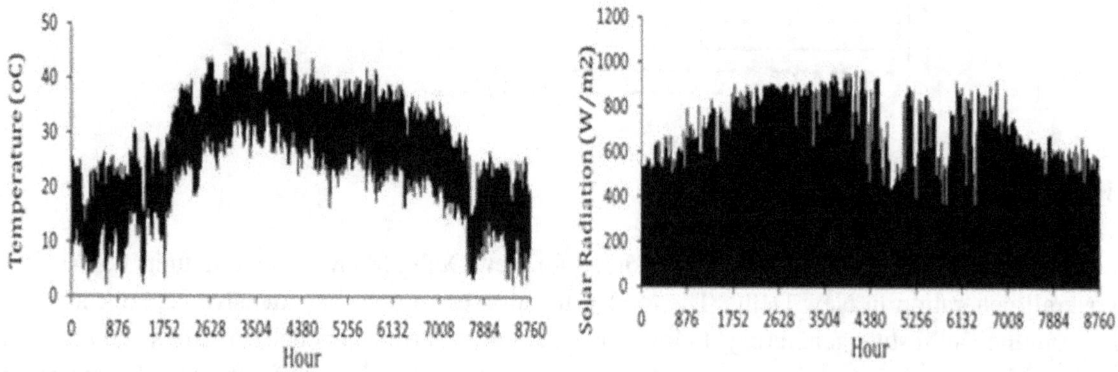

4. MONTE CARLO RELIABILITY

Generating system adequacy assessment is used to evaluate whether the system generation is sufficient to satisfy the load demand or not. This assessment can be conducted using either the Analytical technique or the Monte Carlo technique. For the present study, the Monte Carlo method is considered instead of the Analytical technique as the Monte Carlo can simulate the actual process, including the system's random behaviour better. There are two fundamental simulation approaches – system state sampling and state duration sampling. The state sampling approach is used here for generation system adequacy assessment instead of state duration sampling, as it requires less computing time and storage for an extensive network.

State Sampling Approach: It is seen that many units contribute to the Delhi generation. Generating system adequacy depends upon the possible generating capacity states of the system. Each generating unit can be represented with the uniform distribution Ui between (0- 1). Each generating unit has two states- failure or success. Each unit is assigned a forced outage rate (FOR). Depending upon the FOR of these units, the availability (success state) and unavailability (failure state) of these units can be calculated. Let Si denotes the state of the ith generating unit and FOR_i its failure probability.

$$S_i = \begin{cases} C_i & 0 < U_i < FOR_i \quad \text{Success State} \\ 0 & U_i > FOR_i \quad \text{Failure State} \end{cases}$$

The generation reliability can further be evaluated based upon the aggregation of the available units. Similarly, the annual generation profile is obtained for the available generating units using the same approach. For the computation of the system reliability, many options are available which may be deterministic or probabilistic. For the present study, as there are factors involving weather dependency, it is preferable to adopt the probabilistic approach like the Monte Carlo simulation, which gives a more accurate picture of the stochastic behaviour associated with renewable-based power generation. The generation and load profiles for the desired system are considered on an annual basis. The Monte Carlo technique involves repeated simulation using the database.

For this purpose, a random hour is generated within a year to test if the load is sufficiently met with the available generation. If so, the system is considered reliable for that period. The reliability indices chosen for the study is LOLP.

Loss of Load Probability: Loss of load probability is the measurement of the probability that the generation is not sufficient to meet the load requirement for the given time interval.

$$LOLE = \frac{\sum_{t=1}^{N} LL}{N} \tag{3}$$

Where,

$$LL = \begin{cases} 0 & L(t) < G(t) \\ 1 & L(t) > G(t) \end{cases}$$

L(t)= Load for hour t
G(t)= Generation at hour t
LL= Loss of load

The reliability assessment of the system is done with the Monte Carlo simulation technique, which is discussed in the steps below:

$$\overbrace{\sum_{i=1}^{n} CG + CG_{shutdown}}^{RI^2} \tag{4}$$

$$\overbrace{\sum_{i=1}^{n} CG^{sd} + RG_{install}^{PV}}^{RI^1} \tag{5}$$

$$CC(RG_{install}^{PV}) = CG_{shutdown} \quad \forall RI^1 = RI^2;\ RI \begin{cases} LOEE \\ LOLE \end{cases} \tag{6}$$

Figure 3. Simulation procedure for the Monte Carlo reliability calculation

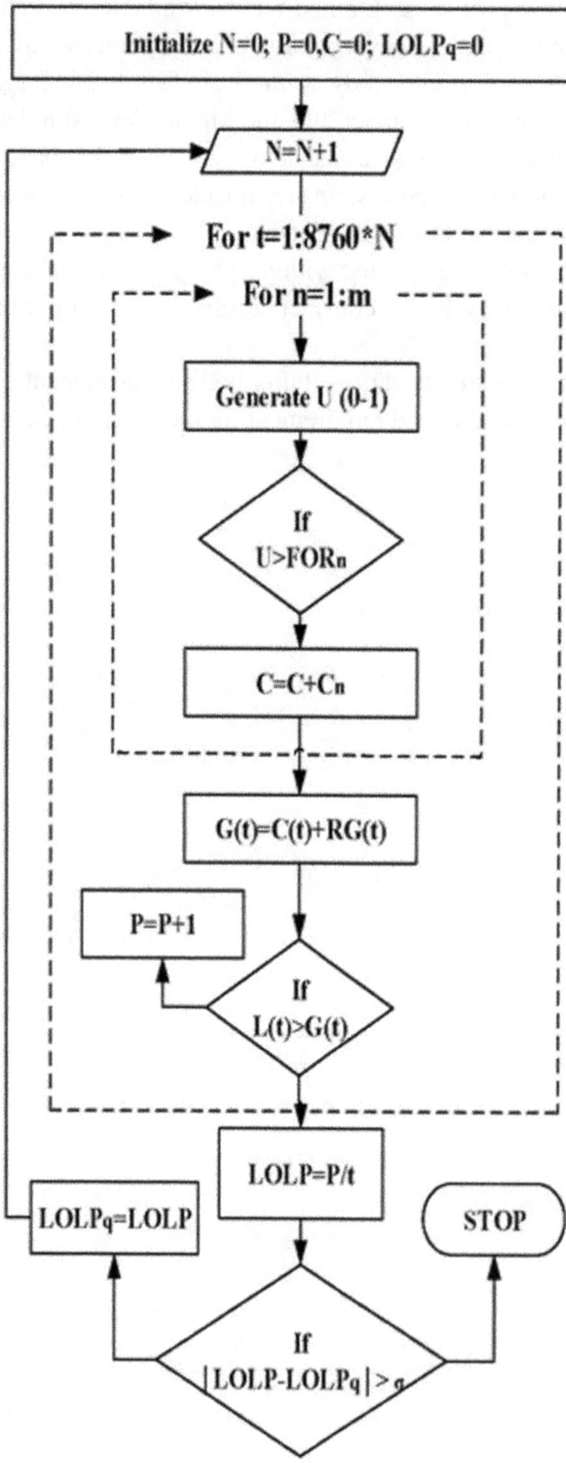

5. METHODOLOGY

To reduce the dependency of the existing generation system on fossil fuels by incorporating solar PV installation, it is necessary to maintain the reliability of the system to the same level. In the present study, we are installing solar PV to fill the gap of the shutdown conventional generating units. To evaluate the replacement amount of conventional shutdown units instead of adding the solar PV for the increased load profile for the same level of reliability. The available generation is considered (after removing shutdown units) for the same load profile. The system with this generation and load profile will be less reliable than the original system, as the generation is reduced for the same load profile. Now to achieve the same reliability level of the original system (including shutdown units) (Eq. (4)), the solar PV installation is added to the remaining generation until the reliability level of the system restored to its previous value as in Eq. (5). The renewable generation corresponding to the same reliability value is known as the capacity credit. As the contribution of renewable generation cannot be achieved without battery storage due to the unavailability of solar power during night time. Thus, the required amount of battery storage is considered along with the solar installation step. As the solar installation increases to achieve the same reliability level, the requirement of battery storage reduces. To make any choice about solar installation, the optimum solution needs to be chosen as the different combination of the battery and solar PV fulfils the requirement of compensating for the conventional shutdown units. The range of the solar PV installation is needed to be chosen carefully, as the cost is a significant barrier for solar PV installation; thus, for the proposed study, the renewable penetration (renewable capacity to the peak load) is restricted to 30% for economic reasons.

battery storage reduces. To make any choice about solar installation, the optimum solution needs to be chosen as the different combination of the battery and solar PV fulfils the requirement of compensating for the conventional shutdown units. The range of the solar PV installation is needed to be chosen carefully, as the cost is a significant barrier for solar PV installation; thus, for the proposed study, the renewable penetration (renewable capacity to the peak load) is restricted to 30% for economic reasons.

The addition of the solar PV units is done in the step size of the 100MW. With the installed capacity of solar PV for each step, the battery storage is added to maintain the original level of system reliability. It will be easy to figure out the trends of capacity credit with the stepwise solar PV installation that will be helpful for making any conclusion.

6. RESULT AND DISCUSSION

To meet the generation of shutdown units in Delhi, solar PV installation is done for a more eco-friendly and sustainable system. The solar PV installation is done for renewable penetration up to 40% with the installation step of 5%. The remaining capacity is achieved with the battery storage for original system reliability. The results are shown in figure 4 for the given solar penetration levels considered. It is seen from the figure that with the increased solar PV installation, the reduction in the required battery storage is not linear. For initial solar installation steps, the slope of battery requirement is low as the solar installation increases the slope becomes steeper. Now the best combination needs to be figure out to choose the optimum solution. As the cost is a major hindrance to the solar installation, the life cycle cost of the overall system is calculated for each step. To do so the capital cost, replacement cost and the annual maintenance cost is considered during the life of the projects. The cost contribution of solar and battery is shown in figure 5 and 6 respectively. The overall cost of the system with these units shows a continuous increment with the increased solar installation as the cost of solar PV quite high as compared to the battery. However, the cost does not increase continuously but has a breakaway point for the solar penetration of 35%. Thus, if battery installation needs to be minimized, the solar installation needs to be chosen above 35%. But the reliability problem is also prominent with increased renewable penetration. Thus, for both the reduced battery installation and restricted penetration level, the solar PV is chosen as 2840.26MW corresponding to the 35% solar penetration. The battery storage need for this installation is 418 MWh to compensate for the conventional shutdown units.

Figure 4. Solar PV and Battery requirement vs renewable penetration

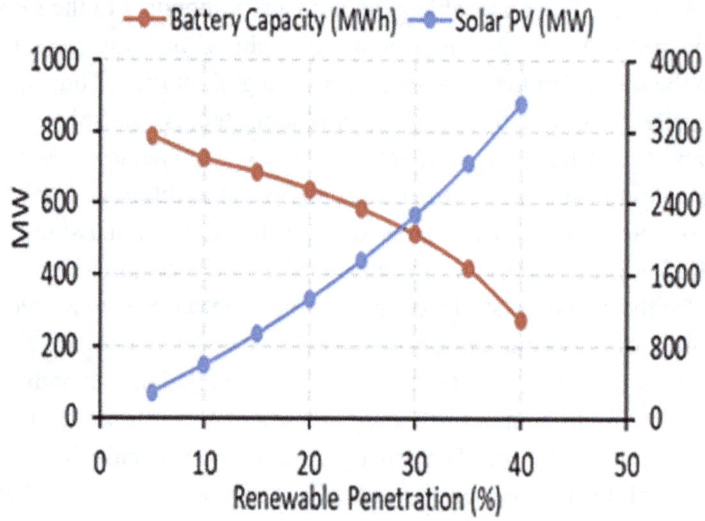

Figure 5. Life cycle cost of Solar PV vs renewable penetration

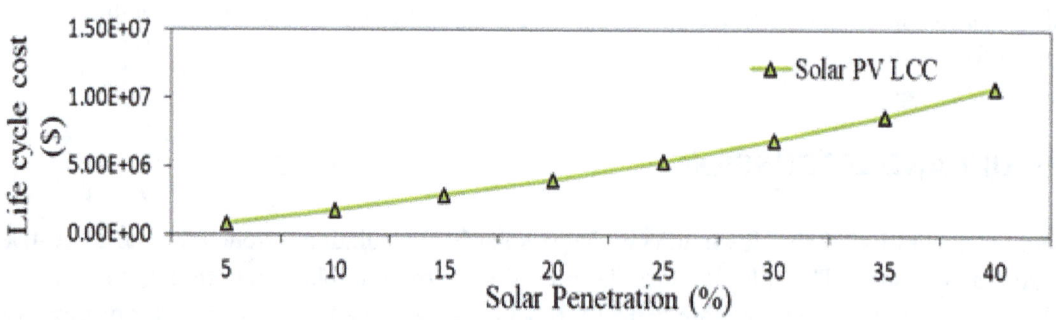

Figure 6. Life cycle cost of battery vs renewable penetration

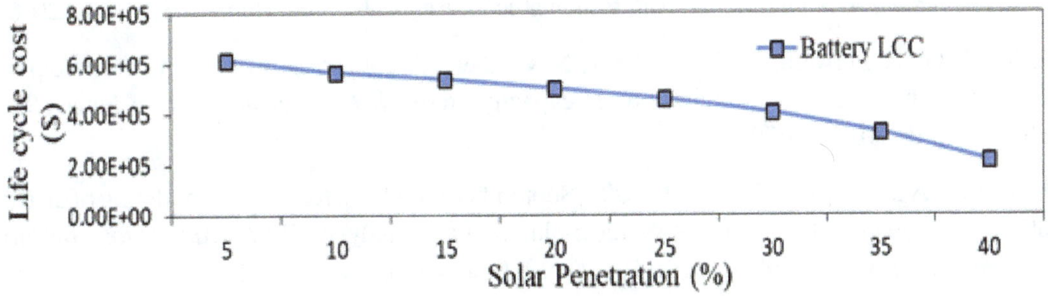

Figure 7. Life cycle cost of system vs renewable penetration

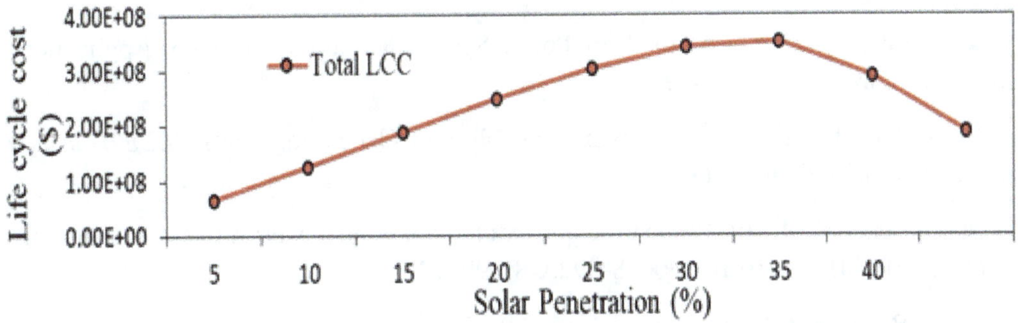

7. CONCLUSION

The conventional generating unit's shutdown at Delhi due to the high pollution level of units than the defined norm. The load burden is imposing on the other state, generating units for satisfactory performance. The paper proposes the solar PV installation to meet the generation gap of the shutdown units. The results show that with the increment in the solar PV (renewable penetration), the battery capacity requirement reduces. Still, the reduction in the battery requirement is not linear and needs to be chosen carefully for the economic system. The breakaway point of the life cycle cost of the solar system occurs at the renewable penetration of 35%.

REFERENCES

Ackerman, T., & Knyazkin, V. (2000). Interaction between distributed generation and the distribution network. *Transmission and Distribution Conference and Exhibition: Asia Pacific IEEE/PES, 2*, 1357–1362.

Aghaei, J., & Alizadeh, M.-I. (2013, February). Demand response in smart electricity grids equipped with renewable energy sources: A review. *Renewable & Sustainable Energy Reviews, 18*, 64–72.

Billinton, R., & Allan, R. N. (1994). *Reliability Evaluation of Power Systems* (2nd ed.). Plenum Press.

Brinkman, G., Jorgenson, J., Ehlen, A., & Caldwell, J. (2016). *Low carbon grid study: Analysis of a 50% emission reduction in California*. Nat. Renewable Energy Lab. Tech. Rep.NREL/TP-6A20-64884.

Castro, R. M. G., & Ferreira, A. F. M. (2001, November). A comparison between chronological and probabilistic methods to estimate wind power capacity credit. *IEEE Transactions on Power Systems, 16*(4), 904–909. doi:10.1109/59.962444

Chowdhury, A., Agarwal, S., & Koval, D. (2003, September/October). Reliability modeling of distributed generation in conventional distribution systems planning and analysis. *IEEE Transactions on Industry Applications, 39*(5), 1493–1498. doi:10.1109/TIA.2003.816554

Ding, M., Xu, Z., Zhao, B., & Bi, R. (2015, September). Solar irradiance model for large-scale photovoltaic generation considering passing cloud shadow effect. *Zhongguo Dianji Gongcheng Xuebao, 35*(17), 4219–4299.

Dragoon & Dvortsov. (2006). Z-Method for Power System Resource Adequacy Applications. *IEEE Transactions on Power Systems, 21*(2).

Garver, L. L. (1966). Efective Load Carrying Capability of Generating Units. *IEEE Transactions on Power Apparatus and Systems, 85*(8).

Haslett, J., & Diesendorf, M. (1981). On the capacity credit of wind power: A theoretical analysis. *Solar Energy, 26*(5), 391–401. doi:10.1016/0038-092X(81)90218-8

Hoff, T., Perez, R., Ross, J. P., & Taylor, M. (2008). *Photovoltaic capacity valuation method*. SEPA, Tech. Rep. 02-08.

Hu, Karki, & Billinton. (2009). Reliability Evaluation of Generating Systems Containing Wind Power and Energy Storage. *Journal of IET Generation, Transmission & Distribution, 3*(8), 783-791.

Hu, P. (2009). *Reliability Evaluation of Electric Power Systems Including Wind Power and Energy Storage* (Thesis). University of Saskatchewan.

Madaeni, Sioshansi, & Denholm. (2012). *Comparison of capacity value methods for photovoltaics in the western US*. NREL, Colorado, Tech. Rep., NREL/TP-6A20-54704.

Marcus, W. (2015). *Low Carbon Grid Study: Comparison of 2030 Fixed Cost of Renewables, Efficiency, Integration With Production Cost Savings*. JBS Energy.

Miller, N. (2015). *Low Carbon Grid Study: Discussion of Dynamic Performance Limitations in WECC*. GE Energy Consulting. Available http://lowcarbongrid2030.org/wpcontent/uploads/2015/08/150429_LCGS-Dynamics-White-Paper_NWMiller.pdf

Milligan, M., & Parsons, B. (1997). A comparison and case study of capacity credit algorithms for intermittent generators. *Proc. Solar'97*. NREL/CP-440-22 591.

Mills, A., Ahlstrom, M., Brower, M., Ellis, A., George, R., Hoff, T., Kroposki, B., Lenox, C., Miller, N., Milligan, M., Stein, J., & Wan, Y. (2011, May/June). Dark shadows: Understanding variability and uncertainty of photovoltaics for integration with the electric power system. *IEEE Power & Energy Magazine, 9*(3), 33–41. doi:10.1109/MPE.2011.940575

Oh, U., Lee, Y., Choi, J., Yoon, Y., Chang, B., & Cha, J.-M. (2016, March). Development of Reliability Contribution Function of Power System including Wind Turbine Generators combined with Battery Energy Storage System. *Journal of KIEE*, *65*(3), 371–381. doi:10.5370/KIEE.2016.65.3.371

Oh, U., Lee, Y., Lim, J., Choi, J., Yoon, Y., Chang, B., & Cho, S. (2015, January). Reliability Evaluation with Wind Turbine Generators and an Energy Storage System for the Jeju Island Power System. *Journal of KIEE*, *64*(1), 1–7. doi:10.5370/KIEE.2015.64.1.001

Pelland, S., & Abboud, I. (2008). Comparing Photovoltaic Capacity Value Metrics: A Case Study for the City of Toronto. *Progress in Photovoltaics: Research and Applications*, *16*(8), 715–724. doi:10.1002/pip.864

Prajapati, S., & Fernandez, E. (2019). Capacity credit estimation for solar PV installations in conventional generation: Impacts with and without battery storage. *Energy Sources. Part A, Recovery, Utilization, and Environmental Effects*, 1–13. Advance online publication. doi:10.1080/15567036.2019.1676326

Richardson, D. B., & Harvey, L. D. D. (2015). Strategies for correlating solar PV array production with electricity demand. *Renewable Energy*, *76*, 432–440. doi:10.1016/j.renene.2014.11.053

Zheng, X., Zeng, B., Wu, G., Zhang, J., Zeng, M., & Shi, J. (2015). Capacity credit assessment of renewable distributed generation in active distribution systems considering demand response impact. *Proc. 5th Int. Conf. Elect. Utility Deregulation Restruct. Power Technol. (DRPT)*, 108–113.

Zhou, Y., Mancarella, P., & Mutale, J. (2016). Framework for capacity credit assessment of electrical energy storage and demand response. *IET Generat. Transmiss. Distrib.*, *10*(9), 2267–2276.

Chapter 17
Thermal Analysis of Realistic Breast Model With Tumor and Validation by Infrared Images

Deepika Singh
Indian Institute of Information Technology, Allahabad, India

Ashutosh Kumar Singh
Indian Institute of Information Technology, Allahabad, India

Sonia Tiwari
Kamla Nehru Memorial Hospital and Regional Cancer Centre Prayagraj, India

ABSTRACT

Breast thermography is an emerging adjunct tool to mammography in early breast cancer detection due to its non-invasiveness and safety. Steady-state infrared imaging proves promising in this field as it is not affected by tissue density. The main aim of the present study is to develop a computational thermal model of breast cancer using real breast surface geometry and internal tumor specification. The model depicting the thermal profile of the subject's aggressive ductal carcinoma is calibrated by variation of blood perfusion and metabolic heat generation rate. The subject's IR image is used for validation of the simulated temperature profile. The thermal breast model presented here may prove useful in monitoring the response of tumor post-chemotherapy for female subjects with similar breast cancer characteristics.

INTRODUCTION

One of the most common cancers among women is breast cancer (BC). In 2018 BC had the second highest rate of mortality among women ("Breast Cancer Facts and Figures 2019-2020," n.d.)("Breast cancer statistics in India," n.d.).Infrared Imaging (IRI) or breast thermography is a comparatively inexpensive diagnostic technique as compared to mammography and has the potential to indicate thermal abnormalities indicating BC. This enables IRI to be used as an adjunct tool to mammography. In IR

DOI: 10.4018/978-1-7998-3327-7.ch017

imaging, an IR camera captures the infrared radiations emitted by a human body and records the variation of surface temperature of that body. Cancer causes patho-physiological changes within the breast including metabolic and vascular changes. The intensity of the radiation emitted is dependent on the temperature of the body, given by Planck's Radiation Law:

$$I(,T)=\frac{2\pi hc^2}{4}\left(e^{\frac{hc}{kT}}-1\right)^{-1} \quad (1)$$

Where I is the intensity of radiation, λ is the wavelength, T is the absolute temperature, h is the Planck's constant, c is the speed of light in vacuum and k is the Boltzmann constant. For human body, most of the radiation is within IR spectrum in the bands between 2 and 20μm. The main purpose of using IR thermography is that it doesn't cause exposure to ionizing radiations and is useful for dense breast tissue. In the present study, thermal simulation of aggressive tumor based on variable blood perfusion rate and metabolic heat generation rate in a realistic three dimensional breast model, is performed. The resulting surface temperature profile is matched with pre-processed and segmented IR images of the subject. The main objective of the present study is to compute the thermal characteristics of aggressive grade 2 breast cancer by calibrating the bioheat model with real clinical data.

BACKGROUND

Cancerous cells produce more heat as compared to normal cells due to the following factors: (1) release of nitric oxide into the blood that results in altered microcirculation (2) Vasodilation (3) Neo-angiogenesis (4) increases metabolic activity of cancerous cell (Kakileti, Manjunath, Madhu, & Ramprakash, 2017) .The rise in temperature of suspicious region can be measured by using advanced infrared cameras. LAWSON (1956) measured the temperature of breast cancer patients and demonstrated that the tumor affected area had temperature higher (max up to 3.5þC), than healthy contra-lateral part of the body. Gautherie (1980, 1983) presented the most exclusive data available till date on metabolic heat generation of cancerous breast tissue by providing empirical relationship between metabolic heat generation rate of breast cancer and its volume doubling time. Based on the computational thermal modelling of breast cancer i.e. solving the Bioheat transfer equation to calculate the surface temperature distribution of the breast embedded with tumor, various studies are presented. Ng and Sudharsan (1999, 2001) validated their steady state numerical model using real IR image from three female subjects; geometry of model was created using the shape of 34 C cup size of mannequin. Mital & Pidaparti (2008) simulated a simplified 2D breast model to determine tumor parameters using evolutionary algorithm. They suggested further investigation in non-uniform 3D breast geometry with chances of increased complexity and computational cost. The authors of Chanmugam (2012) and Hatwar & Herman (2017) performed transient thermal analysis of hemispherical breast model to estimate tumor parameters. González (2011) simulated surface temperature of a hemispherical breast model and validated results for 20 breast cancer patients by variation of metabolic heat generation of tumor in model. Kandlikar et al. (2017) presented a detailed study on current advancements, shortcomings, and future possibilities to improve breast cancer screening methods using numerical simulation, computer-aided diagnosis, and artificial intelligence in the field of dynamic thermography. Singh & Singh (2020) presented a detailed survey on methods used in improving reliability of breast thermography by advanced machine learning techniques and numerical simulation methods. Saniei et al. (2016) estimated breast tumor parameters using gravity defined based on the study by Jiang et al.(2010, 2011). Recently, J. L. Gonzalez-Hernandez et al. (2020) validated surface temperature obtained from Kandlikar et al. (2018) model using IR images of subjects in the prone position. (Lozano et al. (2020) estimated the thermal characteristics of the subject with triple negative invasive ductal carcinoma by calibrating the model using 3D breast scan, magnetic resonance imaging and IR imaging. Augusto et al.(2019) inversely estimated geometric centers of a tumor using a surface temperature of 2D realistic breast model, without prior information of the thermos-physical properties of breast tissues. Based on the detailed literature survey, it can be seen that there is a need to quantify the thermal characteristics-blood perfusion rate and metabolic heat generation rate of tumorous and normal breast by matching computed surface temperature with that of real thermograms of the subject.

METHODS

The present study consists of preprocessing of clinic data obtained from Lozano et al.(2020) using MATLAB to obtain surface temperature parameters from thermograms of subject with invasive ductal carcinoma in left breast and normal right breast. The features are matched with thermal characteristics obtained from 3D model using COMSOL software constructed by using geometrical specifications of the subject and application of Penne's Bioheat Transfer Equation for aggressive triple negative tumor subtype.

THERMAL MODEL

The computational thermal model for representing subject with Grade 2 triple negative receptor status is constructed as three dimensional ellipsoid breast model representing both normal (right) breast and cancerous (left) breast. In right breast model, skin thickness is taken as 1.5mm whereas in left breast model, the thickness is 5.00 mm representing thickening of skin due to advanced cancer stage. Gravity induced deformation Jiang et al. (2010, 2011) is applied by rotating the ellipsoid model 30 degrees around y axis and cut from x-y plane. Dimensions of ellipsoid for both the breasts are based on bounding box dimensions of each CAD model (Lozano et al., 2020) in (Width × Length × Height) format, 8.15 × 15.4 × 4.75 cm (left breast) and 8 × 14.1 × 4.35 cm (right breast).

Right breast model comprises of skin and breast tissue whereas left breast model shown in Figure 1, consists of skin, breast tissue and random shaped tumor formed by merging two spheres. Based on the left breast's thermal profile i.e. location of nipple hotspot and hotspot located approximately 8.0 cm superior to the nipple position at 11 o'clock (CDC, 2001), model the location of tumor is simulated.

Pennes Bioheat Equation for Bioheat transfer

Heat transfer in living tissue (Model & Becker, n.d.) involves multiple mechanisms like heat conduction, heat convection by blood perfusion, and metabolic heat generation. Blood perfusion helps to stabilize tissue temperature relative to the body's core temperature. Mathematically, heat conduction in breast tissue comprising of six layers as well as the tumor, may be expressed by Pennes bio heat equation (Pennes, 1998)

$$\rho_n c_n \frac{\partial T_n}{\partial t} = k_n \nabla^2 T_n + \rho_b \omega_{b,n} c_b \left(T_b - T_n\right) + Q_n \tag{2}$$

n= 1,2,3,4,5,6,7

Figure 1. 3D gravity defined breast model based on bounding box dimensions of cancerous left breast CAD model (Lozano et al., 2020)

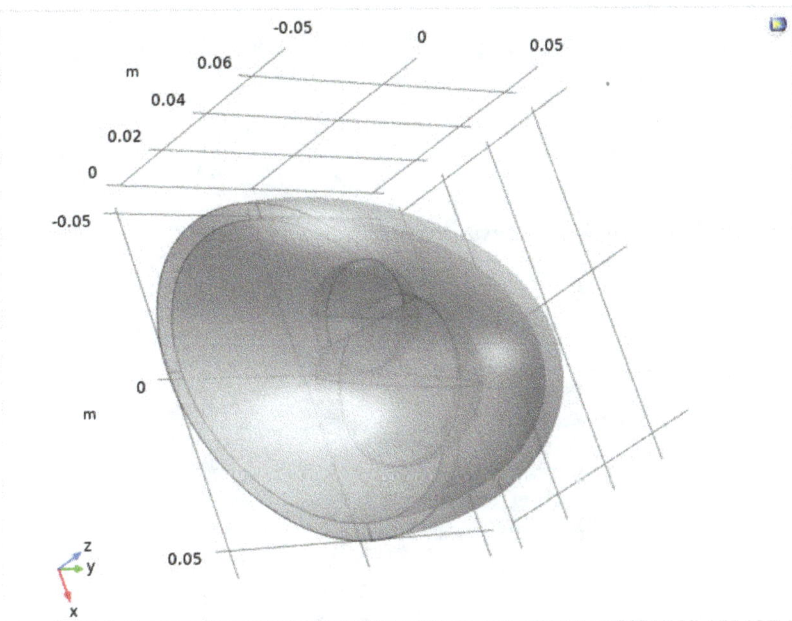

Here, ρ_b, c_b, T_b and $\omega_{b,n}$ represent blood density, specific heat of blood, arterial blood temperature, and blood perfusion rate of the nth tissue layer respectively. Also, ρ_n, c_n, T_n, k_n and Q_n denote the corresponding properties of tissues: density, specific heat, temperature, thermal conductivity, and metabolic heat generation respectively.

Boundary conditions shown in figure 3, for the heat flux continuity and the temperature continuity at the interface of two tissue layers are given by:

$$k_n \frac{\partial T}{\partial \xi} = k_{n+1} \frac{\partial T_{n+1}}{\partial \xi}, \; n = 1,2,3,4,5,6 \tag{3}$$

$$T_n = T_{n+1}, \; n = 1,2.......6 \tag{4}$$

$$k_7 \frac{\partial T_7}{\partial \xi} = k_{adj} \frac{\partial T_{adj}}{\partial \xi} \tag{5}$$

$$T_7 = T_{adj} \tag{6}$$

Where ξ represents a direction perpendicular to the interface between the layers, ka_{dj} and Ta_{dj} are the thermal conductivity and temperature of the tissue layer adjacent to the tumor. The inner boundary condition at the core area of the thoracic wall is defined by the following boundary condition:

$$\frac{\partial T}{\partial r} = 0 \; at \; r = 0 \tag{7}$$

And

$$T_c = 37°C \tag{8}$$

Figure 2. Boundary conditions of 2- Dimensional breast model

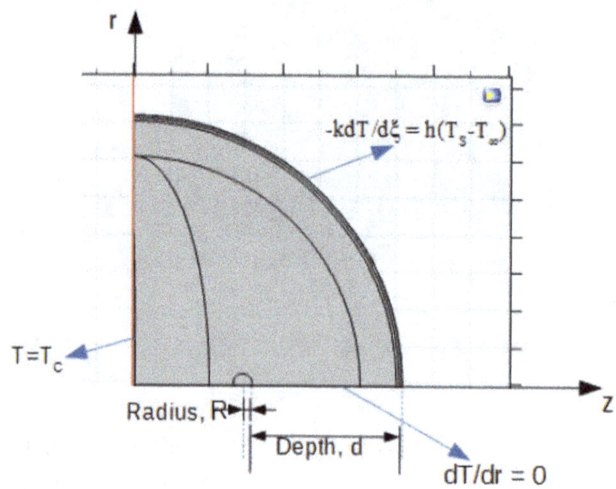

At the breast skin surface, convective boundary condition due to loss of heat to the atmosphere is given as

$$-k_1 \frac{\partial T_1}{\partial \xi} = h\left(T_{(1,s)} - T_\infty\right) \tag{9}$$

where the convective heat transfer coefficient, $h=10Wm^{-2}K^{-1}$, the thermal conductivity of the epidermis, $k_1=0.235Wm^{-1}K^{-1}$, and the temperature of the epidermis at the skin surface interfacing the ambient is $T_{(1,s)}$.

Heat transfer phenomenon occurring in a living tissue involves many complex mechanisms, like heat conduction, heat convection due to blood perfusion and metabolic reactions. Blood flow inside a tissue provides thermal stabilization. Any abnormal change in temperature distribution is normally one of the earliest signs of disease. The irregular and sudden growth of cancer is supported by the formation of new blood vessels, subsequently causing a rise in metabolic activities of the tissues. Pennes Bioheat

Equation is a modified form of the heat transfer mechanism that provides a continuum model to represent the effects of blood flow as a temperature-based heat source term. In the numerical simulation of heat transfer in breast cancer, Pennes model is preferred due to its simple implementation. Moreover, it is suitable for the temperature calculation of capillary vessels of female breast tissue with a diameter of less than 300μm.

THERMOPHYSICAL PROPERTIES AND NUMERICAL SIMULATION

For each breast model, the heat transfer coefficient and arterial blood temperature of skin, breast and cancerous tissue were kept variable. Arterial blood temperature ranged from 33.0 þC to 35.5 þC based on the studies (Gautherie, 1980) for measurement of internal mammary artery temperature. Metabolic heat generation rate and blood perfusion rate of normal and cancerous breast tissue were modelled as adjustable variables to calibrate the computational model with surface temperature obtained in subjects IR image as shown in Figure 2.

Table 1. Thermophysical properties for the thermal breast model (þC)

Thermophysical properties	Skin	Breast Tissue			
		Normal right breast	Non-cancerous left breast	Upper-cancerous	Lower-cancerous
Thermal conductivity, k [W/m.K]	0.266	0.266	0.322	0.564	0.564
Blood density, ρ [kg/m3]	1060	1060	1060	1060	1060
Specific Heat of Blood, C_b [j/kg.K]	3840	3840	3840	3840	3840
Blood perfusion rate, ωb [s-1]	2.22e-3	wb_normal	wb_noncan	wb_uppercan	wb_can
Metabolic heat generation rate, Q_m [W/m^3]	0	Qm_normal	Qm_noncan	Qm_uppercan	Qm_can

Heat transfer coefficient, h was varied from 5 W/ m^2. K to 8 W/m^2.K to match the temperature of left and right breast model with corresponding temperatures room frontal view IR images as shown in table 2.In the steady state analysis, for arterial temperature 308.5K, blood perfusion rates in various tissues in the breast model; wb_noncan varies from 5.98e-3 to 6.668e-3, wb_uppercan varies from 7.356e-3 to 8.56e-3 and wb_can varies from 1.114e-2 to 1.527e-2 (Lozano et al., 2020). Metabolic heat generation rate of breast tissues; Qm_noncan varies from 5.2e4 to 9.7e4, Qm_uppercan varies from 1.03e5 to 1.68e5, and Qm_can varies from 3.05e5 to 5.25e5. For normal right breast model, wb_normal varies from 3.4e- to 1.386e-3 and Qm_normal varies from 2e2 to 2.5e3.

For steady state analysis, tetrahedral meshing with maximum element size 3.3mm is considered and mesh sensitivity analysis is done to ensure that further change in element size does not change temperature more than 0.1%. Each breast model is meshed into predefined finite elements to solve the given set of equations for each tissue layer.

RESULTS

The present study attempts to use clinical data of triple negative grade-2 ductal carcinoma of subject to develop a computational thermal model of breast cancer. High resolution IR image of the subject (Figure 2) were preprocessed and segmented into normal (right) and cancerous (left) breast region using MatlabR2020a version as shown in Figure 3.

Figure 3. IR frontal image of subject with triple negative ductal carcinoma in left breast

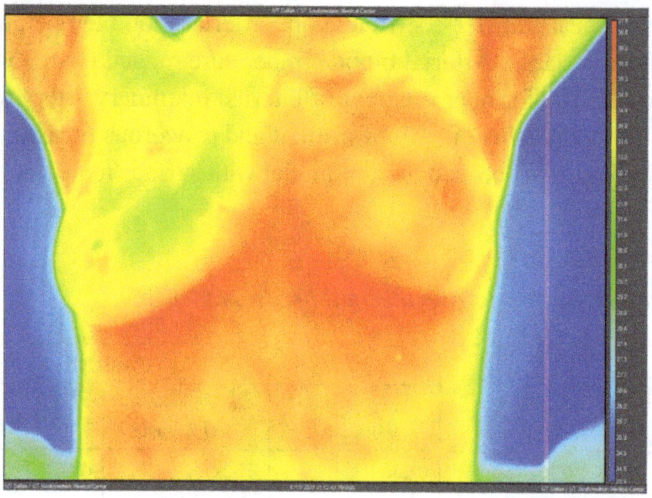

The RGB pixel values of both right and left segmented breast image are converted into corresponding floating point thermal image and histogram as shown in Figure 4 and Figure 5, respectively.

The maximum, minimum and average surface temperature values of both left and right breast are computed from segmented gray scale thermal images. Steady state analysis of computational thermal model of both normal and cancerous breast using PBHT is validated using surface temperatures obtained from subject's IR images as given in Table 2.

Figure 4. Right and left breast region after preprocessing

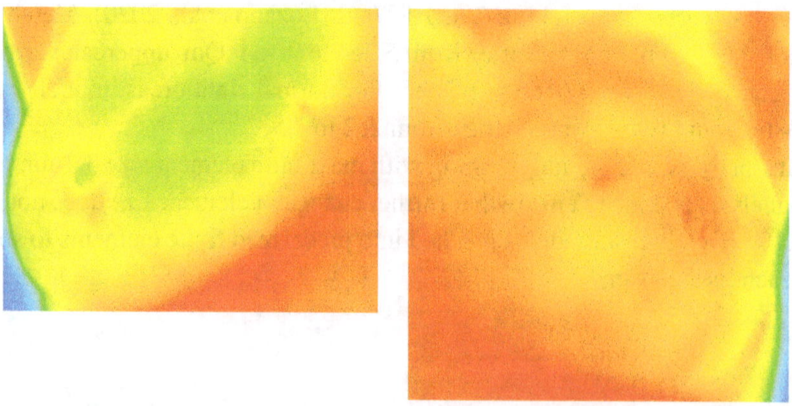

Figure 5. Right breast gray scale image and histogram after conversion from pixel values to temperature in degC

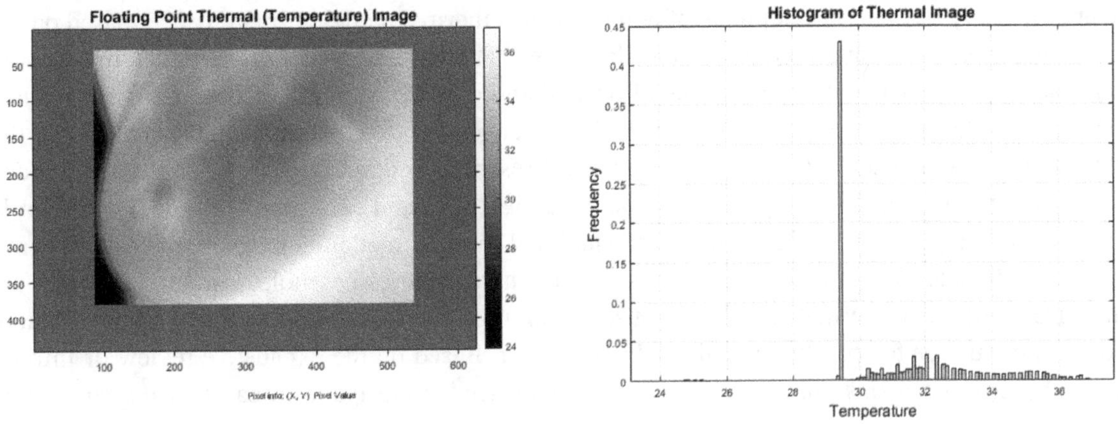

Figure 6. Left breast gray scale image and histogram after conversion from pixel values to temperature in degC

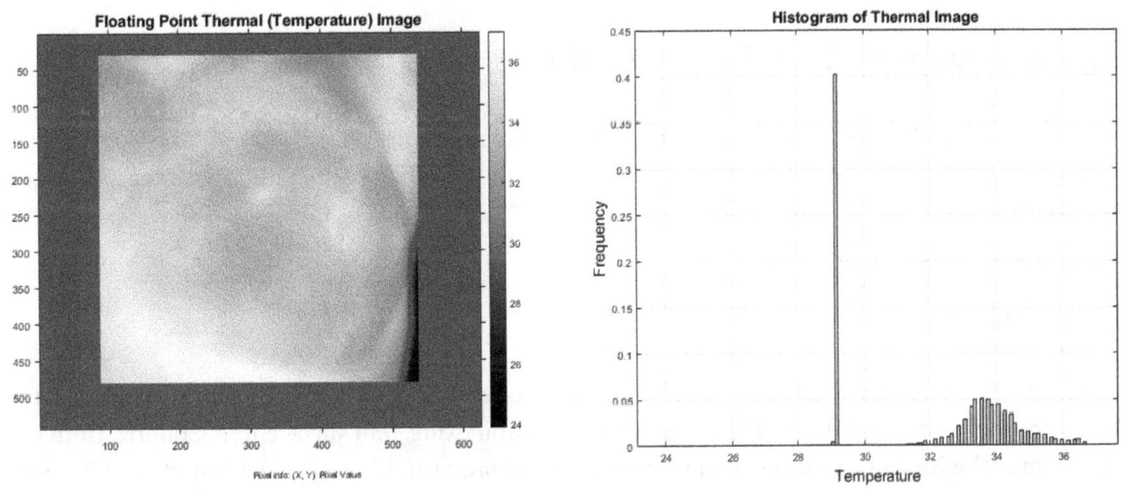

Table 2. IR image temperature and simulated surface temperature (þc)

	Max	*Min*	*Mean*
Left (cancerous) from IR image after preprocessing	36.2	33.4	35
Left (cancerous) simulated	**36.4**	**34.16**	**34.9**
Right (normal) from IR image after preprocessing	35.6	32.6	33.8
Right (normal) simulated	**35.13**	**33.305**	**33.4**

FUTURESCOPE AND CONCLUSION

The present study involves development of computational thermal model of breast cancer based on real breast tumor definitions and high resolution IR images. The thermal model is calibrated by selecting a variable range of thermal characteristics of normal and cancerous breast tissues i.e. metabolic heat generation rate Q_m and blood perfusion rate w_b. These ranges resulted in simulated thermal profile that matched surface temperatures observed on IR images. Present work differs from previous thermal models of breast cancer regarding the application of actual patient data and calibration of model based on IR images. Future application of the present computational thermal model includes monitoring of tumor response for female subject with similar breast cancer characteristics i.e. molecular subtype, stage and tumor size. Further, the thermal model can be used to optimize dynamic thermography for more accurate prediction of surface temperature profile of breast tumor. Based on the exhaustive review of Infrared imaging by (Lozano & Hassanipour, 2019), clinical evidence for and against breast thermography show that it can be used as an adjunct tool for breast cancer screening, especially for women under 40. This enables risk assessment and risk factor for development of breast cancer in future. Future work involves validation of thermal model of grade 1 cancer by clinical data. This may enable early cancer detection thereby assisting physicians in monitoring tumor growth before biopsy.

DECLARATION OF COMPETING INTEREST

The authors declare that there is no conflict of interest regarding the publication of this paper.

This research received no specific grant from any funding agency in the public, commercial, or not-for-profit sectors.

REFERENCES

Augusto, A., Figueiredo, A., Gomes, J., Costa, F., Henrique, L., & Coelho, H. (2019). Computer Methods and Programs in Biomedicine Breast tumor localization using skin surface temperatures from a 2D anatomic model without knowledge of the thermophysical properties. *Computer Methods and Programs in Biomedicine*, *172*, 65–77. doi:10.1016/j.cmpb.2019.02.004 PMID:30902128

Breast Cancer Facts and Figures 2019-2020. (n.d.). Retrieved October 19, 2020, from https://www.cancer.org/content/dam/cancer-org/research/cancer-facts-and-statistics/breast-cancer-facts-and-figures/breast-cancer-facts-and-figures-2019-2020.pdf

Breast cancer statistics in India. (n.d.). Retrieved October 19, 2020, from https://gallery-repo.inshorts.com/gallery/view/796e2a89-eb11-4165-90b7-43bbca6c3738

CDC. (2001). *National Program of Cancer Registries Education and Training Series How to Collect High Quality Cancer Surveillance Data*. Retrieved from https://www.cdc.gov/cancer/npcr/pdf/abstracting/breast.pdf

Chanmugam, A., Hatwar, R., & Herman, C. (2012). Thermal analysis of cancerous breast model. *International Mechanical Engineering Congress and Exposition : International Mechanical Engineering Congress and Exposition, 2012*, 134–143. 10.1115/IMECE2012-88244

Gautherie, M. (1980). Thermopathology of breast cancer: Measurement and analysis of in vivo temperature and blood flow. *Annals of the New York Academy of Sciences, 335*(1), 383–415. doi:10.1111/j.1749-6632.1980.tb50764.x PMID:6931533

Gautherie, M. (1983). Thermobiological assessment of benign and malignant breast diseases. *American Journal of Obstetrics and Gynecology, 147*(8), 861–869. doi:10.1016/0002-9378(83)90236-3 PMID:6650622

González, F. J. (2011). Non-invasive estimation of the metabolic heat production of breast tumors using digital infrared imaging. *Quantitative Infrared Thermography Journal, 8*(2), 139–148. doi:10.3166/qirt.8.139-148

Gonzalez-Hernandez, J.-L., Kandlikar, S. G., Dabydeen, D., Medeiros, L., & Phatak, P. (2018). Generation and Thermal Simulation of a Digital Model of the Female Breast in Prone Position. *Journal of Engineering and Science in Medical Diagnostics and Therapy, 1*(4), 1–34. doi:10.1115/1.4041421

Gonzalez-Hernandez, J. L., Recinella, A. N., Kandlikar, S. G., Dabydeen, D., Medeiros, L., & Phatak, P. (2020). An inverse heat transfer approach for patient-specific breast cancer detection and tumor localization using surface thermal images in the prone position. *Infrared Physics and Technology, 105*(October), 103202. doi:10.1016/j.infrared.2020.103202

Hatwar, R., & Herman, C. (2017). Inverse method for quantitative characterisation of breast tumours from surface temperature data. *International Journal of Hyperthermia, 0*(0), 1–17. doi:10.1080/02656736.2017.1306758 PMID:28540793

Jiang, L., Zhan, W., & Loew, M. H. (2010). Modeling static and dynamic thermography of the human breast under elastic deformation. *Physics in Medicine and Biology, 56*(1), 187–202. doi:10.1088/0031-9155/56/1/012 PMID:21149948

Jiang, L., Zhan, W., Loew, M. H., & Francisco, S. (2011). Toward Understanding the Complex Mechanisms behind Breast Thermography. *An Overview for Comprehensive Numerical Study., 7965*, 1–9. doi:10.1117/12.877839

Kakileti, S. T., Manjunath, G., Madhu, H., & Ramprakash, H. V. (2017). *Advances in Breast Thermography*. New Perspectives in Breast Imaging., doi:10.5772/intechopen.69198

Kandlikar, S. G., Perez-Raya, I., Raghupathi, P. A., Gonzalez-Hernandez, J. L., Dabydeen, D., Medeiros, L., & Phatak, P. (2017). Infrared imaging technology for breast cancer detection – Current status, protocols and new directions. *International Journal of Heat and Mass Transfer, 108*, 2303–2320. doi:10.1016/j.ijheatmasstransfer.2017.01.086

Lawson, R. (1956). Implications of surface temperatures in the diagnosis of breast cancer. *Canadian Medical Association Journal, 75*(4), 309–311. Retrieved from https://pubmed.ncbi.nlm.nih.gov/13343098

Lozano, A., & Hassanipour, F. (2019). Infrared imaging for breast cancer detection: An objective review of foundational studies and its proper role in breast cancer screening. *Infrared Physics and Technology, 97*(December), 244–257. doi:10.1016/j.infrared.2018.12.017

Lozano, A. III, Hayes, J. C., Compton, L. M., Azarnoosh, J., & Hassanipour, F. (2020). Determining the thermal characteristics of breast cancer based on high-resolution infrared imaging, 3D breast scans, and magnetic resonance imaging. *Scientific Reports, 10*(1), 1–14. doi:10.103841598-020-66926-6 PMID:32572125

Mital, M., & Pidaparti, R. M. (2008). Breast Tumor Simulation and Parameters Estimation Using Evolutionary Algorithms. *Modelling and Simulation in Engineering, 756436*, 1–6. Advance online publication. doi:10.1155/2008/756436

Model, P., & Becker, S. M. (n.d.). Analytical Bioheat Transfer : Solution Development of the. In Heat Transfer and Fluid Flow in Biological Processes. doi:10.1016/B978-0-12-408077-5.00004-3

Ng, E. Y. K., & Sudharsan, N. M. (2001). An improved three dimensional direct numerical modelling and thermal analysis of a female breast with tumour. *Proceedings of the Institution of Mechanical Engineers. Part H, Journal of Engineering in Medicine, 215*(1), 25–38. doi:10.1243/0954411011533508 PMID:11323983

Pennes, H. H. (1998). Analysis of tissue and arterial blood temperatures in the resting human forearm. 1948. *Journal of Applied Physiology, 85*(1), 5–34. doi:10.1152/jappl.1998.85.1.5

Saniei, E., Setayeshi, S., Akbari, M. E., & Navid, M. (2016). Parameter estimation of breast tumour using dynamic neural network from thermal pattern. *Journal of Advanced Research, 7*(6), 1045–1055. doi:10.1016/j.jare.2016.05.005 PMID:27857851

Singh, D., & Singh, A. K. (2020). Role of image thermography in early breast cancer detection- Past, present and future. *Computer Methods and Programs in Biomedicine, 183*, 105074. Advance online publication. doi:10.1016/j.cmpb.2019.105074 PMID:31525547

KEY TERMS AND DEFINITIONS

Adjunct: Something that is joined or added to another thing but is not an essential part of it.

Biopsy: The removal of some tissues from somebody's body in order to find out about a disease that a person may have.

Blood Perfusion Rate: Volume of blood per unit time per unit tissue mass.

Continuum: Gradual transitions from condition to another without abrupt change.

Contra-Lateral: Denoting the side of the body opposite to that on which a particular condition occurs.

Ductal Carcinoma: Presence of abnormal cells inside a milk-duct in the breast.

Pre-Processing: Operations on images at the lowest level of abstraction in order to improve the image data.

Thermal Conductivity: Quantity of heat that passes per unit time through a unit area of a particular thickness, where opposite faces differs in temperature by 1 K.

Chapter 18
Metaheuristic Techniques of Parameter Estimation of Solar PV Cell

Nikita Rawat
Graphic Era University (Deemed), India

Padmanabh Thakur
Graphic Era University (Deemed), India

ABSTRACT

The performance and efficiency of a solar PV cell are greatly dependent on the precise estimation of its current-voltage (I-V) characteristic. Usually, it is very difficult to estimate accurate I-V characteristics of solar PV due to the nonlinear relation between current and voltage. Metaheuristic optimization techniques, on the other hand, are very powerful tools to obtain solutions to complex non-linear problems. Hence, this chapter presents two metaheuristic algorithms, namely particle swarm optimization (PSO) and harmony search (HS), to estimate the single-diode model parameters. The feasibility of the metaheuristic algorithms is demonstrated for a solar cell and its extension to a photovoltaic solar module, and the results are compared with the numerical method, namely the Newton Raphson method (NRM), in terms of the solution accuracy, consistency, absolute maximum power error, and computation efficiency. The results show that the metaheuristic algorithms were indeed capable of obtaining higher quality solutions efficiently in the parameter estimation problem.

INTRODUCTION

Over the last few decades, the solar PV has received substantial attentions because of its illimitable availability, non-polluting characteristics, zero maintenance, and good reliability. Further, due to the global warming crises, depletion of the conventional fuel, and its polluting nature, utilization of solar PV received lots of recognition across the globe. According to the renewable global status report (RNE21 2018), the total solar PV capacity has increased from 2.6 GW in 2004 to 505 GW in 2018. Currently,

DOI: 10.4018/978-1-7998-3327-7.ch018

China in 2018 leads the net power generation in solar PV followed by the United States, Japan, and India. India has also set the target off generating 100 GW of renewable energy capacity, from solar energy by 2022 as a part of the Paris agreement according to the Indian renewable energy industry report 2019.

However, due to high installation cost and poor efficiency, the utilization of solar PV has now become one of the most challenging issues. Although, initiative is being taken at the government level to reduce the cost of installation, still its poor efficiency is a major concern among the power producers. In the interest of improved efficiency, several research contributions have also been made. Among the existing research areas, significant efforts have been directed to estimate accurate unknown model parameters, such as, photon current, the diode dark saturation current, series/shunt resistance, and diode ideality factor, of solar PV for obtaining better efficiency. An accurate parameter estimation technique is highly desirable to achieve efficient solar PV model before proceeding to the installation part.

In past, several solar PV modelling techniques, such as, single, double, and triple diode models were analyzed to achieve precise performance and improved efficiency of solar PV (Fariva, 2010). Also, several techniques were established to obtain maximum power point (MPP) and subsequently to achieve better efficiency. Indeed, the performance and efficiency of a solar PV cells are greatly dependent on the precise estimation of its current-voltage (I-V) characteristic (Attivissimo, 2012). Usually, it is very difficult to estimate accurate output characteristics of solar PV due to the nonlinear relation between output I and V. The output characteristics equation of solar PV, i.e., the relation between I and V, has five unknown and five known parameters. The exactness of the PV characteristics curve is largely dependent on the precise estimation of the unknown parameters which are not present in the manufacture's datasheet. Therefore, in recent years, parameter estimation techniques have drawn considerable attention amongst researchers. It has been recommended that a suitable parameter estimation technique is highly desirable for efficient solar modelling of solar PV. In the past decades, numerous methods, such as, analytical, numerical, iterative, and combinations of these methods were developed to obtain accurate unknown model parameters of solar PV either by using the measured illuminated I-V data, available manufacturer data, or from notable points in the output characteristics curve. The analytical method provides results with good accuracy but these analytical methods were not found suitable with large numbers of unknown parameters. Similarly, numerical is found suitable when unknown parameters are less in number. Furthermore, prediction starting points, for the unknown parameters, in the numerical methods is a great challenging task. Any wrong in initial guess results in the solution convergence to local minima instead of global minima (Askarzadeh, 2013, Jadli, 2017). In addition, parameter estimation, through analytical methods, is usually based on some approximations or assumptions on some terms of the system equations. Although, these methods are being reliable and accurate, but the accuracy of this method is highly subjected to the accuracy of the approximations and the appropriateness of the simplifications (Ibrahim, 2017). Also, these methods usually require some preliminary data,, like the values of slope at the notable points on the output characteristics curve, which can involve some intricacy to be accurately measured in practice.

Recently, parameter estimation techniques, based on metaheuristic optimization approach, are successfully addressed in several researches to avoid the limitations of numerical, analytical, and iterative approach. Metaheuristic optimization algorithm notable for their competency in resolving complex transcendental problems viz. PV cells as discussed by Yousri et al. (2019). Metaheuristics optimization algorithms define an objective function, as a difference between the measured values and estimated ones. The success of Metaheuristics algorithms is sustained by the algorithm's search proficiency, while most of the optimization algorithms fall in local minima. In this chapter, two Metaheuristics-based approaches, namely, 'PSO' and 'HS' algorithm, are used to estimate the five unknown parameter solar PV. As an alternative to traditional optimization algorithms, PSO uses direct fitness information instead of function derivatives to guide the search. Also, constructive cooperation between particles participating in the search improvise the optimization process. Whereas, HS is a stochastic search algorithm with fewer mathematical preconditions and generates a new result after accepting the existing results regardless of additional derivative information. This chapter highlights the implementation and shortcomings of PSO and HS along with the comparative analysis of these methods with numerical method, namely, NRM.

The chapter is organized as follows: section 2 summarises the working principle of solar PV cell and presents the layout of solar PV cell and PV system; section 3 presents the classification of solar PV cell, based on material used and types of modelling; section 4 highlights the need for solar PV cell modelling; section 5 presents the significant influence of environmental conditions on solar PV cell parameters and their impact on the performance characteristic. An outline of the various extraction procedures of these parameters is given and their advantage and limitations are highlighted. An objective function is also defined in the above section which will be later used for the parameter estimation process; section 6 discusses the metaheuristic optimization techniques along with the procedures used; section 7 evaluates the metaheuristic optimization techniques by testing it on the solar cell and solar module and discusses the results obtained. Finally, future scope and conclusions of this chapter has been presented.

WORKING PRINCIPLE OF SOLAR PV CELL

PV cell is somewhat similar to p-n diode in construction but differs in working. Combination of p- and n-types semiconductors makes equal Fermi level and that results in creating a band banding region in the p-n semiconductor device. When the light energy, incident on the band-banding region, has photon energy either same or higher than the bandgap energy, the electrons are knocked out from its place and creates electron-hole pairs. These generated free electrons and holes move to their respective side and subsequently creates a high potential difference. If charge collector is placed at the two sides, connected by means of resistance, then the current pass through external load. This current is called photovoltaic current or photon current as it is dependent on incident solar radiations. The mechanism on which solar PV generates photon current is called the photovoltaic effect. Solar PV cell is a current source as the incident solar light produces current. The diagram, as shown in Figure 1, represents a solar PV cell (Goswami, 2015).

Firstly, to enhance the utility of solar cells, numbers of solar cells are wired in series to form solar PV module. Afterward, these modules are connected in series/parallel to form array and to get desired output power. The array of panels along with the power conditioning units, and maximum power point tracking unit represent a PV system. The simplified layout of the stand-alone solar PV system is shown in Figure 2. There are different applications of PV systems, such as hybrid renewable energy system, Hybrid photovoltaic/thermal (PV/T) systems, rooftop, building-integrated PV's (BIPV), ground field, on-grid or stand-alone ones.

Figure 1. Representation of solar PV cell.

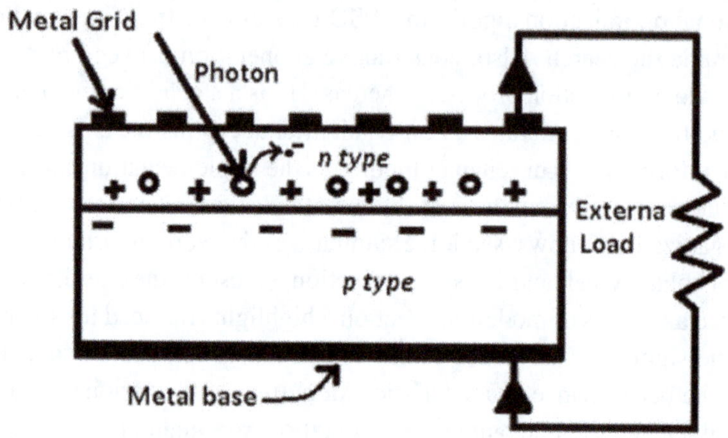

Figure 2. Stand-alone solar PV system.

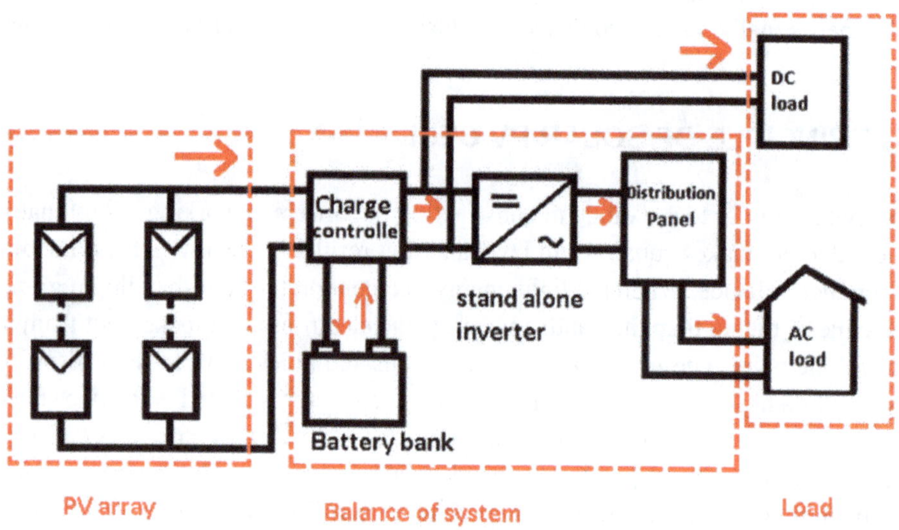

CLASSIFICATION OF SOLAR PV CELL

Classification According to Material

Usually, silicon material is widely used for manufacturing PV cell due to its abundance availability on the earth surface in form of sand (SiO_2). Firstly, the silicon sand converted into metallurgical grade silicon is purified to electronic silicon or solar grade silicon by the Siemens process (or modified Siemens process). Further using various respective technology available single-crystals or multi crystals of silicon block is grown. From the single-crystalline or multi-crystalline silicon wafers, the respective solar cell is fabricated through various steps as mentioned by Green (1991). The various PV technologies are classified in Table 1. Among all the PV technology, I[st] generation solar cell performs well at lower temperature and less unit area whereas IInd generation solar cell has a low manufacturing cost. To date, multijunction PV cell technology (made from the III and V material semiconductor) has the highest efficiency of 44% and they are used for space applications (Hayat, 2019).

Table 1. Classification of the solar PV technology.

PV technology	Solar PV cell types	Efficiency (in %)
I[st] Generation	Mono-crystalline silicon solar cell	25
	Poly-crystalline silicon solar cell	20.4
IInd Generation	Thin-film amorphous silicon (a-Si) solar cell	10.1
	Cadmium telluride (CdTe) solar cell	17
	Copper indium gallium selenide (CIGS) solar cell.	20
IIIrd Generation	Organic solar cell.	10
	Dye sensitized solar cell.	10
III-V PV cell Technology	GaAs (thin film, multi-crystalline)	28
	InP	22.1
Multijunction PV	InGaP/GaAs/InGaAs	37-44
	a-Si/nc-Si/nc-Si	13

Although, it is commonly suspected that the PV technology is the cleanest form of energy, but it has some environmental concern, also. During the fabrication of solar PV cells, heavy metals, such as, cadmium, nickel, and lead are used. These heavy metals are restricted by global environmental protection agencies. Also, PV modules emit CO_2 and other greenhouse gases (GHS) (Aman, 2015). Further, a life cycle assessment (LCA) is often introduced to scrutinize the pernicious influence of PV system over the environment. Also, the energy payback time (EPBT) and life cycle CO_2 emission factor/indicator are calculated to demonstrate the issue. The EPBT for amorphous (thin-film), mono-crystalline (single crystal) and Poly-crystalline solar PV system were estimated as 2.5 to 3.2 years, 3.2 to 15.5 years, and 1.5 to 5.7 years, respectively (Pacca, 2007). Similarly, GHG emission was 15.6-50g-CO_2/kWhe, 44-280 g-CO_2/kWhe, 9.4-104 g-CO_2/kWhe, for amorphous (thin-film), mono-crystalline (single crystal), and Poly-crystalline solar PV system, respectively. The EPBT depends on the electricity conversion efficiency among the other parameters. Additionally, the revision in the performance of the solar cell will cut back GHG emissions.

Classification According to Modelling

For the system design purposes, it is beneficial to comprehend the solar PV output characteristic with respect to the voltage under illumination. To achieve this goal, graphical or pictorial topology is utilized for the model. There are several existing models of PV device, based on the equivalent circuit, ANN, Linear Regression, Fuzzy Rule-Based model, etc. Among these existing models, the equivalent circuit model is highly recommended in most of the studies because of its simplicity. This section presents the various equivalent circuits PV models, namely, the ideal, single-diode, and double diode model. Also, this section highlights the advantages and drawbacks of these PV models.

The Ideal Single Diode Model (ISDM)

As discussed earlier the incident solar radiation produces the photon current and hence solar PV cell is characterize by a current source shorted by a diode in anti-parallel as shown in Figure 3 [18].The current is termed as photon current (I_{ph}). The schematic diagram of the ISDM of the PV cell is shown in Figure 3.

Figure 3. An ideal single diode model.

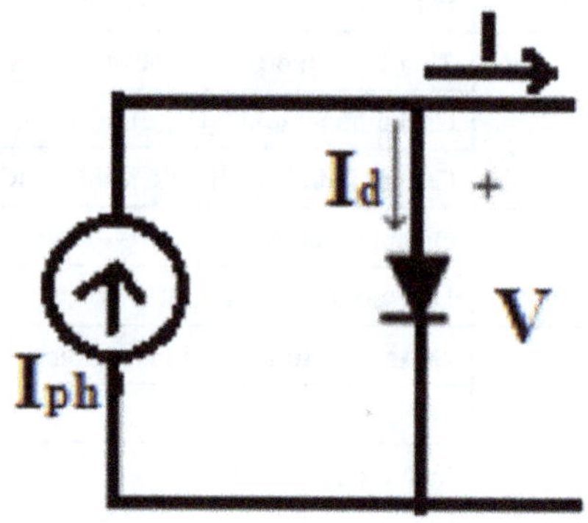

ISDM is the modest model but it is inefficient to predict the actual I-V characteristic curve of PV cell at MPP (Villalva, 2009). This model has only three unknown parameters. These electrical parameters are photon current (I_{ph}), diode dark saturation current (I_o), and ideality factor (A). These parameters can be predicted with the help of the data provided by the manufacturer datasheet, such as, short circuit (SC) current, open circuit (OC) voltage, maximum point output current, maximum point output voltage, maximum power, current temperature coefficient α, voltage temperature coefficient β, at standard temperature condition. The output characteristic of the ISDM of a PV cell is expressed in (2). Where, q is the electron charge, k is the Boltzmann constant.

$$I = I_{ph} - I_d \tag{1}$$

$$I = I_{ph} - I_o\left[\exp\left(\frac{qV}{kAT}\right) - 1\right] \tag{2}$$

The I-V or output curve and P-V curve of the solar cell is shown in Figure 4. The goal of the modelling is to extract accurate output curve at any irradiance and temperature ensuring that these curves cross the three remarkable points i.e. SC point, OC point, and MPP as closely as possible as shown in Figure 4.

Figure 4. The I-V and P-V characteristic curve of solar PV cell.

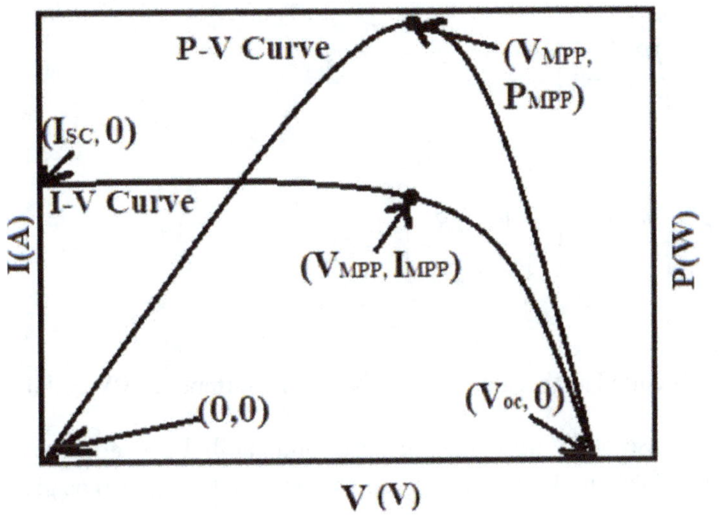

Single Diode Model (SDM)

SDM at standard operational conditions provide a very suitable performance whereas at low irradiance this model fails to give an accurate result. The schematic diagram of R_s-SDM and R_p-SDM are shown in Figure 5. The ISDM was made by assuming that there are no internal losses occurring in the PV cell whereas SDM takes cares of the losses. In the R_s-SDM model, a series resistance (R_s), as shown in Figure 5 (a), signifies the resistance offered by the bulk material in the passage of electrons and hole (Xiao, 2004). Despite, this model is simple and requires only one extra electrical parameter then ISDM namely, R_s, it is unable to demonstrate the accurate electrical behaviour of PV cell at high temperature variations. To make the model more accurate an addition shunt resistance R_p is added which represents the lose due to recombination of electron-hole pair (Liu, 2002). R_p-SDM increases the accuracy but unlike ISDM it has a comparatively large number of electrical parameters that have to be estimated.

Figure 5. (a) R_s-SDM; (b) R_p-SDM.

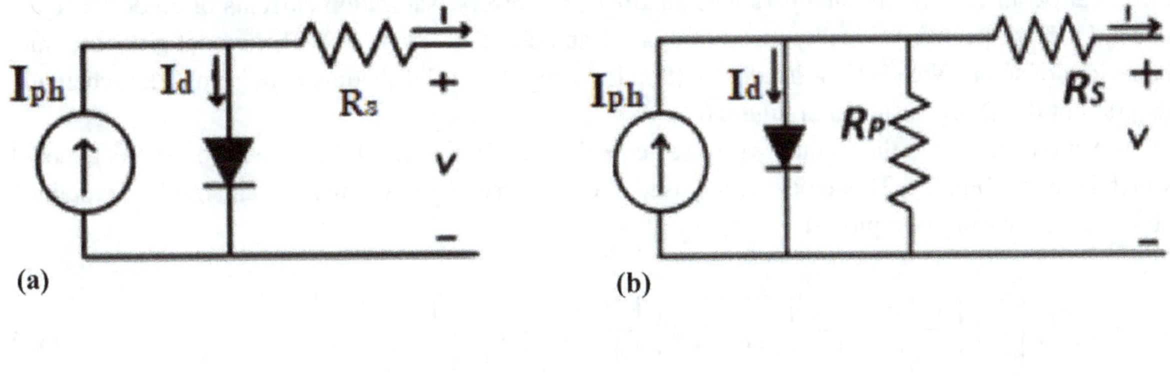

The output current equation for both the R_s-SDM and R_p-SDM is given by equation (3) and (4).

$$I = I_{ph} - I_o\left[\exp\left(\frac{V + IR_s}{AV_T}\right) - 1\right] \tag{3}$$

$$I = I_{ph} - I_o\left[\exp\left(\frac{V + IR_s}{AV_T}\right) - 1\right] - \left(\frac{V + IR_s}{R_p}\right) \tag{4}$$

Where, $V_T = \frac{N_s kT}{q}$, is called as thermal voltage. N_s are the number of PV cell interconnected in series to form PV module. In case of PV array having series and parallel PV modules ($N_{ss} \times N_{pp}$), the output current for both the R_s-SDM and R_p-SDM is given by equation (5) and (6) as given by Liu and Dougal (2002):

$$I = I_{Ph}N_{pp} + I_o N_{pp}\left(\exp\left(\frac{V + IR_s\left(N_{ss}/N_{pp}\right)}{AV_T N_{ss}}\right) - 1\right) \tag{5}$$

$$I = I_{Ph}N_{pp} + I_o N_{pp}\left(\exp\left(\frac{V + IR_s\left(N_{ss}/N_{pp}\right)}{AV_T N_{ss}}\right) - 1\right) - \frac{V + IR_s\left(N_{ss}/N_{pp}\right)}{Rp\left(N_{ss}/N_{pp}\right)} \tag{6}$$

Double Diode Model (DDM)

The recombination loses were not adequately modeled by R_p-SDM, therefore, a need for a new model was felt which can more precisely accumulate the recombination losses (Liu, 2002). The double diode model (DDM) was observed more accurate but relatively lot more complex. Due to the complexity, the challenge faced by DDM was the large computation time needed to estimate all the seven unknown electrical parameters (additional parameters namely, the reverse saturation currents of diode 1 and diode 2, I_{o1}, I_{o2} respectively, ideality factor of diode 1 and diode 2, A_1 and A_2). This model gives accurate behaviour of solar PV cell at low irradiance (that is during the partial shading condition). The schematic diagram of the DDM is shown in Figure 6.

The output current of the module is given by equation (6). Similar to DDM there is a three diode model which is more complex. This model is not used a lot because of its complexity. The models explained above are increasingly employed.

$$I = I_{ph} - I_{o1}\left[\exp\left(\frac{V + IR_s}{A_1 V_{T1}}\right) - 1\right] - I_{o2}\left[\exp\left(\frac{V + IR_s}{A_2 V_{T2}}\right) - 1\right] - \left(\frac{V + IR_s}{R_p}\right) \tag{7}$$

Figure 6. The equivalent circuit of DDM.

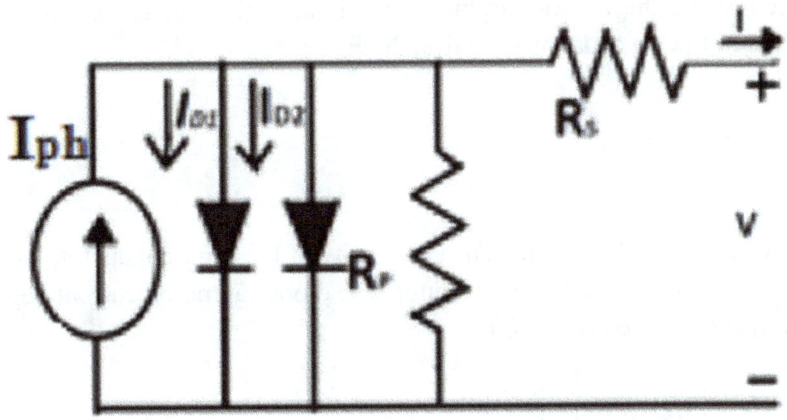

NEED FOR SOLAR PV CELL MODELLING

Accurate solar PV modelling plays very significant role to achieve good understanding of the core physical mechanism as well as the electrical performance of the solar PV. Solar cell modelling primarily involves the formulation of an equivalent circuit, which is called mathematical model formulation of solar PV. The accurate mathematical modelling and unknow electrical parameter estimation of the solar PV cell are significant because of the following reasons:

- It helps in explaining the implicit I-V relationship of a solar cell and to get improved power performance.
- It helps in the adequate designing of the power converter and controller used in a solar PV system, this helps in achieving a desirable dynamic, accurate MPP tracking for the real profile of the environmental condition.
- It helps in estimating the performance parameters. These parameters are helpful in PV system simulation and performance analysis.

Understanding of the electrical characteristics of the solar PV cell is very important to evaluate the unknown electrical parameters of the solar PV cell. Some of the empirical data is provided by the manufacturers for the standard test condition (STC). At STC the irradiation (G_{STC}) of 1000 W/m^2 with a global AM1.5 spectrum (AM stand for air mass) at 25^0 C temperature (T_{STC}) is considered (Villalva, 2009). The empirical data may have the information of the I-V curves obtained experimentally for different operating conditions.

SIGNIFICANCE OF SOLAR PV CELL PARAMETERS

The parameter estimation of R_p-SDM requires the computation of five unknown parameters, namely, I_{ph}, I_o, A, R_s, and R_p. However, these parameters are significantly influenced by the variation in irradiance and temperature. Thus, for accurate modelling of a solar PV, it is pivotal to be able to compute this parameter accurately considering the irradiance and temperature. The I_{ph} is an irradiance and temperature dependent variable because the variation in incident light can affect the generation of charge carriers. This is shown by the following equation (Villalva, 2009):

$$I_{ph} = \frac{G}{G_{STC}}\left(I_{ph,STC} + \alpha(T - T_{STC})\right) \tag{8}$$

The I_o of the PV cell depends upon the current density of the semi-conductor, but this information is not usually provided by the PV cell manufacturer. The diode saturation current dependency on temperature is shown by the equation (Liu, 2002).

$$I_o = I_{o,STC}\left(\frac{T}{T_{STC}}\right)^3 \exp\left[\frac{qE_g}{Ak}\left(\frac{1}{T} - \frac{1}{T_{STC}}\right)\right] \tag{9}$$

Where $I_{o,STC}$ is the value of saturation current at STC, and E_g is the bandgap energy which is 1.12eV for poly-crystalline silicon (De Soto, 2006). However, the diode ideality factor depends upon the PV cell technology (Bellia, 2014), usually, the value of A is arbitrary chosen between 1 and 2. The value of R_p decreases steadily as the light intensity increases whereas the value of R_s increases with an increase in light intensity (Wolf, 1996). The R_s and R_p impact the I-V curve near the MPP. The increase in R_s shifts the MPP to the right of I-V curve, reducing the slope of I-V curve between the MPP and V_{oc}, effectively reducing the maximum power. In contrast, increasing R_p reduces the slope of the curve between I_{sc} and the MPP, causing the output power to increase (Wolf, 1996). R_s and R_p can be calculated using the slope of I-V curve at V_{oc} and I_{sc} respectively. Thus, all the unknown parameters of solar PV affect the I-V characteristic.

These above methods used to determine the parameters of solar PV suffers from two drawbacks. Firstly, these equations turn out to be implicit in nature which is difficult to solve therefore explicit analytical solution using Lambert W-function, Co-content function, Pade´s approximants, special Trans function theory can be used. Secondly, this method needs some prior information which is not given in the manufacturer datasheet like the value of $I_{ph,STC}$, $I_{o,STC}$, $I_{sc,STC}$, $V_{oc,STC}$. These parameters have to be obtained experimentally. Thus a parameter estimation method that can determine the unknown parameter at STC with just the information given in the manufacturer datasheet is required. Usually, by using the information of three remarkable points as shown in Figure 4 the unknown parameters can be estimated without the experimental data (Villalva, 2009). Five equations can be derived from the I-V characteristic equation as shown below (Jadli, 2017 and Rawat, 2019):

The characteristic equation (4) at OC condition (I=0, V=V_{oc}) can be written as

$$0 = I_{Ph} - I_o\left\{\exp\left(V_{oc}/AV_T\right) - 1\right\} - V_{oc}/R_p \tag{10}$$

At SC condition (I=I_{sc}, V=0) the equation (4) is written as

$$I_{sc} = I_{Ph} - I_o\left\{\exp\left(I_{sc}R_s/AV_T\right) - 1\right\} - I_{sc}R_s/R_p \tag{11}$$

For third and fourth equations, (4) is differentiated with respect to voltage and given as

$$\frac{dI}{dV} = -I_o\left\{\frac{1}{AV_T}\left(1+\frac{dI}{dV}R_s\right)\exp\left(\frac{V+IR_s}{AV_T}\right)\right\} - \frac{1}{R_p}\left(1+\frac{dI}{dV}R_s\right) \tag{12}$$

dI/dV at open circuit condition is taken as third equation, whereas dI/dV at short circuit condition is taken as forth equation

$$\left.\frac{dI}{dV}\right|_{I=0} = -I_o\left\{\frac{1}{AV_T}\left(1+\left(\frac{dI}{dV}\right)_{I=0}R_s\right)\exp\left(\frac{V+IR_s}{AV_T}\right)\right\} - \frac{1}{R_p}\left(1+\left(\frac{dI}{dV}\right)_{I=0}R_s\right) \tag{13}$$

$$\left.\frac{dI}{dV}\right|_{V=0} = -I_o\left\{\frac{1}{AV_T}\left(1+\left(\frac{dI}{dV}\right)_{V=0}R_s\right)\exp\left(\frac{V+IR_s}{AV_T}\right)\right\} - \frac{1}{R_p}\left(1+\left(\frac{dI}{dV}\right)_{V=0}R_s\right) \tag{14}$$

The fifth equation is derived at MPP by differentiating P with respect to V and equating it equal to zero Thus, it can be written that

$$dI/dV = -(I_{MPP}/V_{MPP}) \tag{15}$$

On solving this fifth equation can be derived as

$$-\frac{I_{MPP}}{V_{MPP}} = -I_o\left\{\frac{1}{AV_T}\left(1+\frac{I_{MPP}}{V_{MPP}}R_s\right)\exp\left(\frac{V_{MPP}+I_{MPP}R_s}{AV_T}\right)\right\} - \frac{1}{R_p}\left(1+\frac{I_{MPP}}{V_{MPP}}R_s\right) \tag{16}$$

The five equations (10), (11), (13), (14), and (16) could not be solved analytically because of the existence of non-linearity in these equations. Therefore, numerical methods, based on iterative approach, such as, NRM and Gauss Seidel method, were used to estimate five unknown parameters of solar PV. Unfortunately, the correctness of these methods is subjected to the choice of initial values (Easwarakhanthan, 1986). Therefore, another method, based on metaheuristic/evolutionary optimization, such as, PSO, Genetic Algorithm, Chaos Optimization, Simulated Annealing, Artificial Bee Swarm optimization, Multi-Objective Optimization algorithm were recommended by the researchers to estimate more accurate parameters of solar PV than the iterative methods.

In the following section, two Metaheuristics-based approaches, namely the PSO, and HS algorithm, and one classical numerical method, namely, NRM have been used for the estimation of unknown parameters of Rp-SDM. The five equations, as indicated in (10), (11), (13), (14), and (16), have been combined to form an objective function, f(x), which is then minimized by above Metaheuristics optimization to estimate the unknown parameters. The objective function (or fitness function) is given below:

$$f(x) = \{f_1(x), f_2(x), f_3(x), f_4(x), f_5(x)\} \tag{17}$$

Here, x is the array of decision variables $\{x_1, x_2, x_3, x_4, x_5\}$. x_1, x_2, x_3, x_4, and x_5 represents I_{Ph}, I_o, R_s, R_p and V_T, respectively.

METAHEURISTICS /EVOLUTIONARY OPTIMIZATION

Overview of Particle Swarm Optimization

In 1995 James Kennedy, and Russell C. Eberhart developed PSO algorithm inspired by the movement pattern of organisms in a group, namely schools of fish and flocks of birds searching for food (Eberhart, 1995). During the emergence, of Swarm Intelligence, PSO was one of the only two mainstream algorithms in the Swarm Intelligence area. In the last decades, various other swarm intelligence algorithms have been developed, based on the behaviour of fish schools, bees, bacteria, glow worms, fireflies, cockroaches, bats, and cuckoo birds. The remarkable features of swarm intelligence-based algorithms are Self-organization and decentralized control.

PSO has swarm containing the population of potential solutions called particles. All the potential solutions to the problem flow through a hyper dimensional search space. Each particle has a memory that helps in self-organization as a result of which particles can adjust their positions by correlating their own experiences, and the experiences of the adjoining particles. This socio cognitive tendency of individual particles helps in decentralized control. Every individual, i^{th} particle in the swarm has a position x_i, and a velocity v_i, based on the current global best position G_{best} and the present best position P_{best} of the particle the new velocity v_i (t+1) and position x_i(t+1) of each particle is updated at each iteration of optimization process using the following equations.

$$v_i^{t+1} = w^t v_i^t + r_1 c_1^t [P_{best,i}^t - x_i^t] + r_2 c_2^t [G_{best,i}^t - x_i^t] \quad (18)$$

$$x_i^{(t+1)} = v_i^{(t+1)} + x_i^t \quad (19)$$

Where, r_1 and r_2 are arbitrary numbers from 0 to 1. c_1 and c_2 are called the learning factors, commonly the value for c_1 and c_2 is chosen as 2. The inertia weight w is a very important parameter in PSO as it balances the global search and local search. PSO is different from other swarm intelligence algorithm, in a way that PSO has its distinct approach to explore and exploit the search space of the problem, this is possible by dynamically changing the inertia weight to have a more desirable agreement between exploration and exploitation. For a large value of inertia, weight provides a global search whereas a smaller value of inertia weight aids in local search. Secondly, by velocity clamping, since velocities of the particles can readily rise to large values far from the boundaries of the search space. The steps for PSO are given below:

Step 1. Set the iteration index 'i' to zero. Initialize the parameters, such as, population size of the particles (P_s) randomly, inertia weight w (0), and the input data, for the PSO algorithm.

Step 2. For the objective function, as given in (17), calculate the fitness of each particle $f[x_i(j)]$.

Step 3. Compare the current fitness value of each particle with its personal best value and set new $P_{best,i}(j)$ according to the equation given below:

$$P_{best,i}(t) = \begin{cases} P_{best,i}(t-1) & if\ f(x_i(t)) \geq f(P_{best,i}(t-1)) \\ x_i(\mathrm{t}) & if\ (x_i(t)) < f(P_{best,i}(t-1)) \end{cases} \tag{20}$$

This means that the respective value of the position of the particle corresponding to which the fitness function is minimized below the P_{best} should be the new P_{best}.

Step 4. Update the global best $G_{best}(t)$, if P_{best} provides a minimum objective function compared to the fitness function provided by global best, according to the equation

$$G_{best}(t) \in \{P_{best1}(t), P_{best2}(t), \ldots\ldots\} \tag{21}$$

Step 5. Update the new velocity vector for each particle using (18)
Step 6. Calculate the new position for each particle, using (19)
Step 7. Change the inertia weight using (23)
Step 8. Increment the iteration index by one

Repeat the process from step 2 till the termination criteria (i.e. minimum value of objective function) is meat.

b. Overview of Harmonic Search Optimization

Harmony search (HS) algorithm was originally developed by Geem et al. (2001), imitating the refinement mechanism of music by adjusting the pitch to achieve a melodies harmony. Although HS was relatively new to PSO and other optimization algorithms its efficacy and supremacy have been signified in different applications, such as designing the network of water distribution networks, groundwater networks, and designing truss. To understand the working principle behind the HS algorithm let us consider the process of music improvisation by a musician. The musician has three possible choices: (1) performing a musical piece exactly from memory; (2) performing a musical piece close to the aforementioned pitch or (3) playing a random or new pitch. HS algorithm optimizes a problem analogous to the effort of attaining perfect harmony in music, the three corresponding features of HS are: using harmony memory, adjusting the pitch, and randomizing the pitch. Each solution is called harmony, a population of harmony vectors is arbitrary produced with in the potential range and placed in the harmony memory (HM). HM is of great importance as it enables that elite harmony to be included as elements of new solution vectors. The memory is put to use effectually by employing, key parameters like harmony memory considering rate (HMCR), pitch adjusting rate (PAR) and bandwidth of generation (b_w). The value of HMCR is usually chosen between 0.7-0.95, for the reason that for a very low value of HMCR only a few elite harmonies are added in the new solution which eventually slowdowns the convergence rate, whereas for a very high value of HMCR it is observed that the solution is not well explored, resulting in average solution The term pitch adjustment means generating a slightly different value in the HS algorithm. The PAR can be attuned either linearly or nonlinearly, but usually, in practice, linear adjustment is preferred. A low PAR can mitigate the convergence rate due to the restrain in the exploration of a very small sub-space of the entire search space. Contrarily, a very high PAR may cause the solution to search around larger space. The RPA is set in the range of 0.1 to 0.5 in most applications. To increase the diversity of the solution to avoid local minima trapping the third parameter i.e. randomization is employed. Although the PAR has a similar role in comparison with randomization, it is limited to certain areas and thus corresponds to a local search. The PAR and b_w are given as:

$$PAR(t) = PAR_{\max} - \frac{PAR_{\max} - PAR_{\min}}{t_{\max}} \times t \tag{21}$$

$$bw(t) = bw_{\max} \exp\left[Ln(bw_{\min} / bw_{\max}) \times \frac{t}{t_{\max}} \right] \tag{22}$$

The steps for HS algorithm search is given below:

Step 1 Set the number of iteration t_{max}, the objective function, decision variables, harmony memory size (HMS), HMCR, PAR_{max}, PAR_{min}, bw_{max} and bw_{min}.

Step 2. Initialize HM with random harmonies in the harmony search range using equation (23).

$$x_i(j) = l(j) + \text{rand}[u(j) - l(j)] \tag{23}$$

Where, i and j are the harmony's index and decision variable's index respectively, rand denotes the random number drawn uniformly from the interval [0, 1], l(j) is the lower bound value of design variable and u(j) is the upper bounds.

Step 3. Calculate the value of objective function for each harmony.

Step 4: A new harmony, x_{new}, is generate using randomization

$$x_{new}(j) = l(j) + \text{rand}[u(j) - l(j)] \tag{24}$$

Step 5: Calculate the value of objective function value for the new harmonics if it is in the solution space.

Step 6 Compare the fitness value with the fitness value for the worst harmony in the HM, in case the worst harmony gives better fitness function then the new solution is abundant otherwise accept the new one is replaced.

Step 7: Repeat the Step 4, 5 and 6 for the initialized number of iterations.

RESULTS AND DISCUSSION

In this section, the efficiency and feasibility of evolutionary algorithms are verified by testing the algorithm on a commercial silicon solar cell of 57 mm diameter, and a solar PV module Photowatt-PWP201 consisting of 36 poly-crystalline silicon cells connected in series. The experimental data for the solar cell and solar module has been taken from (Easwarakhanthan, 1986).

In this study for the PSO implementation the controlling variables such as learning factors c_1 and c_2 was taken as 2, the population size P_s was taken 60, the number of generation was set to 5000, maximal iteration number i_{max} to 100000 and the inertia weight w was taken as w_{max}=0.9 and w_{min}=0.4. The performance of PSO was improved by using a linearly decreasing inertia weight given by equation (25)

$$w = w_{\max} - \frac{w_{\max} - w_{\min}}{i_{\max}} \times i \tag{25}$$

For the implementation of HS algorithm the controlling parameter such as HMS was taken 30, HMCR was taken 0.95, PAR_{max}= 0.7, PAR_{min} = 0.1, bw_{max} = 1, bw_{min} = 0.0001, and t_{max} = 100000. The range of design variables for both the algorithm is taken the same to conduct a fair comparison and analysis since the range of the parameters can influence the algorithm speed and accuracy. The simulation has been carried out using MATLAB software.

R.T.C Solar Cell

The R.T.C solar cell data given in Table 2 is used to estimate the five unknown parameters via PSO and HS algorithm. A total number of 10 simulation runs for each optimization algorithm were executed and the best results for each algorithm are presented in Table 3. The lower and upper bound range of the unknown parameters are also shown in Table 3.

Table 2. Experimental data of R.T.C solar cell and PWP201 module used as case study.

Datasheet Parameters	RTC solar cell	Photowatt-PWP
Temperature, T	33þC	45þC
Irradiance, G	1000W/m^2	1000W/m^2
V_{oc}(V)	0.5728	16.778
I_{sc} (A)	0.7603	1.030
V_{mp}(A)	0.4507	12.649
I_{mp}(A)	0.6894	0.912
P_{MPP}	0.3105	11.5358

Table 3. Comparison of extracted parameters by HS, PSO, and NRM for R.T.C solar cell along with the search range and initial value used for different parameters.

	I_{ph} (A)	I_o (A)	R_s (Ω)	R_p (Ω)	V_t	A	P(W)	MPP_{error} error	t (sec)
Search-Range	[0 1]	[$0.1e^{-9}$ $0.1e^{-4}$]	[0 1]	[0 100]	[0 1]	-	-	-	-
HS	0.7692	5.6066e-7	0.0286	30.5293	0.0414	1.5693	0.3118	0.0013	52.58
PSO	0.7776	1e-6	0.0251	26.0017	0.0422	1.5996	0.3055	0.005	169.83
NRM	0.7611	2.7802e-7	0.0383	50.1101	0.0387	1.4669	0.3096	0.0009	26.84
Initial value	0.7	0.1e-6	0.01	20	0.03	-	-	-	-

The parameters of the same solar cell are also extracted using NRM and the outcome is compared with the best solutions of PSO and HS as shown in Table 3. Before analysing and comparing the result of Table 3, one more important aspect of algorithms, that is its consistency, should be analyzed and tested. In order to test the consistency of the algorithms, 10 simulation runs for each algorithm are carried out and analysed. Table 4 shows the values of five-parameters extracted using HS along with the fitness values f(x), convergence time t_{conv} and maximum power obtained for 10 number of runs, keeping the control parameters and iteration number the same for each run. The minimum fitness value obtained is 0.003963, and the maximum fitness value is 0.1139, but there is no significant variation observed in the fitness values for different runs. Convergence time and maximum power calculated also remained almost the same. With reference to the result in Table 4, it is clear that the HS algorithm used here is consistent. Since maximum power is one of the remarkable points, the I-V curve should pass through MPP as closely as possible. the I-V curve obtained from the result of the 4th and 10th run is observed to pass through MPP very closely, therefore, the value of these two runs is recorded as the best solution. The I-V curve obtained from the result of the 4th and 10th run is shown in Figure 7 and Figure 8 respectively.

Table 4. Extracted parameter of R.T.C solar cell obtained for back to back 10 number of simulations run to check the consistency and reliability of HS algorithm.

HS	1st	2nd	3rd	4th	5th	6th	7th	8th	9th	10th
I_{ph} (A)	0.7616	0.7618	0.7669	0.7610	0.7604	0.7604	0.7610	0.7617	0.7611	0.7692
I_o (A)	7.5579e-7	4.3265e-7	6.3074e-7	2.7267e-7	2.7123e-7	8.5607e-7	6.0664e-8	7.5388e-7	3.9948e-7	5.6066e-7
R_s (Ω)	0.02	0.0332	0.0301	0.0251	0.0364	0.0130	0.0454	0.0290	0.0338	0.0286
R_p(Ω)	27.5078	40.4873	33.3624	46.7121	50.4575	22.0855	99.2312	30.6079	41.8354	30.5293
A	1.5693	1.5086	1.5503	1.4631	1.4631	1.5844	1.3267	1.5693	1.5010	1.5693
V_t	0.0414	0.0398	0.0409	0.0386	0.0386	0.0418	0.0350	0.0414	0.0396	0.0414
f(x)	0.0622	0.0106	0.0102	0.1139	0.0104	0.108	0.003963	0.0101	0.011	0.009999
t	53sec	52sec	53sec	53.45sec	52.48 sec	53.98 sec	53.55sec	51.63sec	52.9sec	52.58sec
P	0.3011	0.3030	0.3033	0.30915	0.3037	0.3037	0.3078	0.3003	0.3024	0.3118

Figure 7. output characteristic curve of R.T.C solar cell modeled by HS algorithm: 4th simulation run

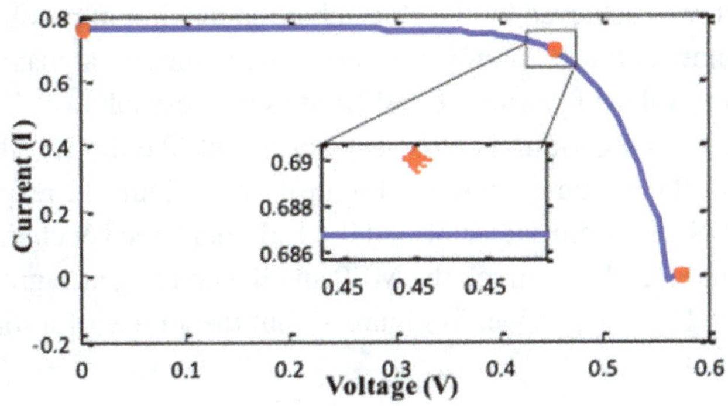

Figure 8. output characteristic curve of R.T.C solar cell modeled by HS algorithm:10th simulation run.

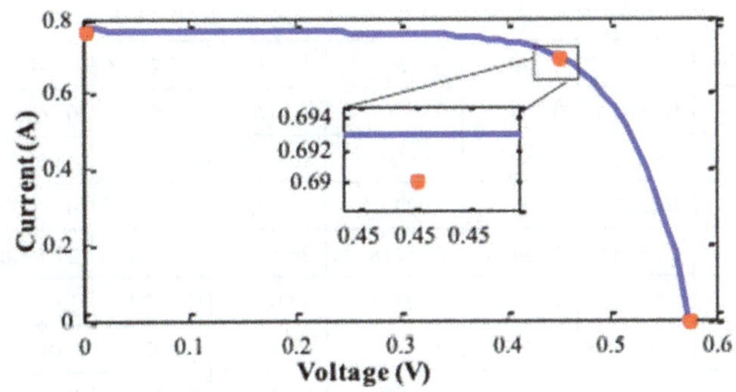

The I-V curve in Figure 7 does not touch the open-circuit voltage point on the x-axis closely whereas the I-V curve in the Figure 8 passes through all the three remarkable points. Thus the I-V curve of Figure 8 is more accurate. It can be concluded that the result obtained in the 10th run is accurate in all aspects such as convergence rate, fitness function, and I-V curve. The convergence graph for the best solution of the HS algorithm is shown in Figure 9.

Figure 9. Convergence process of HS algorithm (for 10th simulation run) during the parameter extraction process of the R.T.C solar cell.

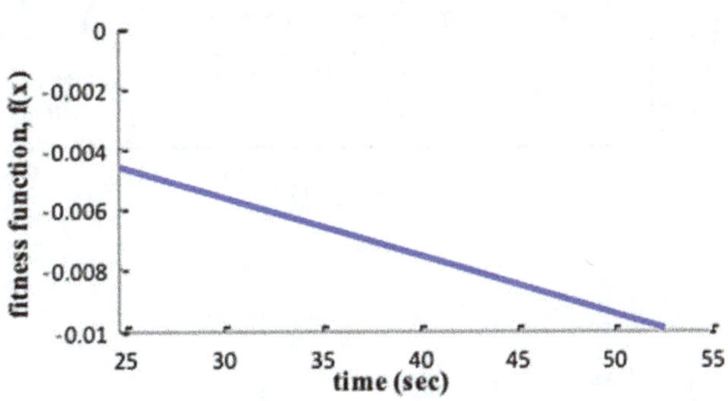

Similar to Table 4, Table 5 shows the outcome for 10 simulation runs of the PSO algorithm. The minimum fitness function was achieved by the 7th run but the maximum power calculated with respect to the estimated model value for the 7th run was not very close to the actual maximum power as obtained by the 9th run, therefore, the value of 9th run is considered as the best solution. It is clear from the results shown in table 6 that the PSO algorithm is also very consistent like the HS algorithm. The I-V curve and convergence graph for the 9th run is shown in Figure 10 and Figure 11 respectively. Out of the two metaheuristic algorithm HS algorithm outperformed PSO, although the I-V characteristic curve obtained by NRM as shown in Figure 12 closely track the MPP and the convergence graph of NRM converge to fitness function zero in 26.84 sec as shown in Figure 13 but the choice of initial value is very close to the solution.

Table 5. Extracted parameter of R.T.C solar cell obtained for back to back 10 number of simulation run to check the consistency and reliability of PSO algorithm.

PSO	1st	2nd	3rd	4th	5th	6th	7th	8th	9th	10th
I_{ph} (A)	0.7636	0.7702	0.7684	0.7611	0.7676	0.7676	0.7536	0.7615	0.7776	0.7757
I_o (A)	1e-6	1e-6	1e-6	1e-6	1e-6	1e-6	1e-6	1e-6	1e-6	1e-6
R_s (Ω)	0.0265	0.0253	0.0278	0.0281	0.0272	0.0272	0.0293	0.0292	0.0251	0.0259
R_p(Ω)	26.4026	26.1692	26.6969	26.8453	26.5418	26.5418	27.1847	27.1240	26.0017	26.2415
A	1.6034	1.5996	1.6034	1.6034	1.6034	1.6034	1.6072	1.6072	1.5996	1.5996
V_t	0.0423	0.0422	0.0423	0.0423	0.0423	0.0423	0.0424	0.0424	0.0422	0.0422
f(x)	0.1755	0.235	0.464	0.2755	0.2545	0.254	0.1243	0.2615	0.233	0.408
t	195.54	183.179	172.23	170.80	177.50	162.824	186.27	161.362	169.83	176.1
P_{max}	0.297	0.3028	0.3006	0.297	0.301	0.301	0.294	0.297	0.3055	0.304

Figure 10. output characteristic curve of R.T.C solar cell modeled by PSO algorithm: 9th simulation run.

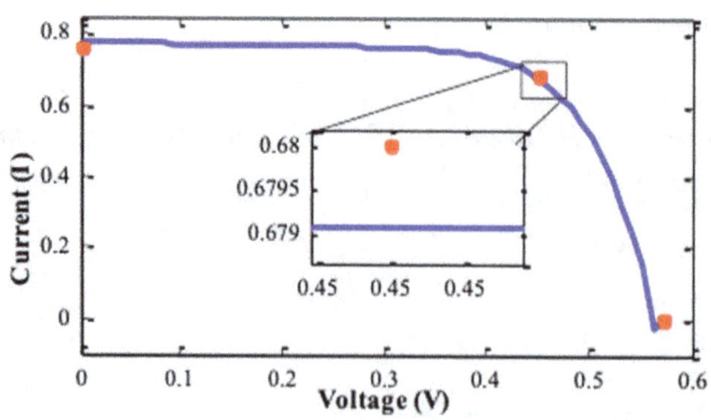

Figure 11. Convergence process of PSO algorithm (for 9th simulation run) during the parameter extraction process of the R.T.C solar cell.

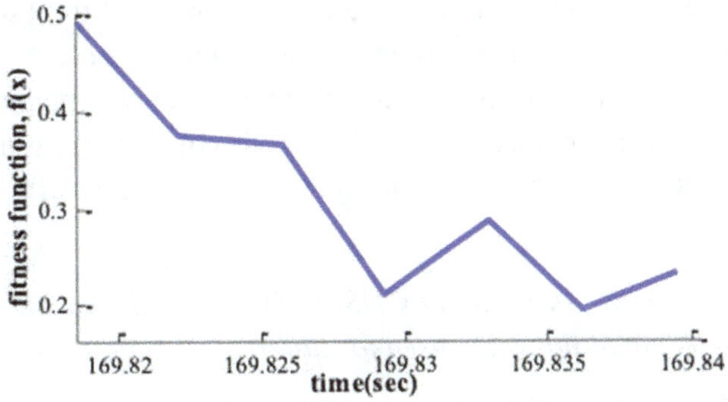

Figure 12. output characteristic curve of R.T.C solar cell modeled by NRM.

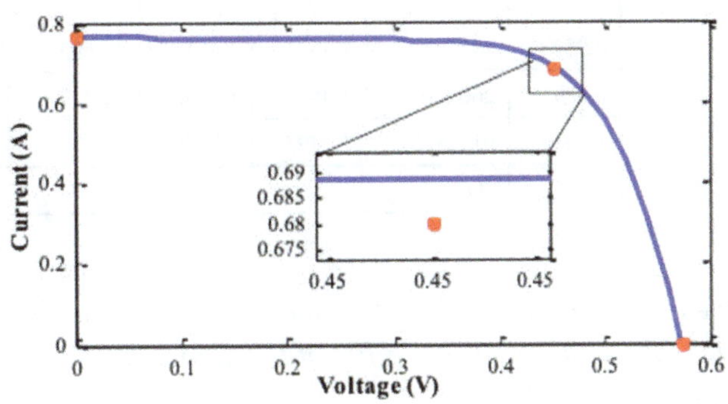

Figure 13. Convergence process of NRM during the parameter extraction process of the R.T.C solar cell.

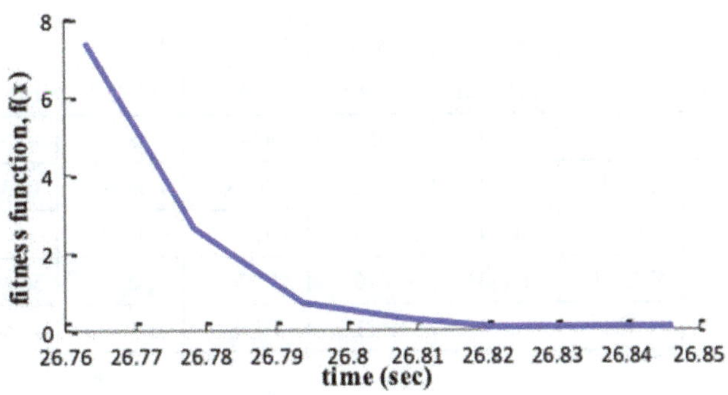

Photowatt-PWP201

The upper and lower search range for the different fitted parameters with respect to the PV module PWP201 is specified in Table 6. Last row of Table 6 indicates initial values for NRM. Here also the HS and PSO algorithm is processed for 10 simulation run to evaluate the consistency of the algorithm and their results are shown in Table 7 and Table 8 respectively. From the result in Table 7 and Table 8, it was clear that both the algorithm is consistent in term of the fitness value. The maximum power value close to the actual maximum power was obtained for the 3rd, 4th and 10th run of HS algorithm but the minimum fitness was observed for the 1st run. Since the I-V curves of the 1st and 3rd simulation run were not smooth therefore the best solution was a tie between the 4th and 10th simulation run. From the I-V curve shown in Figure 14 and Figure 15, it is clear that the I-V curve obtained for the 10th simulation run crosses all the three remarkable points closely. This implies that curve fitting is very important. The convergence graph for the 10th run of the HS algorithm is shown in Figure 16.

Table 6. Comparison of extracted parameters by HS, PSO, and NRM for photowatt PWP201 module along with the search range and initial value used for different parameters.

	I_{ph} (A)	I_o (A)	R_s (Ω)	R_p (Ω)	V_t	A	P_{MPP} (W)	MPP_{error} error	t (sec)
Search-Range	[0 3]	[0.1e^{-9} 0.1e^{-5}]	[0 5]	[0 1000]	[0 3]	-	-	-	-
HS	1.0331	8.338e-6	0.9512	741.3290	1.4321	1.4510	11.505	0.0308	55.11
PSO	1.0628	1e-5	0.8595	260.538	1.4687	1.4496	11.447	0.0888	190.46
NRM	1.0322	1.4799e-6	1.3333	613.0644	1.2494	1.2659	11.472	0.0638	17.87
Initial value	1	1e-6	1	200	1	-	-	-	-

Table 7. Extracted parameter of photowatt PWP201 module obtained for back to back 10 number of simulation run to check the consistency and reliability of HS algorithm.

HS	1st	2nd	3rd	4th	5th	6th	7th	8th	9th	10th
I_{ph} (A)	1.0324	1.0317	1.0257	1.0673	1.0332	1.0368	1.0345	1.0590	1.0338	1.0331
I_o (A)	6.789e-7	8.136e-7	8.781e-7	9.282e-6	7.458e-6	5.522e-6	4.720e-6	6.878e-6	4.865e-6	8.3380e-6
R_s (Ω)	1.4221	1.4506	1.3885	0.8730	0.9874	2.2042	1.0772	0.1835	1.0692	0.9512
R_p(Ω)	844.8815	911.1853	760.8425	269.7621	303.2659	353.6655	260.663	161.8591	367.1322	741.3290
A	1.1944	1.1794	1.2158	1.4589	1.4359	1.4848	1.3824	1.4558	1.3853	1.4510
V_t	1.1788	1.1640	1.2000	1.4399	1.4172	1.4655	1.3644	1.4368	1.3673	1.4321
f(x)	0.0098	0.0064	0.015	0.0279	0.0195	0.513	0.0262	0.1754	0.0261	0.0268
t(sec)	51.96	52	52.2261	51.4503	52.016	52.594	51.225	50.8936	54.736	55.11
P_{max}	11.649	11.214	11.527	11.528	11.181	11.1184	11.1462	11.5472	11.308	11.505

The results obtained for PSO was remarkable in terms of consistency in the value of fitting parameters. The values obtained for the 1st run of the PSO algorithm is considered the best solution. The I-V curve and convergence graph for the best solution of PSO is shown in Figure 17 and 18.

Figure 14. output characteristic curve of PWP201 module modeled by HS algorithm: 4th simulation run

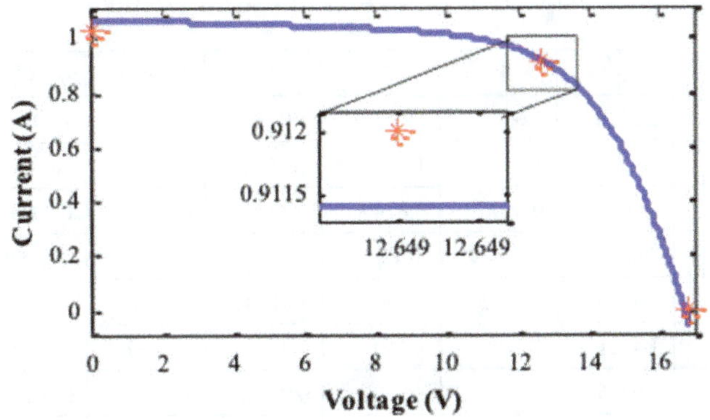

Figure 15. output characteristic curve of PWP201 module modeled by HS algorithm: 10th simulation run.

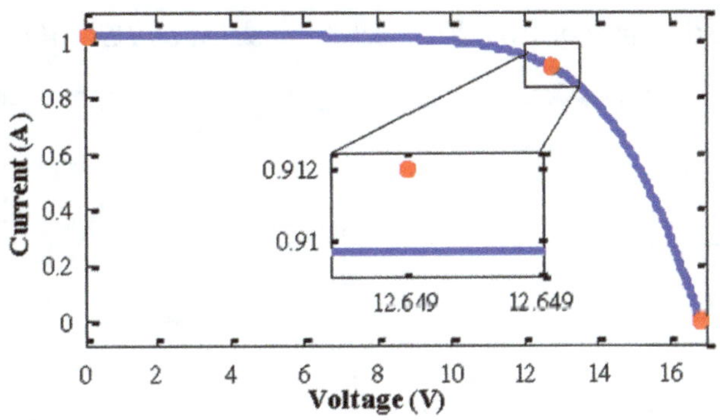

Figure 16. Convergence process of HS algorithm (for 10th simulation run) during the parameter extraction process of the PWP201 module.

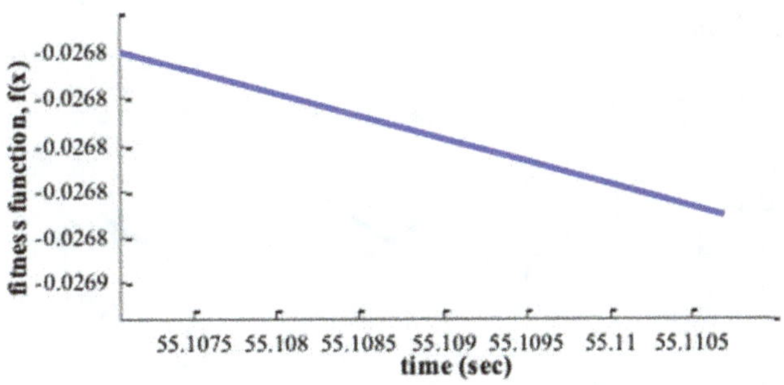

FUTURE RESEARCH DIRECTIONS

There are many metaheuristic optimization techniques for multi-objective optimization problem available in the literature that can be employed for parameter estimation problem. Hybrid techniques consisting of two or more type of metaheuristic optimization can also be designed to increase the computation efficiency. Also, there is a lot of scope of improvisation in solar PV model by designing a more efficient model then SDM.

Table 8. Extracted parameter of photowatt PWP201 module obtained for back to back 10 number of simulation run to check the consistency and reliability of PSO algorithm.

PSO	1st	2nd	3rd	4th	5th	6th	7th	8th	9th	10th
I_{ph} (A)	1.0628	1.0521	1.0426	1.0631	1.0544	1.0488	1.0607	1.0471	1.0606	1.0557
I_o (A)	1e-5	1e-5	1e-5	1e-5	1e-5	1e-5	1e-5	1e-5	1e-5	1e-5
R_s (Ω)	0.8595	0.8787	0.8928	0.8606	0.8749	0.8833	0.8654	0.8864	0.8641	0.8724
R_p(Ω)	260.538	261.8299	262.77	260.589	261.570	262.1633	260.928	262.3395	260.852	261.4130
A	1.4687	1.4700	1.4713	1.4687	1.4698	1.4704	1.4690	1.4708	1.4690	1.4696
V_t	1.4496	1.4509	1.4521	1.4496	1.4507	1.4513	1.4499	1.4516	1.4499	1.4505
f(x)	0.0058	-.0094	0.0266	0.0183	-.0048	0.0008	-.0006	0.0008	-.00144	-0.0055
t(sec)	190.4669	162.136	157.55	154.391	171.90	152.0705	152.044	156.57	225.65	218.11
P_{max}	11.447	11.333	11.207	11.452	11.346	11.282	11.422	11.260	11.422	11.358

Figure 17. I-V characteristic curve of PWP201 module modeled by PSO algorithm: 1th simulation

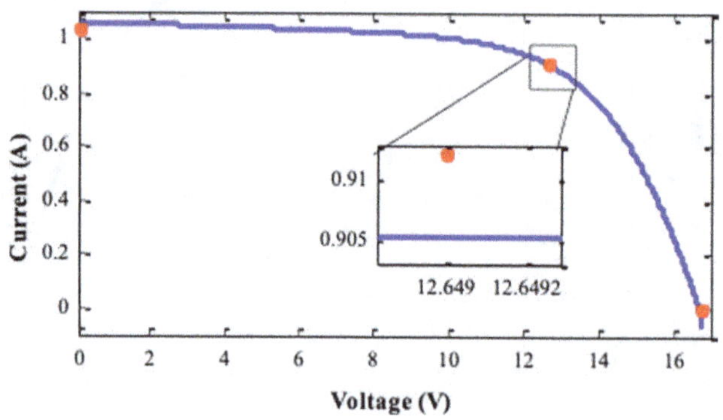

Figure 18. Convergence process of PSO algorithm (for 1th simulation run) during the parameter extraction process of the PWP201 module

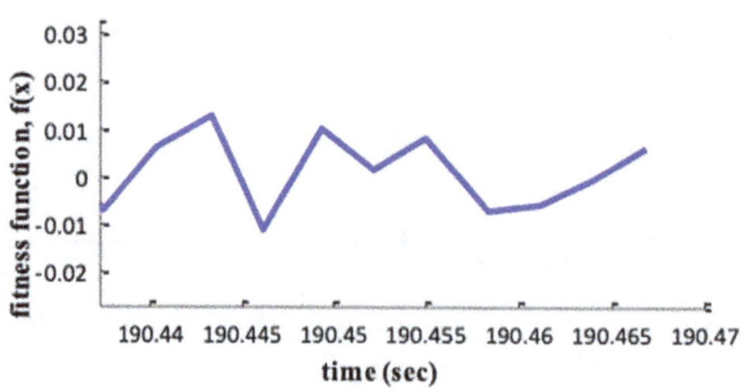

Figure 19. I-V characteristic curve of PWP201 module modeled by NRM

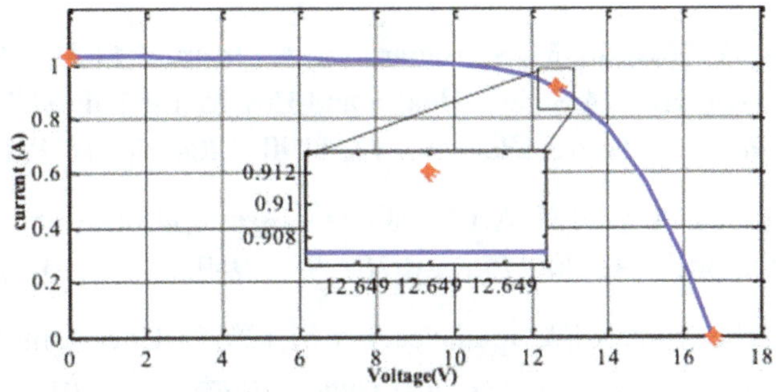

Figure 20. Convergence process of NRM during the parameter extraction process of the PWP201 module

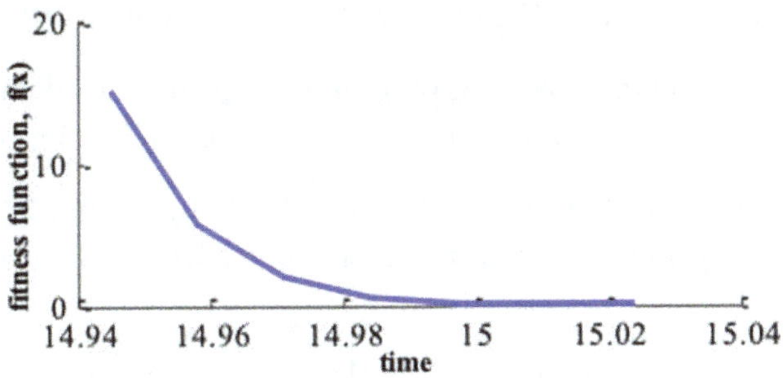

CONCLUSION

In this chapter, the working principle, construction, different PV cell technologies and application of solar PV cells are systematically explained. The various model of PV cell namely, ideal model, Rs-SDM, Rp-SDM, DDM are explained in detail to highlight their advantage and disadvantage. The three major classifications of the parameter estimation techniques were also explained. Their advantages and disadvantages are highlighted so that an accurate method for parameter estimation can be used. A detailed study of two metaheuristic algorithms namely, HS and PSO on two different case studies is performed and compared with a classical numerical method NRM. From the outcome of the study, it was observed that soft computing method like evolutionary method is better compared to other methods because of their ability to converge to a global solution even for a complex problem like parameter estimation of solar PV cell, also they are much easier when DDM and triple diode model is considered for modelling. Another exciting feature of the evolutionary method is its stochastic nature which allows the search process to allow random search. One major recurrent issue faced by numerical and iterative method is the difficulty in choosing or initializing the parameters, like in the case of NRM unknown parameter has to be initialized, any wrong value for even one parameter can converge the solution to local solution, not only this but in case of more complex problem like DDM and triple diode model having seven and more number of unknown parameter to be extracted, the numerical and iterative method fail to converge to a solution. Whereas this limitation is overcome by evolutionary algorithms by providing a large search range of the parameter.

REFERENCES

Aman, M. M., Solangi, K. H., Hossain, M. S., Badarudin, A., Jasmon, G. B., Mokhlis, H., Bakar, A. H. A., & Kazi, S. N. (2015). A review of Safety, Health and Environmental (SHE) issues of solar energy system. *Renewable & Sustainable Energy Reviews, 41*, 1190–1204. doi:10.1016/j.rser.2014.08.086

Askarzadeh, A., & Rezazadeh, A. (2013). Artificial bee swarm optimization algorithm for parameters identification of solar cell models. *Applied Energy, 102*, 943–949. doi:10.1016/j.apenergy.2012.09.052

Attivissimo, F., Di Nisio, A., Savino, M., & Spadavecchia, M. (2012). Uncertainty analysis in photovoltaic cell parameter estimation. *IEEE Transactions on Instrumentation and Measurement, 61*(5), 1334–1342. doi:10.1109/TIM.2012.2183429

Bellia, H., Youcef, R., & Fatima, M. (2014). A detailed modelling of photovoltaic module using MATLAB. *NRIAG Journal of Astronomy and Geophysics, 3*(1), 53–61. doi:10.1016/j.nrjag.2014.04.001

De Soto, W., Klein, S. A., & Beckman, W. A. (2006). Improvement and validation of a model for photovoltaic array performance. *Solar Energy, 80*(1), 78–88. doi:10.1016/j.solener.2005.06.010

Easwarakhanthan, T., Bottin, J., Bouhouch, I., & Boutrit, C. (1986). Nonlinear minimization algorithm for determining the solar cell parameters with microcomputers. *International Journal of Solar Energy, 4*(1), 1-12.

Eberhart, R., & Kennedy, J. (1995, November). Particle swarm optimization. In *Proceedings of the IEEE international conference on neural networks* (Vol. 4, pp. 1942-1948). 10.1109/ICNN.1995.488968

Farivar, G., & Asaei, B. (2010, November). Photovoltaic module single diode model parameters extraction based on manufacturer datasheet parameters. In *2010 IEEE International Conference on Power and Energy* (pp. 929-934). IEEE. 10.1109/PECON.2010.5697712

Geem, Z. W., Kim, J. H., & Loganathan, G. V. (2001). A new heuristic optimization algorithm: harmony search. *Simulation, 76*(2), 60-68.

Goswami, D. Y. (2015). *Principles of solar engineering*. CRC Press. doi:10.1201/b18119

Green, M. A. (1991). Recent advances in silicon solar cell performance. In *Tenth EC Photovoltaic Solar Energy Conference* (pp. 250-253). Springer. 10.1007/978-94-011-3622-8_63

Hayat, M. B., Ali, D., Monyake, K. C., Alagha, L., & Ahmed, N. (2019). Solar energy—A look into power generation, challenges, and a solar-powered future. *International Journal of Energy Research, 43*(3), 1049–1067. doi:10.1002/er.4252

Ibrahim, H., & Anani, N. (2017). Evaluation of analytical methods for parameter extraction of PV modules. *Energy Procedia, 134*, 69–78. doi:10.1016/j.egypro.2017.09.601

Jadli, U., Thakur, P., & Shukla, R. D. (2017). A new parameter estimation method of solar photovoltaic. *IEEE Journal of Photovoltaics, 8*(1), 239–247. doi:10.1109/JPHOTOV.2017.2767602

Liu, S., & Dougal, R. A. (2002). Dynamic multiphysics model for solar array. *IEEE Transactions on Energy Conversion, 17*(2), 285–294. doi:10.1109/TEC.2002.1009482

Pacca, S., Sivaraman, D., & Keoleian, G. A. (2007). Parameters affecting the life cycle performance of PV technologies and systems. *Energy Policy*, *35*(6), 3316–3326. doi:10.1016/j.enpol.2006.10.003

Rawat, N., Thakur, P., & Jadli, U. (2019). Solar PV parameter estimation using multi-objective optimisation. *Bulletin of Electrical Engineering and Informatics*, *8*(4), 1198–1205. doi:10.11591/eei.v8i4.1312

Sulaiman, S. I., Rahman, T. K. A., Musirin, I., & Shaari, S. (2012, May). Artificial neural network versus linear regression for predicting Grid-Connected Photovoltaic system output. In *2012 IEEE International Conference on Cyber Technology in Automation, Control, and Intelligent Systems (CYBER)* (pp. 170-174). IEEE. 10.1109/CYBER.2012.6392548

Villalva, M. G., Gazoli, J. R., & Ruppert Filho, E. (2009). Comprehensive approach to modeling and simulation of photovoltaic arrays. *IEEE Transactions on Power Electronics*, *24*(5), 1198–1208. doi:10.1109/TPEL.2009.2013862

Wolf, M., & Rauschenbach, H. (1963). Series resistance effects on solar cell measurements. *Advanced Energy Conversion*, *3*(2), 455–479. doi:10.1016/0365-1789(63)90063-8

Xiao, W., Dunford, W. G., & Capel, A. (2004, June). A novel modeling method for photovoltaic cells. In *2004 IEEE 35th Annual Power Electronics Specialists Conference* (IEEE Cat. No. 04CH37551) (Vol. 3, pp. 1950-1956). IEEE.

Yousri, D., Allam, D., Eteiba, M. B., & Suganthan, P. N. (2019). Static and dynamic photovoltaic models' parameters identification using Chaotic Heterogeneous Comprehensive Learning Particle Swarm Optimizer variants. *Energy Conversion and Management*, *182*, 546–563. doi:10.1016/j.enconman.2018.12.022

Chapter 19
A Compact Antenna Design With High Gain for Wireless Energy Harvesting

Priya Sharma
Indian Institute of Information Technology, Allahabad, India

Ashutosh Kumar Singh
Indian Institute of Information Technology, Allahabad, India

ABSTRACT

A compact rectangular slotted antenna fed through coplanar waveguide for rectenna system is proposed in the application of radio frequency (RF) energy harvesting at center frequency of 2.45 GHz in the wireless local area network (WLAN) band. Three unequal widths of rectangular slots with equal distance have been created step by step to maximize the peak gain to 3.6 dB of the antenna. Radiation plot of the proposed antenna has been depicted to be omnidirectional for RF energy harvesting with maximum radiation efficiency characteristics. The dimension of the antenna is reduced up to 28 × 17 mm2 with better reflection coefficient of -34.6dB.

INTRODUCTION

With the growth in technology, RF wireless energy harvesting (WEH) has been receiving an extensive research approaches by the researchers for the human welfare. It has the potential to harvest more energy from the various RF energy sources, for example Wi-Fi, WLAN, portable devices etc. Through the use of Internet of Things (IoT) number of users can be connected freely. Battery consuming more power as the number of usages increases by the users then RF energy harvesting come in the environment. The wireless energy harvesting ideas have been prolonged for more than last two decades. Different ambient energy sources are available for energy harvesting such as vibration energy sources (approx. 200μW/cm^2), solar energy (approx. 100mW/cm^2), RF energy (approx. 1μW/cm^2) and thermal energy (approx. 60μW/cm^2).

DOI: 10.4018/978-1-7998-3327-7.ch019

The wireless power transfer (WPT) divided in two parts: transmitter and receiver. The receiving part is the RF WEH and has received extensive growth in the last decades.

RF energy harvesting is useful in the unfavorable applications such as health monitoring, environment monitoring, defense, surveillance etc. in (Bhatt et al., 2019)-(Vu et al., 2020). Rectenna system design is used for RF WEH. It comprises of antenna may be a microstrip antennas or any other, rectifying circuit composed of capacitor, inductor and low threshold diode which is used to change the RF wave to Direct current (DC) and storing devices are introduced in the literature by various energy sources that influenced by various environmental factor but RF energy sources independent of these factors in (Danesh & Long, 2011)-(Bito et al., 2017). RF broaden the range of frequencies in telecommunication area such as industrial, scientific and medical (ISM) band, Universal Mobile Telecommunication Service (UMTS) 2.1GHz, 1.8GHz (GSM) and WLAN (2.4GHz, 5.8GHz) in (Sun et al., 2013),(),(Elwi et al., 2019).

Figure 1. Block diagram of wireless energy harvesting system.

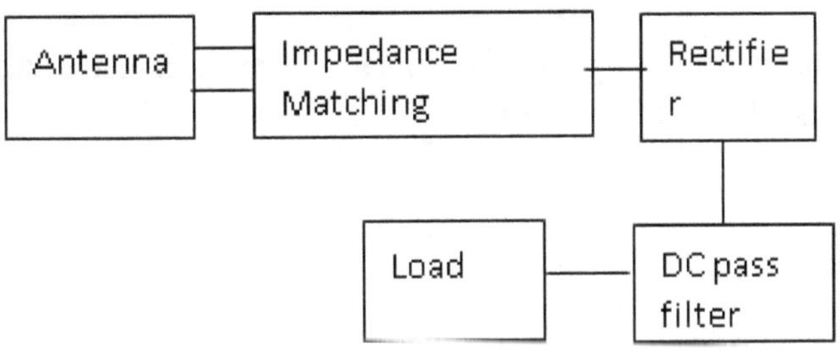

In this, the proposed antenna used to receive energy from various RF sources and rectifying circuit converts it into direct current (DC) from electromagnetic wave with high efficiency at different compact rectenna with single wide band high efficiency is frequencies, DC filter applied after it to smoothen the DC waveform and then some devices is used to store. A being discussed in (Sun et al., 2012)-(Takhedmit, 2010) and slotted microstrip patch antenna with dual band for WLAN has antenna used to receive energy from various RF sources and rectifying circuit converts it into direct current (DC) from electromagnetic wave with high efficiency at different frequencies, DC filter applied after it to smoothen the DC waveform and then some devices is used to store. A compact rectenna with single wide band high efficiency is being discussed in (Sun et al., 2012)-(Takhedmit, 2010) and slotted microstrip patch antenna with dual band for WLAN has reported in (Khemar et al., 2018)-(Heikkinen & Kivikoski, 2003), then for triple band rectenna (Chandravanshi et al., 2018),(Liu & Zhang, 2018), Quadband rectifier (Hsu et al., 2017), six band (Song et al., 2016), multiband planar antenna (Nimo et al., 2012), Frequency tunable energy harvesting rectenna (Song et al., 2017), broadband rectenna (Nie et al., 2015).

In this paper, the proposed antenna has been designed and simulated in section II, results have been discussed in section III and then at last conclusion is proposed.

Antenna Design

Wireless energy harvesting system has been described with the use of block diagram is presented in fig.1. This figure comprises of a receiving antenna work at any band of frequencies, after this impedance matching network is design for getting maximum power transfer, then this RF energy is transforms into direct current (DC) voltage, make this DC signal to pass through dc pass filter and then after to the load and load can be various types of storage devices or directly give to the wireless sensor nodes. Smart devices and power up handheld devices are used by the antenna for the requirement of wireless RF energy harvesting. The proposed antenna design for rectenna system is simulated using commercially available software using Ansoft HFSS v.15 on FR4 epoxy substrate with loss tangent (δ) of 0.02, relative permittivity (ε_r) of 4.4 and substrate thickness (h) of 1.6mm. A coplanar waveguide (CPW) feeding is used to excite the antenna with dimension of substrate 17×28 mm^2. Four vertical rectangular slots are cut to improve the impedance bandwidth in 2.45 GHz WLAN band. Step by step four rectangular slots have been created to improve the complete features of the antenna for example gain, radiation pattern and return loss etc.

For the designing of an antenna, length of the slots has been explained at different frequency bands by the following equations (Bhatt et al., 2019),

$$L = \frac{C}{2f_r\sqrt{(\varepsilon_{reff})}} \tag{1}$$

$$f_r = \frac{C}{4h\sqrt{(\varepsilon_{reff})}} \tag{2}$$

The evolution process for proposed antenna from rectangular patch to slotted rectangular patch antenna is explained in the Fig 2. Coplanar waveguide feeding technique is used in which ground and rectangular patch are on the same plane of dielectric substrate and 50 Ω coplanar waveguide feeding line are designed to excite the antenna. Distance between ground plane and feed line is maximizing to be 1.25mm. The complete dimensions of proposed antenna are displayed in the Fig. 3 and their values in Table 1. Slots are introduced to reduce the dimension of an antenna. Increasing width of three rectangular slots are created to amplify the radiation efficiency of the antenna in particular direction through enhancing the reflection coefficients and antenna gain.

Figure 2. Step by step process of the proposed antenna

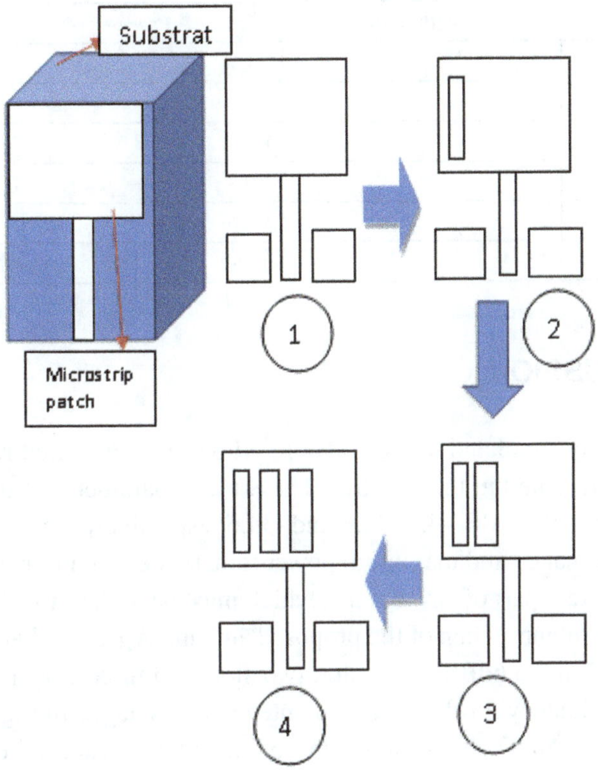

Figure 3. Complete dimensions of the proposed antenna

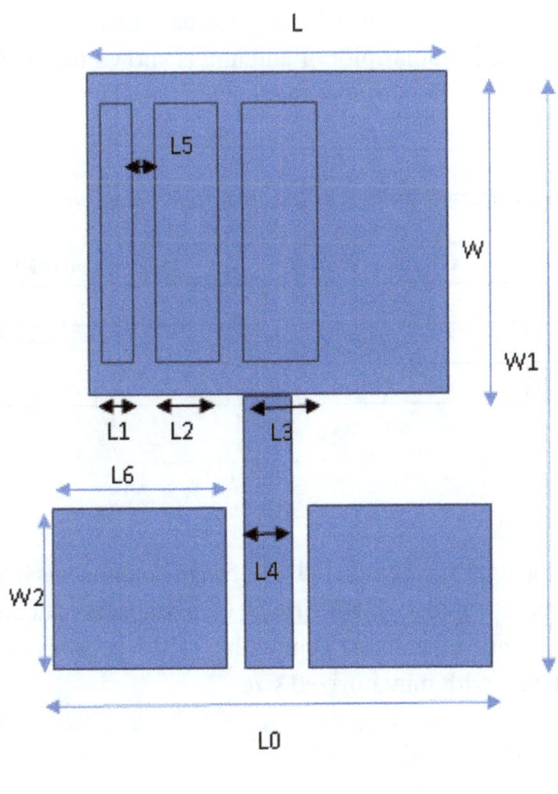

Table 1. Dimensions in mm of fig. 3.

Parameters	Length (mm)	Parameters	Length (mm)
L	15	L4	2
W	15	L5	0.5
L0	17	L6	6.5
L1	1.5	W1	28
L2	2	W2	8
L3	2.5		

RESULT AND DISCUSSION

The design has successfully simulated in the software HFSS and Simulated result of S(1,1) parameter for each design steps is presented in the Fig 4(a). The S (1, 1) parameter of the antenna with one slot, two slots and three slots are -35.17 dB, -42.03 dB and -34.6 respectively. Return loss with larger than -10 dB result in fruitful performance and maximum power transfer between transmitter and receiver of the antenna. Gain is the important part of any antenna which must be high as much as possible with respect to other parameters of the antenna. Gain of the proposed antenna is greater than other reference antenna, which is 0.34 dB, 2.68 dB and 3.6dB for one slot, two slots and three slots respectively is depicted in Fig 4(b). The radiation efficiency for the proposed antenna is displayed in Fig.5. The real part of Input Impedance for the antenna is 50.57 ohm which is approximate 50 ohm and 1.3 ohm is the imaginary part which is approximately -0.67 ohm and it is shown in the Fig. 6. Radiation pattern for antenna E- radiated field (XZ plane, $\phi=0^{o}$) and H- radiated field (YZ plane, $\phi=90^{o}$) is displayed in the Fig. 7(a) and Fig. 7(b) respectively. E- Field has omni-directional radiation pattern and H- field has isotropic radiation pattern in this proposed antenna. 3D Polar plot of antenna is shown in Fig.8. Comparison between the different paper's antenna design and the proposed antenna is tabulated in Table 2.

Table 2. Comparison between different antennas and proposed antenna.

Frequency (GHz)	Antenna size (mm^2)	Antenna Bandwidth (GHz)	Gain (dB)
2.45 [1]	1408	-	1.48
2.45 (Elwi et al., 2019)	1225	2.4 - 2.5	3.24
2.45 [Proposed work]	476	1.9 – 8	3.6

CONCLUSION

A compact rectangular slot coplanar waveguide fed microstrip antenna has been demonstrated and simulated in this paper with proper return loss. A wide range of frequencies at center frequency 2.45 GHz in WLAN band is represented. Improved antenna gain with 3.6dB is attained in this paper for appropriate functioning of proposed antenna with miniaturized size.

Figure 4. (a) Return loss (b) gain for of the proposed antenna at three different stages.

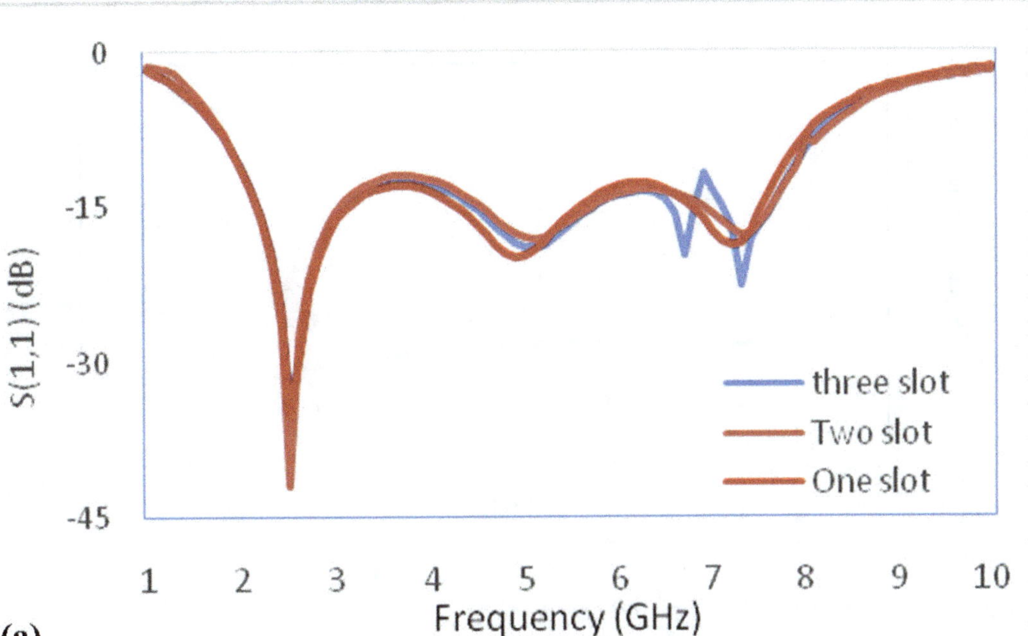

(a)

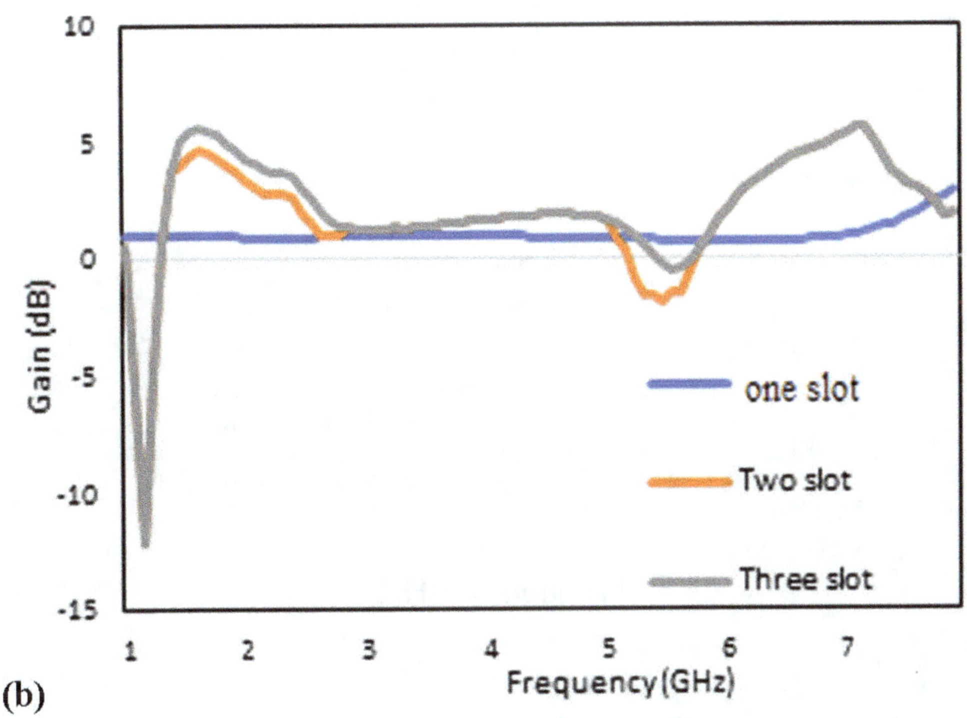

(b)

Figure 5. Radiation efficiency of the proposed antenna

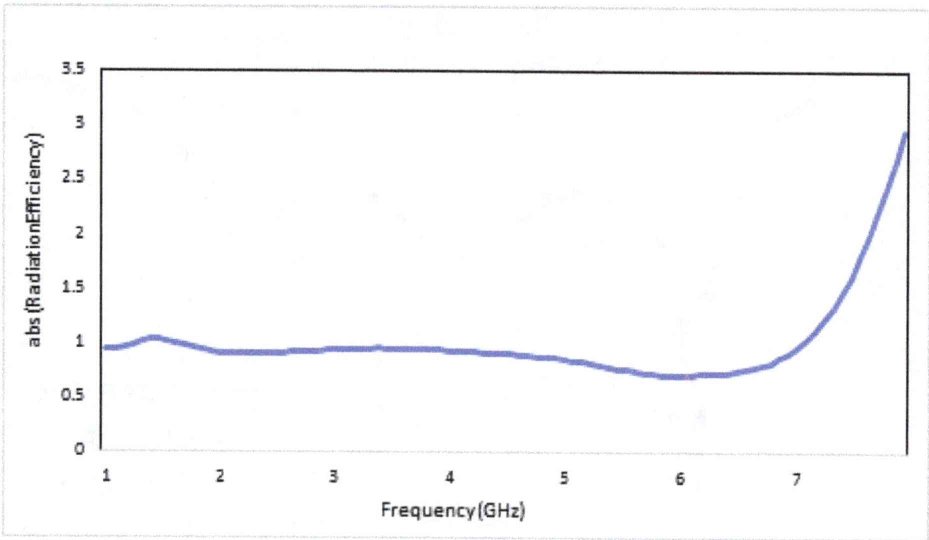

Figure 6. Z (1,1) parameter of the proposed antenna.

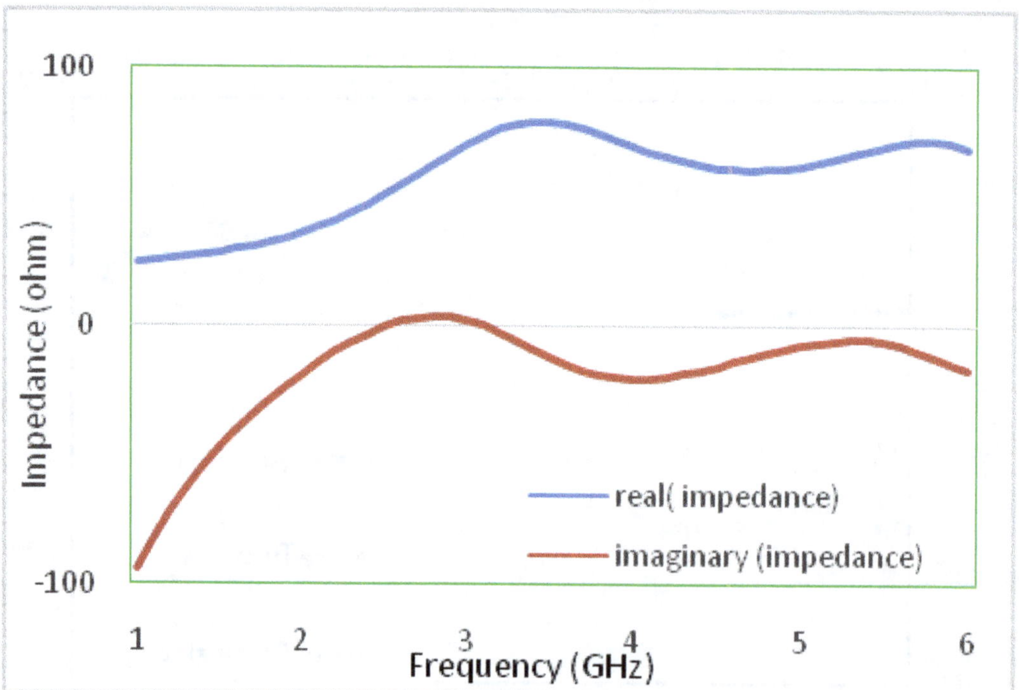

Figure 7. Radiation pattern in (a) E- plane and (b) H- plane at 2.45 GHz.

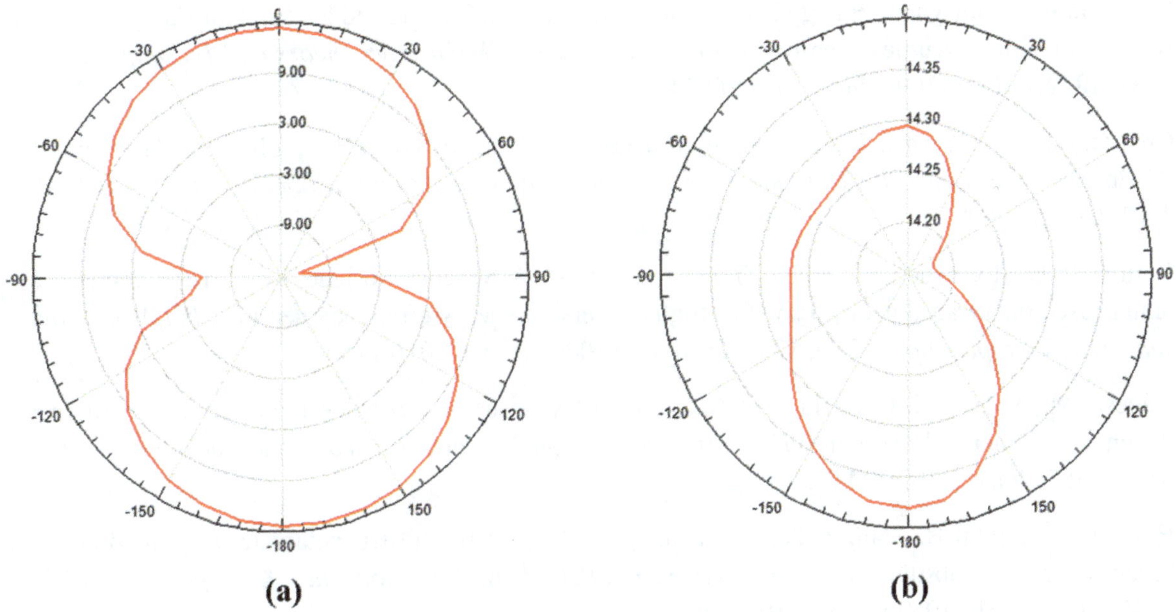

Figure 8. 3D- polar plot of the proposed antenna.

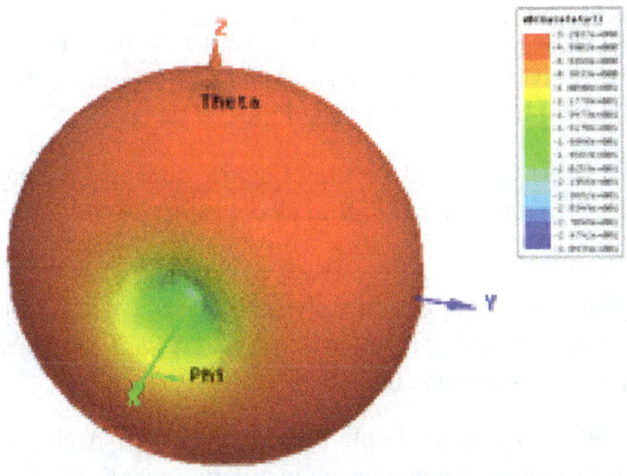

REFERENCES

Bhatt, K., Kumar, S., Kumar, P., & Tripathi, C. C. (2019). Highly Efficient 2.4 and 5.8 GHz Dual-Band Rectenna for Energy Harvesting Applications. *IEEE Antennas and Wireless Propagation Letters*, *18*(12), 2637–2641. doi:10.1109/LAWP.2019.2946911

Bito, J., Bahr, R., Hester, J. G., Nauroze, S. A., Georgiadis, A., & Tentzeris, M. M. (2017). A Novel Solar and Electromagnetic Energy Harvesting System With a 3-D Printed Package for Energy Efficient Internet-of-Things Wireless Sensors. *IEEE Transactions on Microwave Theory and Techniques*, *65*(5), 1831–1842. doi:10.1109/TMTT.2017.2660487

Chandravanshi, S., Sarma, S. S., & Akhtar, M. J. (2018). Design of Triple Band Differential Rectenna for RF Energy Harvesting. *IEEE Transactions on Antennas and Propagation*, *66*(6), 2716–2726. doi:10.1109/TAP.2018.2819699

Chuma, E. L., Rodríguez, L. T., Iano, Y., Roger, L. L. B., & Sanchez-Soriano, M. (2018). Compact rectenna based on a fractal geometry with a high conversion energy efficiency per area. *IET Microwaves, Antennas & Propagation*, *12*(2), 173–178. doi:10.1049/iet-map.2016.1150

Danesh, M., & Long, J. R. (2011). An Autonomous Wireless Sensor Node Incorporating a Solar Cell Antenna for Energy Harvesting. *IEEE Transactions on Microwave Theory and Techniques*, *59*(12), 3546–3555. doi:10.1109/TMTT.2011.2171043

Elwi, T. A., Abdul Hassain, Z. A., & Tawfeeq, O. A. (2019). Hilbert metamaterial printed antenna based on organic substrates for energy harvesting. *IET Microwaves, Antennas & Propagation*, *13*(12), 2185–2192. doi:10.1049/iet-map.2018.5948

Heikkinen, J., & Kivikoski, M. (2003). A novel dual-frequency circularly polarized rectenna. *IEEE Antennas and Wireless Propagation Letters*, *2*, 330–333. doi:10.1109/LAWP.2004.824166

Hoang, M. H., Phan, H. P., Hoang, T. Q. V., & Vuong, T. (2014). Efficient compact dual-band antennas for GSM and Wi-Fi energy harvesting. *2014 International Conference on Advanced Technologies for Communications*, 401-404. 10.1109/ATC.2014.7043420

Hsu, C., Lin, S., & Tsai, Z. (2017). Quadband Rectifier Using Resonant Matching Networks for Enhanced Harvesting Capability. *IEEE Microwave and Wireless Components Letters*, *27*(7), 669–671. doi:10.1109/LMWC.2017.2711578

Khemar, A., Kacha, A., Takhedmit, H., & Abib, G. (2018). Design and experiments of a dual-band rectenna for ambient RF energy harvesting in urban environments. *IET Microwaves, Antennas & Propagation*, *12*(1), 49–55. doi:10.1049/iet-map.2016.1040

Liu, J., & Zhang, X. Y. (2018). Compact Triple-Band Rectifier for Ambient RF Energy Harvesting Application. *IEEE Access: Practical Innovations, Open Solutions*, *6*, 19018–1902. doi:10.1109/ACCESS.2018.2820143

Lyu, H., Liu, X., Sun, Y., Jian, Z., & Babakhani, A. (2019). A 915-MHz Far-Field Energy Harvester With −22-dBm Sensitivity and 3-V Output Voltage Based on Antenna-and- Rectifier Codesign. *IEEE Microwave and Wireless Components Letters*, *29*(8), 557–559. doi:10.1109/LMWC.2019.2923685

Nie, M., Yang, X., Tan, G., & Han, B. (2015). A Compact 2.45-GHz Broadband Rectenna Using Grounded Coplanar Waveguide. *IEEE Antennas and Wireless Propagation Letters*, *14*, 986–989. doi:10.1109/LAWP.2015.2388789

Nimo, Grgić, & Reindl. (2012). Ambient Electromagnetic wireless energy harvesting using multiband planar antenna. *International Multi-Conference on Systems, Signals & Devices*, 1-6.

Niotaki, K., Collado, A., Georgiadis, A., Kim, S., & Tentzeris, M. M. (2014). Solar/Electromagnetic Energy Harvesting and Wireless Power Transmission. *Proceedings of the IEEE*. 1712-1722. 10.1109/JPROC.2014.2358646

Peter, T., Rahman, T. A., Cheung, S. W., Nilavalan, R., Abutarboush, H. F., & Vilches, A. (2014). A Novel Transparent UWB Antenna for Photovoltaic Solar Panel Integration and RF Energy Harvesting. *IEEE Transactions on Antennas and Propagation*, *62*(4), 1844–1853. doi:10.1109/TAP.2014.2298044

Song, C., Huang, Y., Carter, P., Zhou, J., Yuan, S., Xu, Q., & Kod, M. (2016). A Novel Six-Band Dual CP Rectenna Using Improved Impedance Matching Technique for Ambient RF Energy Harvesting. *IEEE Transactions on Antennas and Propagation*, *64*(7), 3160–3171. doi:10.1109/TAP.2016.2565697

Song, C., Huang, Y., Zhou, J., & Carter, P. (2017). A novel compact and frequency-tunable rectenna for wireless energy harvesting. *2017 IEEE Wireless Power Transfer Conference (WPTC)*, 1-3. 10.1109/WPT.2017.7953841

Sun, H., Guo, Y., He, M., & Zhong, Z. (2012). Design of a High-Efficiency 2.45-GHz Rectenna for Low-Input-Power Energy Harvesting. *IEEE Antennas and Wireless Propagation Letters*, 929–932.

Sun, H., Guo, Y., He, M., & Zhong, Z. (2013). A Dual-Band Rectenna Using Broadband Yagi Antenna Array for Ambient RF Power Harvesting. *IEEE Antennas and Wireless Propagation Letters*, *12*, 918–921. doi:10.1109/LAWP.2013.2272873

Takhedmit. (2010). Efficient 2.45 GHz rectenna design includingharmonic rejecting rectifier device. *Electron. Lett.*, 811-812.

Vu, H. S., Nguyen, N., Ha-Van, N., Seo, C., & Thuy Le, M. (2020). Multiband Ambient RF Energy Harvesting for Autonomous IoT Devices. *IEEE Microwave and Wireless Components Letters*, *30*(12), 1189–1192. doi:10.1109/LMWC.2020.3029869

Compilation of References

Abid, G., Shaikh, S. A., Rajput, S. H., & Mazid, U. A. (2020). IoT based smart industrial panel for controlling three phase induction motor. IEEE iCoMET, 1-8.

Ackerman, T., & Knyazkin, V. (2000). Interaction between distributed generation and the distribution network. *Transmission and Distribution Conference and Exhibition: Asia Pacific IEEE/PES, 2*, 1357–1362.

Aghaei, J., & Alizadeh, M.-I. (2013, February). Demand response in smart electricity grids equipped with renewable energy sources: A review. *Renewable & Sustainable Energy Reviews, 18*, 64–72.

Aharchi, M. (2019). A Review on 3D Reconstruction Techniques from 2D Images. *Springer Third International Conference on Smart City Applications (SCA).*

Ahmad, E. D., Kusrini, K., & Sudarmawan, S. (2017). Classification of intrusion detection system (IDS) based on computer network. In *Proceedings of 2017 2nd International conferences on Information Technology, Information Systems and Electrical Engineering (ICITISEE)*. IEEE.

Ajam, N. (2015). Heart Disease Diagnoses using Artificial Neural Network. *The International Insitute of Science, Technology and Education, 5*(4), 7–11.

Akella, Saini, & Sharma. (2009). Social, economical and environmental impacts of renewable energy systems. *Renewable Energy, 34*(2), 390–396.

Alabdulwahab, A. S. (2017). *Teaching reliability assessment of smart grid protection systems.* Paper presented at the 2017 IEEE Global Engineering Education Conference (EDUCON). 10.1109/EDUCON.2017.7943036

Al-Abdulwahab, A. S., Winter, K. M., & Winter, N. (2011). *Reliability assessment of distribution system with innovative Smart Grid technology implementation.* Paper presented at the 2011 IEEE PES Conference on Innovative Smart Grid Technologies-Middle East. 10.1109/ISGT-MidEast.2011.6220780

Albasrawi, M. N., Jarus, N., Joshi, K. A., & Sarvestani, S. S. (2014). *Analysis of reliability and resilience for smart grids.* Paper presented at the 2014 IEEE 38th Annual Computer Software and Applications Conference. 10.1109/COMPSAC.2014.75

Aldous, D. (1991, January). The continuum random tree. I. *Annals of Probability, 19*(1), 1–28.

Alizadeh Shabestary, S. M., Saeedmanesh, M., Rahimi Kian, A., & Jalalabadi, E. (2015). Real-time frequency and voltage control of an islanded mode microgrid. *J. Iran. Assoc. Electr. Electron. Eng., 12*(3), 9–14.

Almoghathawi, Y., & Barker, K. (2019). Component importance measures for interdependent infrastructure network resilience. *Computers and Industrial Engineering, 133*(September), 153–164.

Al-Wafi, E., Al-Subhi, A., & Al-Muhaini, M. (2015). *Reliability assessment of a practical power system using Monte Carlo simulation.* Paper presented at the 2015 Saudi Arabia Smart Grid (SASG). 10.1109/SASG.2015.7449276

Amal, H., Khalid, C., & Rajae, T. (2016). Intrusion detection system using PCA and Fuzzy PCA techniques. In *Proceedings of International Conference on Advanced Communication Systems and Information Security (ACOSIS)*. IEEE.

Aman, M. M., Solangi, K. H., Hossain, M. S., Badarudin, A., Jasmon, G. B., Mokhlis, H., Bakar, A. H. A., & Kazi, S. N. (2015). A review of Safety, Health and Environmental (SHE) issues of solar energy system. *Renewable & Sustainable Energy Reviews*, *41*, 1190–1204. doi:10.1016/j.rser.2014.08.086

Amer, S. H., & Hamilton, J. (2010). Intrusion detection systems (IDS) taxonomy-a short review. *Defense Cyber Security*, *13*(2), 23–30.

Amrutkar, K. P., & Kamalja, K. K. (2017). An overview of various importance measures of reliability system. *International Journal of Mathematical. Engineering and Management Sciences*, *2*(3), 150–171.

Anastasiadis, A. G., Poulimenos, G., Polyzakis, A., Manousakis, N., & Vokas, G. (2019). Algorithms development for the energy management of a micro combined heat and power unit in an AC-DC microgrid. *AIP Conference Proceedings*. 10.1063/1.5117020

Anbarasi, M., Anupriya, E., & Iyengar, N. C. S. N. (2010). Enhanced prediction of heart disease with feature subset selection using genetic algorithm. *International Journal of Engineering Science and Technology*, *2*(10), 5370–5376.

Aneetha, A. S., & Bose, S. (2012). The combined approach for anomaly detection using neural networks and clustering techniques. *Computing in Science & Engineering*, *2*(4), 37–46.

Aneetha, A. S., & Bose, S. (2012, August). The combined approach for anomaly detection using neural networks and clustering techniques. *Computer Science & Engineering: An International Journal*, *2*(4), 37–46. doi:10.5121/cseij.2012.2404

Ardagna, C. A., & Damiani, E. (2014). *Business Intelligence meets Big Data: An Overview on Security and Privacy.* NSF Workshop on Big Data Security and Privacy, Dallas, TX.

Asghar Ghadimi, A., Rastegar, H., & Keyhani, A. (2007). Development of average model for control of a full bridge PWM DC-DC converter. *J. Iran. Assoc. Electr. Electron. Eng.*, *4*(2), 52–59.

Askarzadeh, A., & Rezazadeh, A. (2013). Artificial bee swarm optimization algorithm for parameters identification of solar cell models. *Applied Energy*, *102*, 943–949. doi:10.1016/j.apenergy.2012.09.052

Atick, J. J., Griffin, P. A., & Redlich, N. A. (1996). Statistical approach to shape from shading: Reconstruction of 3d face surfaces from single 2d images. *Neural Computation*, *8*, 1321–1340. doi:10.1162/neco.1996.8.6.1321 PMID:8768397

Attivissimo, F., Di Nisio, A., Savino, M., & Spadavecchia, M. (2012). Uncertainty analysis in photovoltaic cell parameter estimation. *IEEE Transactions on Instrumentation and Measurement*, *61*(5), 1334–1342. doi:10.1109/TIM.2012.2183429

Atzori, A. L., Iera, A., & Morabito, G. (2010). The internet of things: A survey. *Computer Networks*, *54*(15), 2787–2805. doi:10.1016/j.comnet.2010.05.010

Augusto, A., Figueiredo, A., Gomes, J., Costa, F., Henrique, L., & Coelho, H. (2019). Computer Methods and Programs in Biomedicine Breast tumor localization using skin surface temperatures from a 2D anatomic model without knowledge of the thermophysical properties. *Computer Methods and Programs in Biomedicine*, *172*, 65–77. doi:10.1016/j.cmpb.2019.02.004 PMID:30902128

Avatefipour, O., & Sadry, F. (2018). Traffic Management System Using IoT Technology - A Comparative Review. *IEEE International Conference on Electro/Information Technology (EIT)*, 1041-1047. 10.1109/EIT.2018.8500246

Awadallah, S. K. E., Milanovic, J. V., Wang, Z., & Jarman, P. N. (2015). Assessment of suitability of different reliability importance measures for prioritising replacement of transmission system components. *2015 IEEE Eindhoven PowerTech. PowerTech*, *2015*, 1–6.

Badamasi, Y. A. (2014). The working principle of an arduino. *2014 11th International Conference on Electronics, Computer and Computation (ICECCO)*, 1–4. 10.1109/ICECCO.2014.6997578

Banaei, M. R., Ardi, H., & Farakhor, A. (2014). Analysis and implementation of a new single-switch buck–boost DC-DC converter. *IET Power Electronics*, *7*(7), 1906–1914. doi:10.1049/iet-pel.2013.0762

Banzi, M., & Shiloh, M. (2014). *Getting started with Arduino: the open source electronics prototyping platform*. Maker Media, Inc.

Basant, S., Santosh, B., & Sushanta, K. (2016). Enhancing performance of anomaly-based intrusion detection systems through dimensionality reduction using principal component analysis. In *Proceedings of IEEE International Conference on Advanced Networks and Telecommunications Systems (ANTS)*. IEEE.

Bellekens, Tachtatzis, Atkinson, Renfrew, & Kirkham. (2014). Glop: Enabling massively parallel incident response through gpu log processing. In *Proceedings of the 7th International Conference on Security of Information and Networks*. ACM.

Bellia, H., Youcef, R., & Fatima, M. (2014). A detailed modelling of photovoltaic module using MATLAB. *NRIAG Journal of Astronomy and Geophysics*, *3*(1), 53–61. doi:10.1016/j.nrjag.2014.04.001

Best First Search. (n.d.). Available at: http://weka.sourceforge.net/doc.dev/weka/attributeSelection/BestFirst.html

Beyene, . (2018). Survey on Prediction and Analysis the Occurrence of Heart Disease Using Data Mining Techniques. *International Journal of Pure and Applied Mathematics*.

Bhargava, A., & Kumar, A. (2017). Arduino controlled robotic arm. In *2017 International conference of Electronics, Communica- tion and Aerospace Technology (ICECA)*, *2*, 376–380. doi: 10.1109/ICECA.2017.8212837

Bhattacharya & Jana. (2009). Renewable energy in India: Historical developments and prospects. *Energy*, *34*(8), 981–991.

Big Data Security – Issues, Challenges, Tech & Concerns. (n.d.). *Research data alliance*. https://www.rd-alliance.org/group/big-data-ig-data-security-and-trust-wg/wiki/big-data-security-issues-challenges-tech-concerns

Big Data Security, Challenges and Solutions. (n.d.). https://www.dataversity.net/big-data-security-challenges-and-solutions/

Big Data, Preliminary Report 2014, ISO/IEC JTC 1. (2014). *Information technology*. www.iso.org/iso/home/about/iso_members.htm,Preliminary

Billinton, R., & Allan, R. N. (1994). *Reliability Evaluation of Power Systems* (2nd ed.). Plenum Press.

Birnbaum, Z. W. (1969). On the importance of different components in a multicomponent system. *Multivariate Analysis*, *2*, 581–592.

Blanz, V., & Vetter, T. (1999). A morphable model for the synthesis of 3d faces. *International Conference on Computer Graphics and Interactive Techniques*, 187–194. 10.1145/311535.311556

Brahmi, B., & Shirvani, M. H. (2015, February). Prediction and Diagnosis of Heart Disease by Data Mining Techniques. *Journals of Multidisciplinary Engineering Science and Technology*, *2*(2), 164–168.

Breast Cancer Facts and Figures 2019-2020. (n.d.). Retrieved October 19, 2020, from https://www.cancer.org/content/dam/cancer-org/research/cancer-facts-and-statistics/breast-cancer-facts-and-figures/breast-cancer-facts-and-figures-2019-2020.pdf

Breast cancer statistics in India. (n.d.). Retrieved October 19, 2020, from https://gallery-repo.inshorts.com/gallery/view/796e2a89-eb11-4165-90b7-43bbca6c3738

Bressan, N., Bazzaco, L., Bui, N., Casari, P., Vangelista, L., & Zorzi, M. (2010). The deployment of a smart monitoring system using wireless sensor and actuator networks. *Proc. IEEE Smart Grid. Comm.*, 49–54.

Brinkman, G., Jorgenson, J., Ehlen, A., & Caldwell, J. (2016). *Low carbon grid study: Analysis of a 50% emission reduction in California*. Nat. Renewable Energy Lab. Tech. Rep.NREL/TP-6A20-64884.

Bulat, A., & Tzimiropoulos, G. (2017). How far are we from solving the 2d and 3d face alignment problem? (and a dataset of 230,000 3d facial landmarks). *2017 IEEE International Conference on Computer Vision (ICCV)*. arXiv:1703.07332

Butun, Morgera, & Sankar. (2014). A Survey of Intrusion Detection Systems in Wireless Sensor Networks. *IEEE Communications Surveys & Tutorials, 16*(1), 266–282. .2013.050113.00191 doi:10.1109/SURV

Calderaro, V., Galdi, V., Graber, G., Graditi, G., & Lamberti, F. (2014). *Impact assessment of energy storage and electric vehicles on smart grids*. Paper presented at the 2014 Electric Power Quality and Supply Reliability Conference (PQ). 10.1109/PQ.2014.6866775

Cardiovascular disease. (n.d.). Available at: https://www.who.int/cardiovascular_diseases/en/

Casari, P., Castellani, A., Cenedese, A., Lora, C., Rossi, M., Schenato, L., & Zorzi, M. (2009, June). The WIreless SEnsor networks for city-Wide Ambient Intelligence (WISE-WAI) project. *MDPI J. Sensors, 9*(6), 4056–4082. doi:10.339090604056 PMID:22408513

Castellani, A. P., Bui, N., Casari, P., Rossi, M., Shelby, Z., & Zorzi, M. (2010). Architecture and protocols for the Internet of Things: A case study. *Proc. 8th IEEE Int. Conf. Pervasive comput. Commun. WorkshopS (PERCOM Workshops)*, 678–683.

Castellani, A. P., Dissegna, M., Bui, N., & Zorzi, M. (2012). WebIoT: A web application framework for the internet of things. *Proc. IEEE Wireless Commun. Netw. Conf. Workshops*. 10.1109/WCNCW.2012.6215491

Castro, R. M. G., & Ferreira, A. F. M. (2001, November). A comparison between chronological and probabilistic methods to estimate wind power capacity credit. *IEEE Transactions on Power Systems, 16*(4), 904–909. doi:10.1109/59.962444

CDC. (2001). *National Program of Cancer Registries Education and Training Series How to Collect High Quality Cancer Surveillance Data*. Retrieved from https://www.cdc.gov/cancer/npcr/pdf/abstracting/breast.pdf

Central Electricity Authority of India. (2019). *Executive Summary on Power Sector Jan-2019*. https://cea.nic.in/reports/monthly/executivesummary/2019/exe_summary-01.pdf

Čepin, M. (2011a). Event tree analysis. In *Assessment of Power System Reliability* (pp. 89–99). Springer. doi:10.1007/978-0-85729-688-7_6

Čepin, M. (2011b). Reliability block diagram. In *Assessment of Power System Reliability* (pp. 119–123). Springer. doi:10.1007/978-0-85729-688-7_9

Chamandoust, H., Bahramara, S., & Derakhshan, G. (2020). Multi-objective operation of smart stand-alone microgrid with the optimal performance of customers to improve economic and technical indices. *Journal of Energy Storage, 31*, 101738. doi:10.1016/j.est.2020.101738

Chandrashekhar, A. M., & Raghuveer, K. (2013, January). Fortification of hybrid intrusion detection system using variants of neural networks and support vector machines. *International Journal of Network Security and Its Applications, 5*(1), 71–90. doi:10.5121/ijnsa.2013.5106

Chanmugam, A., Hatwar, R., & Herman, C. (2012). Thermal analysis of cancerous breast model. *International Mechanical Engineering Congress and Exposition : International Mechanical Engineering Congress and Exposition, 2012*, 134–143. 10.1115/IMECE2012-88244

Chaudhary & Srivastava. (2016). Big data security issues and challenges. *International Conference on Computing, Communication and Automation (ICCCA2016).*

Chauhan, U., Pahuja, G. L., Singh, V., & Rani, A. (2016). Reliability analysis of wind turbine system using importance measures. *12th IEEE International Conference Electronics, Energy, Environment, Communication, Computer, Control: (E3-C3), INDICON 2015.*

Chen, B., Lu, Z., & Zhou, H. (2018). *Reliability assessment of distribution network considering cyber attacks.* Paper presented at the 2018 2nd IEEE Conference on Energy Internet and Energy System Integration (EI2). 10.1109/EI2.2018.8582610

Chen, X., & Lin, C. (2009). Tetris game system design based on at89s52 single chip microcomputer. In *2009 Third International Symposium on Intelligent Information Technology Application, Shanghai,* (pp. 256–259). IEEE.

Chernyshev, M., Baig, Z., Bello, O., & Zeadally, S. (2018, June). Internet of Things (IoT): Research, Simulators, and Testbeds. *IEEE Internet of Things Journal, 5*(3), 1637–1647. doi:10.1109/JIOT.2017.2786639

Choi, S., Agelidis, V. G., Yang, J., Coutellier, D., & Marabeas, P. (2011). Analysis, design and experimental results of a floating-output interleaved-input boost-derived dc–dc high-gain transformer-less converter. *IET Power Electronics, 4*(1), 168–180. doi:10.1049/iet-pel.2009.0339

Chowdhury, A., Agarwal, S., & Koval, D. (2003, September/October). Reliability modeling of distributed generation in conventional distribution systems planning and analysis. *IEEE Transactions on Industry Applications, 39*(5), 1493–1498. doi:10.1109/TIA.2003.816554

Cisco. (2017). *Cisco Visual Networking Index: Forecast and Methodology, 2016-2021.* https://www.cisco.com/c/en/us/solutions/collateral/service-provider/visual-networking-index-vni/complete-white-paper-c11-481360.html

Classifier Subset Evaluation. (n.d.). Available at: http://weka.sourceforge.net/doc.dev/weka/attributeSelection/ClassifierSubsetEval.html

Components101. (n.d.). *Servo motor pin config.* Available: https://components101.com/servo-motor-basics-pinout-datasheet.

Correlation based Feature Selection. (n.d.). Available at: https://machinelearningmastery.com/perform-feature-selection-machine-learning-data-weka/

Correlation Based Feature Selection. (n.d.). Available at: https://machinelearningmastery.com/perform-feature-selection-machine-learning-data-weka/

Crowder, R. M. (1991). Local Actuation of Multijointed Robotic Fingers. *International Conference on Control.*

Dahlia, A. A. Z., & Hanapi, Z. M. (2013). Hybrid of fuzzy clustering neural network over NSL dataset for intrusion detection system. *Journal of Computational Science, 9*(3), 391–403. doi:10.3844/jcssp.2013.391.403

Das, R., Turkoglu, I., & Sengur, A. (2009). Effective diagnosis of heart disease through neural networks ensembles. *Expert Systems with Applications, 36*(4), 7675–7680. doi:10.1016/j.eswa.2008.09.013

De Sanctis, M., Cianca, E., Araniti, G., Bisio, I., & Prasad, R. (2016, February). Satellite Communications Supporting Internet of Remote Things. *IEEE Internet of Things Journal, 3*(1), 113–123. doi:10.1109/JIOT.2015.2487046

De Soto, W., Klein, S. A., & Beckman, W. A. (2006). Improvement and validation of a model for photovoltaic array performance. *Solar Energy, 80*(1), 78–88. doi:10.1016/j.solener.2005.06.010

Debar, H., Dacier, M., & Wespi, A. (1999). Towards a taxonomy of intrusion-detection systems. *Computer Networks, 31*(8), 805–822.

Debnath, B., Dey, R., & Roy, S. (2019). Smart switching system using bluetooth technology. 2019 *Amity International Conference on Artificial Intelligence (AICAI)*, 760–763. doi: 10.1109/AICAI.2019.8701298

Dey, A., Singh, J., & Singh, N. (2016). Analysis of Supervised Machine Learning Algorithms for Heart Disease Prediction with Reduced Number of Attributes using Principal Component Analysis. *Analysis, 140*(2), 27–31.

Dharmaraja, S., Vinayak, R., & Trivedi, K. S. (2016). Reliability and survivability of vehicular ad hoc networks: An analytical approach. *Reliability Engineering & System Safety, 153*, 28–38. doi:10.1016/j.ress.2016.04.004

Di He, X. C., Zou, D., Pei, L., & Jiang, L. (2018). *An Improved Kernel Clustering Algorithm Used in Computer Network Intrusion Detection.* Advance online publication. doi:10.1109/ISCAS.2018.8350994

Ding, M., Xu, Z., Zhao, B., & Bi, R. (2015, September). Solar irradiance model for large-scale photovoltaic generation considering passing cloud shadow effect. *Zhongguo Dianji Gongcheng Xuebao, 35*(17), 4219–4299.

Distefano, S., Puliafito, A., & Computing, S. (2009). *Dependability evaluation with dynamic reliability block diagrams and dynamic fault trees.* Academic Press.

Dohler, M., Vilajosana, I., Vilajosana, X., & Llosa, J. (2011). Smart Cities: An action plan. *Proc. Barcelona Smart Cities Congress*, 1–6.

Dragoon & Dvortsov. (2006). Z-Method for Power System Resource Adequacy Applications. *IEEE Transactions on Power Systems, 21*(2).

Duch, P., & Jaworski, T. (2018). Enriching computer science programming classes with Arduino game development. *2018 11th International Conference on Human System Interaction (HSI)*, 148–154.

Dui, H., Si, S., & Yam, R. C. M. (2017). A cost-based integrated importance measure of system components for preventive maintenance. *Reliability Engineering & System Safety, 168*(May), 98–104. doi:10.1016/j.ress.2017.05.025

Easwarakhanthan, T., Bottin, J., Bouhouch, I., & Boutrit, C. (1986). Nonlinear minimization algorithm for determining the solar cell parameters with microcomputers. *International Journal of Solar Energy, 4*(1), 1-12.

EasyEDA. (n.d.). *Easyeda circuit editor.* Available: https://easyeda.com/editor

Eberhart, R., & Kennedy, J. (1995, November). Particle swarm optimization. In *Proceedings of the IEEE international conference on neural networks* (Vol. 4, pp. 1942-1948). 10.1109/ICNN.1995.488968

Eesa, A. S., Orman, Z., & Adnan, M. A. B. (2015). A novel feature-selection approach based on the cuttlefish optimization algorithm for intrusion detection systems. Academic Press.

Effendy, D. A., Kusrini, K., & Sudarmawan, S. (2017). Classification of intrusion detection system (IDS) based on computer network. In *Proceedings of 2017 2nd International conferences on Information Technology, Information Systems and Electrical Engineering (ICITISEE).* IEEE.

Elavarasan, Shafiullah, Padmanaban, & Kumar. (2020). A Comprehensive Review on Renewable Energy Development, Challenges, and Policies of Leading Indian States With an International Perspective. *IEEE Access : Practical Innovations, Open Solutions.*

electronicaestudio. (n.d.). *Bluetooth module hc-5 datasheet.* Available: http://www.electronicaestudio.com/docs/istd016a.pdf

electrosome. (n.d.). *hc-05-serial-bluetooth-module.* Available: https://electrosome.com/hc-05-serial-bluetooth- module/

Elshazly, H. I., Elkorany, A. M., & Hassanien, A. E. (2014). Lymph diseases diagnosis approach based on support vector machines with different kernel functions. *Computer Engineering & Systems 9th International Conference (ICCES)*, 198–203.

Elshazly, H. I., Elkorany, A. M., & Hassanien, A. E. (2014). Lymph diseases diagnosis approach based on support vector machines with different kernel functions. *Computer Engineering & Systems 9th International Conference (ICCES)*, 198–203. 10.1109/ICCES.2014.7030956

Espiritu, J. F., Coit, D. W., & Prakash, U. (2007). Component criticality importance measures for the power industry. *Electric Power Systems Research, 77*(5–6), 407–420. doi:10.1016/j.epsr.2006.04.003

Facchinetti, T., & Della Vedova, M. (2011). *Real-time modeling for direct load control in cyber-physical power systems.* Academic Press.

Faheem, M., Shah, S. B. H., Butt, R. A., Raza, B., Anwar, M., Ashraf, M. W., Ngadi, M. A., & Gungor, V. C. (2018). Smart grid communication and information technologies in the perspective of Industry 4.0: Opportunities and challenges. *Computer Science Review, 30*, 1–30. doi:10.1016/j.cosrev.2018.08.001

Falahati, B., & Fu, Y. (2014). *Reliability assessment of smart grids considering indirect cyber-power interdependencies.* Academic Press.

Falahati, B., Fu, Y., & Wu, L. (2012). *Reliability assessment of smart grid considering direct cyber-power interdependencies.* Academic Press.

Fang, Y.-P., Pedroni, N., & Zio, E. (2016). Resilience-Based Component Importance Measures for Critical Infrastructure Network Systems. *IEEE Transactions on Reliability, 65*(2), 502–512. doi:10.1109/TR.2016.2521761

Farivar, G., & Asaei, B. (2010, November). Photovoltaic module single diode model parameters extraction based on manufacturer datasheet parameters. In *2010 IEEE International Conference on Power and Energy* (pp. 929-934). IEEE. 10.1109/PECON.2010.5697712

Feng, Y., Wu, F., Shao, X., Wang, Y., & Zhou, X. (2018). *Joint 3D Face Reconstruction and Dense Alignment with Position Map Regression Network.* arXiv:1803.07835,2018.

fritzing. (n.d.). *fritzing circuit editor*. Available: https://fritzing.org/building-circuit/

Gain Ratio. (n.d.). *Hong Kong University of Science and Technology*. Available at: www.cse.ust.hk/~qyang/537/PPT/dtrees1.ppt

Garcia, A. M., Lipo, T. A., & Novotny, D. W. (1998). A new induction motor V/f control method capable of high-performance regulation at low speeds. *IEEE Transactions on Industry Applications, 34*(4), 813–821. doi:10.1109/28.703982

Garcia, F. S., Pomilio, J. A., & Spiazzi, G. (2010). Modeling and control design of the six-phase interleaved double dual boost. *Proc. 9th IEEE Int. Conf. Ind. Appl., 2010*, 1–6.

Garcia, F. S., Pomilio, J. A., & Spiazzi, G. (2013). Modeling and control design of the interleaved double dual boost converter. *IEEE Transactions on Industrial Electronics, 60*(8), 3283–3290. doi:10.1109/TIE.2012.2203770

Garver, L. L. (1966). Efective Load Carrying Capability of Generating Units. *IEEE Transactions on Power Apparatus and Systems, 85*(8).

Gautherie, M. (1980). Thermopathology of breast cancer: Measurement and analysis of in vivo temperature and blood flow. *Annals of the New York Academy of Sciences, 335*(1), 383–415. doi:10.1111/j.1749-6632.1980.tb50764.x PMID:6931533

Gautherie, M. (1983). Thermobiological assessment of benign and malignant breast diseases. *American Journal of Obstetrics and Gynecology, 147*(8), 861–869. doi:10.1016/0002-9378(83)90236-3 PMID:6650622

Geem, Z. W., Kim, J. H., & Loganathan, G. V. (2001). A new heuristic optimization algorithm: harmony search. *Simulation, 76*(2), 60-68.

Ghadge, Girme, Kokane, & Deshmukh. (2015). Intelligent Heart Disease Prediction System using Big Data. *International Journal of Recent Research in Mathematics Computer Science and Information Technology, 2*, 73-77.

Giacinto, G., Perdisci, R., Del Rio, M., & Roli, F. (2008). *Intrusion detection in computer networks by a modular ensemble of one-class classifiers*. Academic Press.

Global Alliance on Health and Pollution. (2019). *Pollution and Metrics*. Available: http://gahp.net/wp-content/uploads/2019/12/PollutionandHealthMetrics-final-12_18_2019.pdf

Gnana Priya, B., & Arulselvi, M. (2019). 3d Image Generation from Single 2d Image using Monocular Depth Cues. *International Journal of Engineering and Advanced Technology.*

Gomathi, K. (2016). Multi Disease Prediction using Data Mining Techniques. *International Journal of System and Software Engineering*, (December), 12–14.

González, F. J. (2011). Non-invasive estimation of the metabolic heat production of breast tumors using digital infrared imaging. *Quantitative Infrared Thermography Journal, 8*(2), 139–148. doi:10.3166/qirt.8.139-148

Gonzalez-Hernandez, J. L., Recinella, A. N., Kandlikar, S. G., Dabydeen, D., Medeiros, L., & Phatak, P. (2020). An inverse heat transfer approach for patient-specific breast cancer detection and tumor localization using surface thermal images in the prone position. *Infrared Physics and Technology, 105*(October), 103202. doi:10.1016/j.infrared.2020.103202

Gonzalez-Hernandez, J.-L., Kandlikar, S. G., Dabydeen, D., Medeiros, L., & Phatak, P. (2018). Generation and Thermal Simulation of a Digital Model of the Female Breast in Prone Position. *Journal of Engineering and Science in Medical Diagnostics and Therapy, 1*(4), 1–34. doi:10.1115/1.4041421

Goswami, D. Y. (2015). *Principles of solar engineering*. CRC Press. doi:10.1201/b18119

Goundar, S. S., Pillai, M. R., Mamun, K. A., Islam, F. R., & Deo, R. (2015). Real time condition monitoring system for industrial motors. *2nd Asia-Pacific World Congress on Computer Science and Engineering (APWC on CSE)*, 1-9.

Green, M. A. (1991). Recent advances in silicon solar cell performance. In *Tenth EC Photovoltaic Solar Energy Conference* (pp. 250-253). Springer. 10.1007/978-94-011-3622-8_63

Gupta, P. K., Singh, N. K., & Mahajan, V. (2020). *Monitoring of Cyber Intrusion in Wireless Smart Grid Network Using Weight Reduction Technique.* Paper presented at the 2020 International Conference on Electrical and Electronics Engineering (ICE3). 10.1109/ICE348803.2020.9122981

Hadri, A., Chougdali, K., & Touahni, R. (2016). Intrusion detection system using PCA and Fuzzy PCA techniques. In *Advanced Communication Systems and Information Security (ACOSIS), International Conference on*. IEEE. 10.1109/ACOSIS.2016.7843930

Haghifam, M. R., & Manbachi, M. (2011). Reliability and availability modelling of combined heat and power (CHP) systems. *International Journal of Electrical Power & Energy Systems, 33*(3), 385–393. doi:10.1016/j.ijepes.2010.08.035

Hajian-Hoseinabadi, H. (2013). Reliability and component importance analysis of substation automation systems. *International Journal of Electrical Power & Energy Systems, 49*, 455–463. doi:10.1016/j.ijepes.2010.06.012

Hamed, Ernst, & Kremer. (2018). A Survey and Taxonomy of Classifiers of Intrusion Detection Systems. In Computer and Network Security Essentials. Springer.

Harish Kumar, B. (2017). WSN based Automatic Irrigation and Security System using Raspberry Pi Board. *International Conference on Current Trends in Computer, Electrical, Electronics and Communication (CTCEEC)*, 1097-1103. 10.1109/CTCEEC.2017.8455140

Hashemi-Dezaki, H., Askarian-Abyaneh, H., Haeri-Khiavi, H. (2016). *Impacts of direct cyber-power interdependencies on smart grid reliability under various penetration levels of microturbine/wind/solar distributed generations*. Academic Press.

Haslett, J., & Diesendorf, M. (1981). On the capacity credit of wind power: A theoretical analysis. *Solar Energy, 26*(5), 391–401. doi:10.1016/0038-092X(81)90218-8

Hatwar, R., & Herman, C. (2017). Inverse method for quantitative characterisation of breast tumours from surface temperature data. *International Journal of Hyperthermia, 0*(0), 1–17. doi:10.1080/02656736.2017.1306758 PMID:28540793

Hayat, M. B., Ali, D., Monyake, K. C., Alagha, L., & Ahmed, N. (2019). Solar energy—A look into power generation, challenges, and a solar-powered future. *International Journal of Energy Research, 43*(3), 1049–1067. doi:10.1002/er.4252

Hindy, Brosset, Bayne, Seeam, Tachtatzis, Atkinson, & Bellekens. (2018). A Taxonomy and Survey of Intrusion Detection System Design Techniques. *Network Threats and Datasets, 1*, 1. doi:10.1145/nnnnnnn.nnnnnnn

Hodo, E., Bellekens, X., Hamilton, A., Tachtatzis, C., & Atkinson, R. (2017). *Shallow and deep networks intrusion detection system: A taxonomy and survey.* arXiv preprint arXiv:1701.02145.

Hoff, T., Perez, R., Ross, J. P., & Taylor, M. (2008). *Photovoltaic capacity valuation method.* SEPA, Tech. Rep. 02-08.

Honarmand, M. E., Ghazizadeh, M. S., Hosseinnezhad, V., Siano, P., & Systems, E. (2019). *Reliability modeling of process-oriented smart monitoring in the distribution systems.* Academic Press.

Hu, Hu, & Maybank. (2008). AdaBoost-based algorithm for network intrusion detection. *IEEE Transactions on Systems, Man, and Cybernetics, Part B: Cybernetics, 38*(2), 577–583. doi:10.1109/TSMCB.2007.914695

Hu, Karki, & Billinton. (2009). Reliability Evaluation of Generating Systems Containing Wind Power and Energy Storage. *Journal of IET Generation, Transmission & Distribution, 3*(8), 783-791.

Hu, P. (2009). *Reliability Evaluation of Electric Power Systems Including Wind Power and Energy Storage* (Thesis). University of Saskatchewan.

Hu. (2019). Research on a three-dimensional reconstruction method based on the feature matching algorithm of a scale-invariant feature transform. In *Elsevier Mathematical and Computer Modelling Book.* Academic Press.

Hwu, K. I., & Jiang, W. Z. (2014). Voltage gain enhancement for a step-up converter constructed by KY and buckboost converters. *IEEE Transactions on Industrial Electronics, 61*(4), 1758–1768. doi:10.1109/TIE.2013.2263779

Ibrahim, H., & Anani, N. (2017). Evaluation of analytical methods for parameter extraction of PV modules. *Energy Procedia, 134*, 69–78. doi:10.1016/j.egypro.2017.09.601

IEEE S. A Standard Board. (2016). *Recommended practice for motor protection in industrial power system.* https://mentor.ieee.org/3000-stds/dcn/18/stds-18-0003-00-PUBS-3004-8-2016.pdf

India's Population. (2020). http://worldpopulationreview.com/countries/india-population/

Indore Population Estimation. (2020). https://indiapopulation2020.in/population-of-indore-2020.html

Information About Growth of IoT. (2019). https://www.thalesgroup.com/en/markets/digital-identity-and-security/iot/magazine/infographic-growth-iot

Information Gain. (n.d.). Available at: https://www.saedsayad.com/decision_tree.htm

International Renewable Energy Agency. (2017). *Renewable Energy Prospects for India.* Available: http://www.irena.org/-/media/Files/IRENA/Agency/Publication/2017/May/IRENA_REmap_India_paper_2017.pdf

Internet of Things, Market Forecast. (2019). https://www.researchandmarkets.com/reports/3395405/india-internet-of-things-iot-market-forecast

Inventor, A. (n.d.). *MIT app inventor.* Available: https://appinventor.mit.edu/

Ismail, E. H., Al-Saffar, M. A., & Sabzali, A. J. (2008). High conversion ratio DC–DC converters with reduced switch stress. *IEEE Trans. Circuits Syst. I*, *55*(7), 2139–2151. doi:10.1109/TCSI.2008.918195

Jackson, A., Bulat, A., & Tzimiropoulos, Y. (2017). Large Pose 3D Face Reconstruction from a Single Image via Direct Volumetric CNN Regression. *2017 IEEE International Conference on Computer Vision (ICCV).* arxiv:1703.07834

Jackson, A. S., Bulat, A., Argyiou, V., & Tzimiropoulos, G. (2017). Large pose 3d face reconstruction from a single image via direct volumetric CNN regression. *2017 IEEE International Conference on Computer Vision (ICCV)*, 1031-1039. 10.1109/ICCV.2017.117

Jadli, U., Thakur, P., & Shukla, R. D. (2017). A new parameter estimation method of solar photovoltaic. *IEEE Journal of Photovoltaics*, *8*(1), 239–247. doi:10.1109/JPHOTOV.2017.2767602

Jang, Y., & Jovanovic, M. M. (2007). Interleaved boost converter with intrinsic voltage-doubler characteristic for universal-line PFC front end. *IEEE Transactions on Power Electronics*, *22*(4), 1394–1401. doi:10.1109/TPEL.2007.900502

Javaid, S., Sufian, A., Pervaiz, S., & Tanveer, M. (2018). Smart traffic management system using Internet of Things. *20th International Conference on Advanced Communication Technology (ICACT)*, 393-398.

Jiang, L., Zhan, W., & Loew, M. H. (2010). Modeling static and dynamic thermography of the human breast under elastic deformation. *Physics in Medicine and Biology*, *56*(1), 187–202. doi:10.1088/0031-9155/56/1/012 PMID:21149948

Jiang, L., Zhan, W., Loew, M. H., & Francisco, S. (2011). Toward Understanding the Complex Mechanisms behind Breast Thermography. *An Overview for Comprehensive Numerical Study.*, *7965*, 1–9. doi:10.1117/12.877839

Jiao, J. (2020). *Application and prospect of artificial intelligence in smart grid.* Paper presented at the IOP Conference Series: Earth and Environmental Science.

Joga, Sree, Spandana, & Pushpa. (2019). 3D reconstruction of regular objects from multiple 2D images using a reference object. International Journal of Advance Research, Ideas and Innovations in Technology.

Jourabloo, A., & Liu, X. (2016). Large-pose face alignment via CNN-based dense 3d model fitting. Computer Vision and Pattern Recognition.

Kakileti, S. T., Manjunath, G., Madhu, H., & Ramprakash, H. V. (2017). *Advances in Breast Thermography.* New Perspectives in Breast Imaging., doi:10.5772/intechopen.69198

Kanabar, M. G., & Sidhu, T. S. (2009). Reliability and availability analysis of IEC 61850 based substation communication architectures. *2009 IEEE Power & Energy Society General Meeting*, 1–8.

Kandlikar, S. G., Perez-Raya, I., Raghupathi, P. A., Gonzalez-Hernandez, J. L., Dabydeen, D., Medeiros, L., & Phatak, P. (2017). Infrared imaging technology for breast cancer detection – Current status, protocols and new directions. *International Journal of Heat and Mass Transfer*, *108*, 2303–2320. doi:10.1016/j.ijheatmasstransfer.2017.01.086

Kerler, M., Burda, P., Baumann, M., & Lienkamp, M. (2004). A Concept of a High-Energy, Low-Voltage EV Battery Pack. *2004 IEEE International Electric Vehicle Conference (IEVC).*

Kim, H., Lee, D., Cho, J., & Park, D. (2018). Software execution freeze-safe microcontroller using power profile tracking for IoT-driven connected services. *IEEE 4th World Forum on Internet of Things*, 237-240.

Kimani, K., Oduol, V., & Langat, K. J. (2019). *Cyber security challenges for IoT-based smart grid networks*. Academic Press.

Kim, I., James, J.-A., & Crittenden, J. (2017). The case study of combined cooling heat and power and photovoltaic systems for building customers using HOMER software. *Electric Power Systems Research*, *143*, 490–502. doi:10.1016/j.epsr.2016.10.061

K-Nearest Neighbour. (n.d.). Available at: https://en.wikipedia.org/wiki/K-nearest_neighbors_algorithm

Kothari. (2000). Renewable energy scenario in India. *2000 IEEE Power Engineering Society Winter Meeting. Conference Proceedings.*

Kounev, V., Lévesque, M., Tipper, D., & Gomes, T. (2016). Reliable Communication Networks for Smart Grid Transmission Systems. *Journal of Network and Systems Management*, *24*(3), 629–652. doi:10.100710922-016-9375-y

Kumar, N., & Mahajan, V. (2018). *Reconfiguration of Distribution Network For Power Loss Minimization & Reliability Improvement using Binary Particle Swarm Optimization.* Paper presented at the IEEE 8th Power India International Conference (PIICON), Kurukshetra, India. 10.1109/POWERI.2018.8704466

Kumar, S. A., & Kumar, D. A. (2014, August). An ensemble model for classification of attacks with feature selection based on KDD99 and NSL-KDD data set. *International Journal of Computers and Applications*, *99*(15), 8–13. doi:10.5120/15560-4109

Kumar, S. M., & Gursel, S. (2004). *Why Machine Learning Algorithms Fail in Misuse Detection on KDD Intrusion Detection Data Set. ACM Transactions on Intelligent Data Analysis.*

Kuo, W., & Zhu, X. (2012a). *Importance Measures in Reliability, Risk, and Optimization.* John Wiley & Sons. doi:10.1002/9781118314593

Kuo, W., & Zhu, X. (2012b). Some recent advances on importance measures in reliability. *IEEE Transactions on Reliability*, *61*(2), 344–360. doi:10.1109/TR.2012.2194196

Kushner, D. (2011). The making of arduino. *IEEE Spectrum*, 26.

Kuthong, J., Sapaklom, T., Konghirun, M., Prapanavarat, C., Ayudhya, P. N., Mujjalinvimut, E., & Boonjeed, S. (2017). *IoT based traction motor drive condition monitoring in electric vehicles- Part 1.* IEEE PEDS.

Lakshminarasimhan, M. (2016). *IoT Based Traffic Management System.* Academic Press.

Lawson, R. (1956). Implications of surface temperatures in the diagnosis of breast cancer. *Canadian Medical Association Journal, 75*(4), 309–311. Retrieved from https://pubmed.ncbi.nlm.nih.gov/13343098

Lee, S., Yoon, D., & Ghosh, A. (2008). Intelligent parking lot application using wireless sensor networks. *Proc. Int. Symp. Collab. Technol. Syst.* 10.1109/CTS.2008.4543911

Leo, B., Friedman, J.H., Olshen, R.A., & Stone, C.J. (1984). Classification and regression trees. Monterey, CA: Wadsworth & Brooks/ColeAdvanced Books & Software.

Levent, K., Mazzuchi, T. A., & Shahram, S. (2012). A network intrusion detection system based on a Hidden Naïve Bayes multiclass classifier. *Expert Systems with Applications*, *39*, 13492–13500.

Li, L., Yang, D.-Z., & Shen, F.-C. (2010). A Novel Rule-based Intrusion detection System Using Data Mining. *Proc. of 3rd IEEE International Conference on Computer Science and Information Technology*, 169–172.

Li, Q., Meng, S., Zhang, S., Wu, M., Zhang, J., Ahvanooey, M. T., & Aslam, M. S. (2019). *Safety risk monitoring of cyber-physical power systems based on ensemble learning algorithm*. Academic Press.

Lin, S.-W., Ying, K.-C., Lee, C.-Y., & Lee, Z.-J. (2012). *An intelligent algorithm with feature selection and decision rules applied to anomaly intrusion detection*. Academic Press.

Lin, W.-C., Ke, S.-W., & Tsai, C.-F. (2015). *CANN: An intrusion detection system based on combining cluster centers and nearest neighbors*. Academic Press.

Lin, W.-C., Shih-Wen, K., & Chih-Fong, T. (2015). CANN: An intrusion detection system based on combining cluster centers and nearest neighbors. *Knowledge-Based Systems*, 13–21.

Linear Forward Selection. (n.d.). Available at: http://weka.sourceforge.net/doc.stable/weka/attributeSelection/LinearForwardSelection.html

Lisehroodi, M. M., Muda, Z., & Yassin, W. (2013). A hybrid framework based on neural network MLP and K-means Clustering for Intrusion Detection System. In *Proceedings of 4th International Conference on Computing and Informatics*. ICOCI.

Liu, Y., Deng, L., Gao, N., & Sun, X. (2019). *A reliability assessment method of cyber physical distribution system*. Academic Press.

Liu, S., & Dougal, R. A. (2002). Dynamic multiphysics model for solar array. *IEEE Transactions on Energy Conversion*, *17*(2), 285–294. doi:10.1109/TEC.2002.1009482

Li, W., & He, X. (2011). Review of non-isolated high-step-up DC-DC converters in photovoltaic grid-connected applications. *IEEE Transactions on Industrial Electronics*, *58*(4), 1239–1250. doi:10.1109/TIE.2010.2049715

Lozano, A., & Hassanipour, F. (2019). Infrared imaging for breast cancer detection: An objective review of foundational studies and its proper role in breast cancer screening. *Infrared Physics and Technology*, *97*(December), 244–257. doi:10.1016/j.infrared.2018.12.017

Lozano, A. III, Hayes, J. C., Compton, L. M., Azarnoosh, J., & Hassanipour, F. (2020). Determining the thermal characteristics of breast cancer based on high-resolution infrared imaging, 3D breast scans, and magnetic resonance imaging. *Scientific Reports*, *10*(1), 1–14. doi:10.103841598-020-66926-6 PMID:32572125

Lüdtke, I., & Jayne, M. G. (1995). A comparative study of high performance speed control strategies for voltage source PWM inverter fed I.M drives. *Electrical Machines and Drives, Conference Publications No. 412*.

Machine Learning RepositoryU. C. I. (n.d.). Available at: https://archive.ics.uci.edu/ml/machine-learning-databases/heart-disease/

Madaeni, Sioshansi, & Denholm. (2012). *Comparison of capacity value methods for photovoltaics in the western US*. NREL, Colorado, Tech. Rep., NREL/TP-6A20-54704.

Manohar, L. P. (2009). Reliability Assessment of a Power Grid with Customer Operated CHP Systems Using Monte Carlo Simulation. *Masters Theses*, 348.

Marcus, W. (2015). *Low Carbon Grid Study: Comparison of 2030 Fixed Cost of Renewables, Efficiency, Integration With Production Cost Savings*. JBS Energy.

Masethe, H. D., & Masethe, M. A. (2014). Prediction of heart disease using classification algorithms. *World Congress on Engineering and Computer Science 2014, 2.*

Max, B. (2013). *Principles of Data Mining* (2nd ed.). Springer.

McRoberts, M. (2011). *Beginning Arduino.* Apress.

Miller, N. (2015). *Low Carbon Grid Study: Discussion of Dynamic Performance Limitations in WECC.* GE Energy Consulting. Available http://lowcarbongrid2030.org/wpcontent/uploads/2015/08/150429_LCGS-Dynamics-White-Paper_NWMiller.pdf

Milligan, M., & Parsons, B. (1997). A comparison and case study of capacity credit algorithms for intermittent generators. *Proc. Solar'97.* NREL/CP-440-22 591.

Mills, A., Ahlstrom, M., Brower, M., Ellis, A., George, R., Hoff, T., Kroposki, B., Lenox, C., Miller, N., Milligan, M., Stein, J., & Wan, Y. (2011, May/June). Dark shadows: Understanding variability and uncertainty of photovoltaics for integration with the electric power system. *IEEE Power & Energy Magazine, 9*(3), 33–41. doi:10.1109/MPE.2011.940575

Ministry of New and Renewable Energy Government of India. (n.d.). *Vision.and.Mission.* Available: http://mnre.gov.in/mission-andvision-2/mission-and-vision/

Ministry of New and Renewable Energy. (2020). *Annual Reports 2019-2020.* Available: http:// mnre.gov.in/img/documents/uploads/file_f-1585710569965.pdf

Misbahuddin, S., Zubairi, J. A., Saggaf, A., & Basuni, J. (2015). IoT based dynamic road traffic management for smart cities. *12th International Conference on High-capacity Optical Networks and Enabling/Emerging Technologies (HONET),* 1-5. 10.1109/HONET.2015.7395434

Mital, M., & Pidaparti, R. M. (2008). Breast Tumor Simulation and Parameters Estimation Using Evolutionary Algorithms. *Modelling and Simulation in Engineering, 756436,* 1–6. Advance online publication. doi:10.1155/2008/756436

Model, P., & Becker, S. M. (n.d.). Analytical Bioheat Transfer : Solution Development of the. In Heat Transfer and Fluid Flow in Biological Processes. doi:10.1016/B978-0-12-408077-5.00004-3

Mok, M. S., Sohn, S. Y., & Yong, H. J. (2010). *Random effects logistic regression model for anomaly detection.* Academic Press.

Monica, S. (2016, February). Analysis of CardioVasular Disease Prediction using Data Mining Techniques. *International Journal of Modern Computer Science, 4*(1), 55–58.

Moreno, J., Serrano, M. A., Fernandez-Medina, E., & Fernandez, E. B. (2018). *Towards a Security Reference Architecture for Big Data.* http://ceur-ws.org/Vol-2062/paper04.pdf

Moreno, M. V., Terroso-Saenz, F., Gonzalez-Vidal, A., Valdes-Vela, M., Skarmeta, A. F., Zamora, M. A., & Chang, V. (2017, April). Applicability of Big Data Techniques to Smart Cities Deployments. *IEEE Transactions on Industrial Informatics, 13*(2), 800–809. doi:10.1109/TII.2016.2605581

Moslehi, K., & Kumar, R. (2010). *A reliability perspective of the smart grid.* Academic Press.

Mudgal, S. M., Yadav, A. K., & Mahajan, V. (2019). *Reliability Evaluation Of Power System Network With Solar Energy.* Paper presented at the 8th International Conference on Power Systems (ICPS), Jaipur. 10.1109/ICPS48983.2019.9067364

Mudgal, S., & Mahajan, V. (2019). *Reliability and Active Power Loss Assessment Of Power System Network With Wind Energy.* Paper presented at the IEEE Student Conference on Research and Development (SCOReD), Bandar Seri Iskandar, Malaysia. 10.1109/SCORED.2019.8896327

Mujawar & Devale. (2015). Prediction of Heart Disease using Modified K-means and by using Naïve Bayes. *International Journal of Innovative research in Computer and Communication Engineering, 3*, 10265-10273.

Mujawar & Devale. (2015). Prediction of Heart Disease using Modified K-means and by using Naïve Bayes. *International Journal of Innovative Research in Computer and Communication Engineering, 3*, 10265-10273.

Mulligan, C. E. A., & Olsson, M. (2013, June). Architectural implications of smart city business models: An evolutionary perspective. *IEEE Communications Magazine, 51*(6), 80–85. doi:10.1109/MCOM.2013.6525599

Muna Mhammad, T. J., & Mehrotra, M. (2010). Design network intrusion detection system using hybrid fuzzy-neural network. *International Journal of Computer Science and Security, 4*(3), 285–294.

Naderipour, A., Abdul-Malek, Z., Nowdeh, S. A., Ramachandaramurthy, V. K., Kalam, A., & Guerrero, J. M. (2020). Optimal allocation for combined heat and power system with respect to maximum allowable capacity for reduced losses and improved voltage profile and reliability of microgrids considering loading condition. *Energy, 196*, 117124. doi:10.1016/j.energy.2020.117124

Naïve Bayes Classifier. (n.d.). Available at: https://en.wikipedia.org/wiki/Naive_Bayes_classifier

National Institute of Bio-Energy. (n.d.). Available: http://nibe.res.in/

National Institute of Solar Energy. (n.d.). Available: https://nise.res.in/

National Institute of Wind Energy. (n.d.). Available: http://niwe.res.in/

National Portal of India. (2019). https://www.india.gov.in/my-government

Nazeem Basha, Jilani, & Arun. (2016). An Intelligent Door System using Raspberry Pi and Amazon Web Services IoT. *International Journal of Engineering Trends and Technology, 33*, 84-89.). doi:10.14445/22315381/IJETT-V33P217

Negm, M. M., Bakhashwain, J. M., & Shwehdi, M. H. (2006). Speed control of three-phase induction motor based on optimal preview control theory. *IEEE Transactions on Energy Conversion, 21*(1), 77–84.

Ng, E. Y. K., & Sudharsan, N. M. (2001). An improved three dimensional direct numerical modelling and thermal analysis of a female breast with tumour. *Proceedings of the Institution of Mechanical Engineers. Part H, Journal of Engineering in Medicine, 215*(1), 25–38. doi:10.1243/0954411011533508 PMID:11323983

Nomura, H., Fujiwara, K., & Yoshida, M. (2006). A new DCDC converter circuit with larger step-up/down ratio. *Proc. 37th IEEE Power Electron. Spec. Conf.*, 1 - 7.

Nskh, P., Varma, M. N., & Naik, R. R. (2016). Principle component analysis based intrusion detection system using support vector machine. In *Recent Trends in Electronics, Information and Communication Technology (RTEICT), IEEE International Conference on*. IEEE. 10.1109/RTEICT.2016.7808050

Nsl-kdd data set for network-based intrusion detection systems. (2014). Available at: http://nsl.cs.unb.ca/NSL-KDD/

Obi, M., Slay, T., & Bass, R. (2020). Distributed energy resource aggregation using customer-owned equipment: A review of literature and standards. *Energy Reports, 6*, 2358–2369. doi:10.1016/j.egyr.2020.08.035

Oh, U., Lee, Y., Choi, J., Yoon, Y., Chang, B., & Cha, J.-M. (2016, March). Development of Reliability Contribution Function of Power System including Wind Turbine Generators combined with Battery Energy Storage System. *Journal of KIEE, 65*(3), 371–381. doi:10.5370/KIEE.2016.65.3.371

Oh, U., Lee, Y., Lim, J., Choi, J., Yoon, Y., Chang, B., & Cho, S. (2015, January). Reliability Evaluation with Wind Turbine Generators and an Energy Storage System for the Jeju Island Power System. *Journal of KIEE, 64*(1), 1–7. doi:10.5370/KIEE.2015.64.1.001

Pacca, S., Sivaraman, D., & Keoleian, G. A. (2007). Parameters affecting the life cycle performance of PV technologies and systems. *Energy Policy, 35*(6), 3316–3326. doi:10.1016/j.enpol.2006.10.003

Paliwal, P. (2020). Reliability constrained planning and sensitivity analysis for Solar-Wind-Battery based Isolated Power System. *International Journal of Sustainable Energy Planning and Management, 29*, 109–126.

Patel, Upadhay, & Patel. (2017). *Heart disease Prediction using Machine Learning and Data mining Technique*. Academic Press.

Patel, Upadhay, & Patel. (2017). *Heart disease Prediction using Machine Learning and Data mining Technique*. UCI Machine Learning Repository. Available at: http://archive.ics.uci.edu/ml/about.html

Patel, Upadhyay, & Patel. (2015). Heart Disease Prediction using Machine Learning and Data Mining Technique. *International Journal of Computer Science and Communication*, 129-137.

Paul, Sivan, & Balachandran. (2013). Energy sector in India: Challenges and solutions. *2013 International Conference on Green Computing,Communication and Conservation of Energy (ICGCE)*.

Pazouki, S., Mohsenzadeh, A., Ardalan, S., & Haghifam, M.-R. (2016). Optimal place, size, and operation of combined heat and power in multi carrier energy networks considering network reliability, power loss, and voltage profile. *IET Generation, Transmission & Distribution, 10*(7), 1615–1621. doi:10.1049/iet-gtd.2015.0888

Pelland, S., & Abboud, I. (2008). Comparing Photovoltaic Capacity Value Metrics: A Case Study for the City of Toronto. *Progress in Photovoltaics: Research and Applications, 16*(8), 715–724. doi:10.1002/pip.864

Peña, J. M., & Diaz, E. V. (2016). *Implementation of V/f Scalar Control for speed regulation of three phase induction motor*. IEEE ANDESCON Arequipa.

Pennes, H. H. (1998). Analysis of tissue and arterial blood temperatures in the resting human forearm. 1948. *Journal of Applied Physiology, 85*(1), 5–34. doi:10.1152/jappl.1998.85.1.5

Petrov, N., Dobrilovic, D., Kavalić, M., & Stanisavljev, S. (2016). *Examples of Raspberry Pi usage in Internet of Things.*). doi:10.20544/AIIT2016.15

Polaraju, K., & Durga Prasad, D. (2017). Prediction of Heart Disease using Multiple Linear Regression Model. *International Journal of Engineering Development and Research Development.*

Prabhavathi & Chitra. (2016). Analysis and Prediction of Various Heart Diseases using DNFS Techniques. *International Journal of Innovations in Scientific and Engineering Research, 2*(1), 1-7.

Prajapati, S., & Fernandez, E. (2019). Capacity credit estimation for solar PV installations in conventional generation: Impacts with and without battery storage. *Energy Sources. Part A, Recovery, Utilization, and Environmental Effects*, 1–13. Advance online publication. doi:10.1080/15567036.2019.1676326

Prerana, T. H. M., Shivaprakash, N. C., & Swetha, N. (2015). Prediction of Heart Disease Using Machine Learning Algorithms- Naïve Bayes, Introduction to PAC Algorithm, Comparison of Algorithms and HDPS. *International Journal of Science and Engineering, 3*(2), 90-99.

Press Information Bureau, Government of India, Ministry of Power. (2019). *47.86 GW of Renewable Energy Capacity Installed in Last Six Years*. Available: https://pib.gov.in/Pressreleaseshare.aspx?PRID=1596217

Priyanka & Kavita. (2016). Feature Selection Using Genetic Algorithm and Classification using Weka for Ovarian Cancer. *International Journal of Computer Science and Information Technologies, 7*(1), 194–196.

Purushottam, P. (2016). Efficient Heart Disease Prediction System. Academic Press.

Purushottam, S. (2015). Heart Disease Prediction System Evaluation using C4.5 Rules and Partial Tree. *Computational Intelligence in Data Mining, 2*, 285–294.

Quiles, E., & Rold, Ã. (2020). Accurate sizing of residential stand-alone photovoltaic systems considering system reliability. *Sustainability, 12*(3), 1274. doi:10.3390u12031274

Rahman, C. M., Farid, D. M., & Rahman, M. Z. (2011). Adaptive intrusion detection based on boosting and naive bayesian classifier. *International Journal of Computers and Applications, 24*(3), 11–19.

Rai, A. K. (2020). Advances in Reliability of Solar PV Systems. In *Advances in Energy and Built Environment* (pp. 13–21). Springer. doi:10.1007/978-981-13-7557-6_2

Ramana, M. V. (2007). *Nuclear power in India: Failed past, dubious future*. Available at www. npec-web. org/Frameset. asp

Rama, S. A., & Windu, G. (2017). Intrusion detection system using hybrid binary PSO and K-nearest neighborhood algorithm. In *Proceedings of 11th International Conference on Information & Communication Technology and System (ICTS)*. IEEE.

Ramirez-Marquez, J. E., Rocco, C. M., Gebre, B. A., Coit, D. W., & Tortorella, M. (2006). New insights on multi-state component criticality and importance. *Reliability Engineering & System Safety, 91*(8), 894–904. doi:10.1016/j.ress.2005.08.009

Ranjan, R., & Sahoo, G. (2014). *A new clustering approach for anomaly intrusion detection*. arXiv preprint arXiv:1404.2772.

Rath, M. (2018). Smart Traffic Management System for Traffic Control using Automated Mechanical and Electronic Devices. *IOP Conference Series. Materials Science and Engineering, 377*, 012201. doi:10.1088/1757-899X/377/1/012201

Ravi Kiran Varma, P., Valli Kumari, V., & Srinivas Kumar, S. (2018). A Survey of Feature Selection Techniques in Intrusion Detection System: A Soft Computing Perspective. In Progress in Computing, Analytics and Networking. Springer.

Rawat, N., Thakur, P., & Jadli, U. (2019). Solar PV parameter estimation using multi-objective optimisation. *Bulletin of Electrical Engineering and Informatics, 8*(4), 1198–1205. doi:10.11591/eei.v8i4.1312

Refaat, S. S., Abu-Rub, H., Trabelsi, M., & Mohamed, A. (2018). *Reliability evaluation of smart grid system with large penetration of distributed energy resources*. Paper presented at the 2018 IEEE International Conference on Industrial Technology (ICIT). 10.1109/ICIT.2018.8352362

Richardson, D. B., & Harvey, L. D. D. (2015). Strategies for correlating solar PV array production with electricity demand. *Renewable Energy, 76*, 432–440. doi:10.1016/j.renene.2014.11.053

Ridzuan, M. I. M., Roslan, N. N. R., Fauzi, N. F. M., & Rusli, M. A. Z. (2020). Reliability-based DG location using Monte-Carlo simulation technique. *SN Applied Sciences, 2*(2), 1–11.

Sai, Reddy, Palagi, & Jaya. (2017). Heart Disease Prediction using ANN Algorithm in Data Mining. *International Journal of Computer Science and Mobile Computing*, (April), 168–172.

Salary, Banaei, & Ajami. (2017). Design of novel stepup boost DC-DC converter. *Iran. J. Sci. Technol. Trans. Electr. Eng., 41,* 13-22.

Salary, E., Banaei, M. R., & Ajami, A. (2015). Step-up DC-DC converter based on partial power processing. *Gazi Univ. J. Sci., 28*(4), 599–607.

Salary, E., Banaei, M. R., & Ajami, A. (2016). Multi-stage DC-AC converter based on new DC-DC converter for energy conversion. *J. Oper. Autom. Power Eng., 4*, 42–53.

Salazar, J. C., Nejjari, F., Sarrate, R., Weber, P., & Theilliol, D. (2016). Reliability importance measures for a health-aware control of drinking water networks. *Conference on Control and Fault-Tolerant Systems, SysTol*, 572–578. 10.1109/SYSTOL.2016.7739810

Salih & Omer. (2018). *Raspberry pi as a Video Server*. .8515817 doi:10.1109/ICCCEEE.2018

Saniei, E., Setayeshi, S., Akbari, M. E., & Navid, M. (2016). Parameter estimation of breast tumour using dynamic neural network from thermal pattern. *Journal of Advanced Research*, *7*(6), 1045–1055. doi:10.1016/j.jare.2016.05.005 PMID:27857851

Sarkar. (2010). *Global Climate Change and Sustainable Energy Development: Focus on Emerging Issues and Strategies for the Asia- Pacific Region*. Strategic Planning for Energy and the Environment.

Sarwat, A. I., Domijan, A., Amini, M. H., Damnjanovic, A., & Moghadasi, A. (2015). *Smart Grid reliability assessment utilizing Boolean Driven Markov Process and variable weather conditions*. Paper presented at the 2015 North American Power Symposium (NAPS). 10.1109/NAPS.2015.7335101

Satwik, C. V. A., Kumar, L. P., Vineeth, K., & Pillai, K. N. (2018). Intelligent Road Management System for Daily Transit. International Conference on Communication and Signal Processing (ICCSP), 523-526.

Selinger-Lutz, O., & Gro, Ã. Ÿ. (2020). Dynamic feed-in tariffs with reduced complexity and their impact on the optimal operation of a combined heat and power plant. *International Journal of Electrical Power & Energy Systems*, *118*, 105770. doi:10.1016/j.ijepes.2019.105770

Shafi, K., & Abbass, H. A. (2009). *An adaptive genetic-based signature learning system for intrusion detection*. Academic Press.

Shahi, M., & Gurm, R. K. (2017). Heart Disease Prediction System using Data Mining Techniques. *Orient J. Computer Science Technology*, *6*, 457–466.

Shalev-Shwartz & Ben-David. (2014). *Understanding Machine Learning*. Cambridge University Press.

Sharma, K., Saini, L., & Reviews, S. E. (2015). *Performance analysis of smart metering for smart grid: An overview*. Academic Press.

Sharmila, R., & Chellammal, S. (2018, May). A conceptual method to enhance the prediction of heart diseases using the data techniques. *International Journal on Computer Science and Engineering*.

Sharp, M. A., & Vagapov, Y. (2017). *Comparative analysis and practical implementation of the ESP32 Microcontroller Module of the internet of things*. Internet Technologies and Applications.

Shedole & Deepika. (2016). *Predictive analytics to prevent and control chronic disease*. https://www.researchgate.net/punlication/316530782

Shone, N., Ngoc, T. N., Phai, V. D., & Shi, Q. (2018). A deep learning approach to network intrusion detection. *IEEE Transactions on Emerging Topics in Computational Intelligence*, *2*(1), 41–50. doi:10.1109/TETCI.2017.2772792

Shu-Kong, K., & Lu, D. D. C. (2013). A high step-down transformer-less single-stage single-switch AC/DC converter. *IEEE Transactions on Power Electronics*, *28*(4), 36–45.

Shyam, B., & Kanakasabapathy, P. (2017). Renewable energy utilization in India—policies, opportunities and challenges. In *2017 International Conference on Technological Advancements in Power and Energy (TAP Energy)*. IEEE. 10.1109/TAPENERGY.2017.8397311

Shyu, M., Chen, S., Sarinnapakorn, K., & Chang, L. (2003). A novel anomaly detection scheme based on principal component classifier. *Proceedings of the IEEE Foundations and New Directions of Data Mining Workshop, in conjunction with the Third IEEE International Conference on Data Mining (ICDM03)*, 172–179.

Singh & Sood. (2016). Technoeconomic development of renewable energy sector in India. *2016 IEEE Region 10 Humanitarian Technology Conference (R10-HTC)*.

Singh, N. K., & Mahajan, V. (2020). *Detection of cyber cascade failure in smart grid substation using advance grey wolf optimization*. Academic Press.

Singh, D., Saini, J. K., & Sood, Y. R. (2016). The development and potential of wind power sector in India. In *2016 International Conference on Electrical Power and Energy Systems (ICEPES)*. IEEE.

Singh, D., & Singh, A. K. (2020). Role of image thermography in early breast cancer detection- Past, present and future. *Computer Methods and Programs in Biomedicine, 183*, 105074. Advance online publication. doi:10.1016/j.cmpb.2019.105074 PMID:31525547

Singh, J. (2016). Management of the agricultural biomass on decentralized basis for producing sustainable power in India. *Journal of Cleaner Production*.

Singh, N. K., Gupta, P. K., & Mahajan, V. (2020). *Intrusion Detection in Wireless Network of Smart Grid Using Intelligent Trust-Weight Method*. Academic Press.

Singh, N. K., Gupta, P. K., Mahajan, V., Yadav, A. K., & Mudgal, S. Monitoring Cyber-Physical Layer of Smart Grid Using Graph Theory Approach. In *Control Applications in Modern Power System* (pp. 519–525). Springer.

Singh, V. K., Sahu, A., Beg, A., Khan, B., & Kumar, S. (2018). Speed direction control of dc motor through bluetooth hc-05 using Arduino. *2018 International Conference on Advanced Computation and Telecommunication (ICACAT)*, 1–3.

Si, S., Liu, M., Jiang, Z., Jin, T., & Cai, Z. (2019). System Reliability Allocation and Optimization Based on Generalized Birnbaum Importance Measure. *IEEE Transactions on Reliability, 68*(3), 831–843. doi:10.1109/TR.2019.2897026

Sivasankari, B. (2017). IOT based Indoor Air Pollution Monitoring using Raspberry PI. *International Journal of Innovations in Engineering and Technology, 9*(2), 16–2.

Smart Mobility Solution in Indore. (2019). http://urbanmobilityindia.in/Upload/Conference/16d93b11-5067-48da-b854-1bff507a99c2.pdf

Su, M.-Y. (2011). *Real-time anomaly detection systems for Denial-of-Service attacks by weighted k-nearest-neighbor classifiers*. Academic Press.

Su, M.-Y. (2011). Real-time anomaly detection systems for Denial-of-Service attacks by weighted k-nearest-neighbor classifiers. *Expert Systems with Applications, 38*, 3492–3498.

Subcommittee, P. M. (1979). IEEE reliability test system. *IEEE Transactions on Power Apparatus and Systems, PAS-98*(6), 2047–2054. doi:10.1109/TPAS.1979.319398

Suetake, M., Da Silva, I. N., & Goedtel, A. (2011). Embedded DSP-Based Compact Fuzzy System and Its Application for Induction-Motor V/fV/f Speed Control. *IEEE Transactions on Industrial Electronics*, 750–760.

Sulaiman, S. I., Rahman, T. K. A., Musirin, I., & Shaari, S. (2012, May). Artificial neural network versus linear regression for predicting Grid-Connected Photovoltaic system output. In *2012 IEEE International Conference on Cyber Technology in Automation, Control, and Intelligent Systems (CYBER)* (pp. 170-174). IEEE. 10.1109/CYBER.2012.6392548

Sultana & Haider. (2017). *Heart Disease Prediction using WEKA tool and 10-Fold cross-validation.* The Institute of Electrical and Electronics Engineers.

Support Vector Machine. (n.d.). Available at: https://docs.opencv.org/2.4.13.7/doc/tutorials/ml/introduction_to_svm/introduction_to_svm.html

Sutar, P. P., & Panchade, V. M. (2017). Induction motor faults mitigation using microcontroller. *International Conference on Energy, Communication, Data Analytics and Soft Computing (ICECDS)*, 489-493.

Swachh Survekshan. (2020). http://www.swachhsurvekshan2020.org/

Swarna Priya, R. M., Aarthy, S. L., Gunavathi, C., & Venkatesh, P. (2017). 3D Reconstruction of a Scene from Multiple 2D Images. *International Journal of Civil Engineering and Technology*, *8*(12), 324–331.

Syarif, A. R., & Gata, W. (2017). Intrusion detection system using hybrid binary PSO and K-nearest neighborhood algorithm. In *Information and Communication Technology and System (ICTS), 2017 11th International Conference on*. IEEE.

Sykora, P., Kamencay, P., & Hudec, R. (2014). Comparison of SIFT and SURF Methods for Use on Hand Gesture Recognition based on Depth Map. *AASRI Procedia*, *9*, 19–24. doi:10.1016/j.aasri.2014.09.005

Thomas, M. S., & Ali, I. (2010). Reliable, Fast, and Deterministic Substation Communication Network Architecture and its Performance Simulation. *IEEE Transactions on Power Delivery*, *25*(4), 2364–2370. doi:10.1109/TPWRD.2010.2042824

Thomson, W. T., & Fenger, M. (2001). Current signature analysis to detect induction motor faults. *IEEE Industry Applications Magazine*, *7*(4), 26–3.

Tsuji, T., Fukuda, O., Shigeyoshi, H., & Kaneko, M. (2000). Bio-mimetic Impedance Control of an EMG- controlled Prosthetic Hand. *IEEE/RSJ International Conference on Intelligent Robots and Systems*, 377-382. 10.1109/IROS.2000.894634

Ul-Haq, Jalal, Sindi, & Ahmad. (2020). Energy Scenario in South Asia: Analytical Assessment and Policy Implications. *IEEE Access : Practical Innovations, Open Solutions*.

Umera, A. (2017). *IOT Based Theft Detection using Raspberry Pi*. Academic Press.

Vadicherla, D., & Sonawane, S. (2013). Decision Support System for Heart Disease Based on Sequential Minimal Optimization in Support. *International Journal of Engineering Sciences and Emerging Technologies*, *4*(2), 19–26.

Vanisree, K. (2011). Decision Support System for Congenital Heart Disease Diagnosis based on Signs and Symptoms using Neural Networks. *International Journal of Computers and Applications*, *19*(6), 6–12. doi:10.5120/2368-3115

Villalva, M. G., Gazoli, J. R., & Ruppert Filho, E. (2009). Comprehensive approach to modeling and simulation of photovoltaic arrays. *IEEE Transactions on Power Electronics*, *24*(5), 1198–1208. doi:10.1109/TPEL.2009.2013862

Vipin, K., Jaideep, S., & Aleksandar, L. (2006). *Managing cyber threats: Issues, approaches, and challenges* (Vol. 5). Springer.

Wadi, M., Baysal, M., Shobole, A., & Tur, M. R. (2018). *Reliability Evaluation in Smart Grids via Modified Monte Carlo Simulation Method.* Paper presented at the 2018 7th International Conference on Renewable Energy Research and Applications (ICRERA). 10.1109/ICRERA.2018.8566982

Waikato environment for knowledge analysis (weka) version 3.6.9. (2013). Available at: https://download.cnet.com/Weka-32-bit/3000-10254_4-75852610.html

Walravens, N., & Ballon, P. (2013, June). Platform business models for smart cities: From control and value to governance and public value. *IEEE Communications Magazine*, *51*(6), 72–79. doi:10.1109/MCOM.2013.6525598

Wang, Q., Cai, X., Tang, Y., Ni, M., & Systems, E. (n.d.). *Methods of cyber-attack identification for power systems based on bilateral cyber-physical information*. Academic Press.

Wang, D., Chen, D., Song, B., Guizani, N., Yu, X., & Du, X. (2018, October). From IoT to 5G I-IoT: The Next Generation IoT-Based Intelligent Algorithms and 5G Technologies. *IEEE Communications Magazine*, *56*(10), 114–120. doi:10.1109/MCOM.2018.1701310

Weng, S. F., Reps, J., Kai, J., Garibaldi, J. M., & Qureshi, N. (2017). Can machine-learning improve cardiovascular risk prediction using routine clinical data? *PLoS One*, *12*(4), e0174944. doi:10.1371/journal.pone.0174944 PMID:28376093

Wolf, M., & Rauschenbach, H. (1963). Series resistance effects on solar cell measurements. *Advanced Energy Conversion*, *3*(2), 455–479. doi:10.1016/0365-1789(63)90063-8

Worachai, S., & Silada, I. (2015). Classification model of network intrusion using Weighted Extreme Learning Machine. In *Proceedings of 12th International Joint Conference on Computer Science and Software Engineering (JCSSE)*. IEEE.

Wu, S.-Y., & Yen, E. (2009). *Data mining-based intrusion detectors*. Academic Press.

Wu, & Kumar, Quinlan, Joydeep, Yang, Hiroshi, McLachlan, Ng, Liu, Yu, Zhou, Steinbach, Hand, & Steinberg. (2007). Top Ten Data Mining Algorithms. Knowledge and Information Systems Journal. *Springer-Verlag London*, *14*(1), 1–37.

Wu, S., Chen, Y., Wu, Q., & Wang, Z. (2016). Linking component importance to optimisation of preventive maintenance policy. *Reliability Engineering & System Safety*, *146*, 26–32. doi:10.1016/j.ress.2015.10.008

Xiang, Y., Ding, Z., Zhang, Y., & Wang, L. (2016). *Power system reliability evaluation considering load redistribution attacks*. Academic Press.

Xiao, W., Dunford, W. G., & Capel, A. (2004, June). A novel modeling method for photovoltaic cells. In *2004 IEEE 35th Annual Power Electronics Specialists Conference* (IEEE Cat. No. 04CH37551) (Vol. 3, pp. 1950-1956). IEEE.

Xu, Y., & Singh, C. (2012). Adequacy and economy analysis of distribution systems integrated with electric energy storage and renewable energy resources. *IEEE Transactions on Power Systems*, *27*(4), 2332–2341. doi:10.1109/TPWRS.2012.2186830

Yadav, A. K., & Mahajan, V. (2019a). *Reliability Improvement of Power System Network With Optimal Transmission Switching*. Paper presented at the IEEE 1st International Conference on Energy, Systems and Information Processing (ICESIP), Kancheepuram. 10.1109/ICESIP46348.2019.8938283

Yadav, A. K., & Mahajan, V. (2019b). *Transmission System Reliability Evaluation by Incorporating STATCOM in the System Network*. Paper presented at the IEEE Student Conference on Research and Development (SCOReD), Perak, Malaysia. 10.1109/SCORED.2019.8896263

Yadav, A. K., Mudgal, S., & Mahajan, V. (2020b). *Transmission Switching Based Available Transfer Capability Assessment to Make System Network Reliable*. Paper presented at the International Conference on Electrical and Electronics Engineering (ICE3), Gorakhpur.

Yadav, E. P., Mittal, E. A., & Yadav, D. H. (2018). IoT: Challenges and Issues in Indian Perspective. *2018 3rd International Conference on Internet Of Things: Smart Innovation And Usages (IoT-SIU)*, 1-5. 10.1109/IoT-SIU.2018.8519869

Yadav, A. K., Mudgal, S., & Mahajan, V. (2020a). Monte Carlo Simulation Application in Composite Power System Reliability Analysis. In *Control Applications in Modern Power System* (pp. 379–385). Springer.

Yang, H., Park, M., Cho, M., Song, M., & Kim, S. (2014). A System Architecture for Manufacturing Process Analysis based on Big Data and Process Mining Techniques. *2014 IEEE International Conference on Big Data*.

Yousri, D., Allam, D., Eteiba, M. B., & Suganthan, P. N. (2019). Static and dynamic photovoltaic models' parameters identification using Chaotic Heterogeneous Comprehensive Learning Particle Swarm Optimizer variants. *Energy Conversion and Management, 182*, 546–563. doi:10.1016/j.enconman.2018.12.022

Yu, Saito, Li, & Ceylan. (2017). *Learning dense facial correspondences in unconstrained images.* arXiv:1709.00536.

Yuan, H., Li, G., Bie, Z., & Arif, M. (2019). *Distribution system reliability assessment considering cyber-physical integration.* Academic Press.

Yu, D., Wang, S., & Deng, L. (2010). Sequential labeling using deep-structured conditional random fields. *Journal of Selected Topics in Signal Processing, 4*(6), 965–973. doi:10.1109/JSTSP.2010.2075990

Zanella, A., Bui, N., Castellani, A., Vangelista, L., & Zorzi, M. (2014, February). Internet of Things for Smart Cities. *IEEE Internet of Things Journal, 1*(1), 22–32. doi:10.1109/JIOT.2014.2306328

Zhang, R., Zhao, Z., & Chen, X. (2010). *An overall reliability and security assessment architecture for electric power communication network in smart grid.* Paper presented at the 2010 International Conference on Power System Technology.

Zhang, X., Bie, Z., & Li, G. (2011). *Reliability assessment of distribution networks with distributed generations using Monte Carlo method.* Academic Press.

Zhang, Y. (2012). Studies on application of Support Vector Machine in diagnose of coronary heart disease. In *Electromagnetic Field Problems and Applications 2012 Sixth International Conference (ICEF).* IEEE.

Zhang, Y. (2012). Studies on application of Support Vector Machine in diagnose of coronary heart disease. In *Electromagnetic Field Problems and Applications 2012 Sixth International Conference (ICEF).* IEEE. 10.1109/ICEF.2012.6310380

Zhang, P., Portillo, L., & Kezunovic, M. (2006). Reliability and Component Importance Analysis of All-Digital Protection Systems. *2006 IEEE PES Power Systems Conference and Exposition*, 1380–1387. 10.1109/PSCE.2006.296504

Zhang, Y., Liu, J., & Zhang, C. (2013). Improved pulse-width modulation of diode-assisted buck-boost voltage source inverter. *IEEE Transactions on Power Electronics, 28*(8), 3675–3699. doi:10.1109/TPEL.2012.2227816

Zhao, C.W., Jegatheesan, J., & Loon, S.C. (2015). *Exploring IOT Application Using Raspberry Pi.* Academic Press.

Zheng, J., Okamura, H., & Dohi, T. (2018). Component Importance Analysis of Mobile Cloud Computing System in the Presence of Common-Cause Failures. *IEEE Access: Practical Innovations, Open Solutions, 6*, 18630–18642. doi:10.1109/ACCESS.2018.2822338

Zheng, X., Zeng, B., Wu, G., Zhang, J., Zeng, M., & Shi, J. (2015). Capacity credit assessment of renewable distributed generation in active distribution systems considering demand response impact. *Proc. 5th Int. Conf. Elect. Utility Deregulation Restruct. Power Technol. (DRPT)*, 108–113.

Zhou, Y., Mancarella, P., & Mutale, J. (2016). Framework for capacity credit assessment of electrical energy storage and demand response. *IET Generat. Transmiss. Distrib., 10*(9), 2267–2276.

Zhu, W., Wu, H. T., & Chen, Z. (2020). ReDA: Reinforced Differentiable Attribute for 3D Face Reconstruction. *Proceedings of the IEEE/CVF Conference on Computer Vision and Pattern Recognition (CVPR)*, 4958-4967.

About the Contributors

Rajiv Singh is with the Department of Electrical Engineering, College of Technology, G.B. Pant University of Agriculture & Technology, Pantnagar, Uttarakhand, India. His area of interest includes control and instrumentation, wind and solar power, energy markets and policies, energy management, and audit. He has over 15 years of experience in teaching and research at various levels of technical education. He has also vast administrative experience of serving in various capacities at technical institutions and universities. He has organised many international short-term courses funded by the Ministry of human resource & development govt of India in his affiliating department. He has also served as a reviewer for many international journals in IEEE and Elsevier and attended many international conferences and training in India.

Ashutosh Kumar Singh is currently associated with the Department of Electronics & Communication, Indian Institute of Information Technology Allahabad. He has 14 years' experience in teaching and research at various levels of technical education. His teaching areas are Electrical Engineering, Control Engineering & wireless sensor network. He was also associated with the department of electronics engineering CSJM, University Kanpur prior to this institute. He is also the principal investigator of two sponsored projects. He has published more than 60 papers in reputed journals & international conferences.

Ajay Kumar Dwivedi is associated with the Department of Electronics & Communication, SIET, Dr. A.P.J. Abdul Kalam Technical University, Lucknow, India. He has seven years of teaching experience with UG and PG students. He has authored or co-authored more than 30 research papers in international/national journal/conference proceedings. His research interests include RF & microwave, microstrip patch antennas, wireless communication, antenna theory, Dielectric resonator antenna, MIMO DRA, metamaterial-based dielectric resonator antenna.

P. Nagabhushan is currently holding the position of Director at the Indian Institute of Information Technology Allahabad. He has been a professor in the Department of Computer Science Engineering, University of Mysore, Mysore. His research area includes Artificial Intelligence, Remote sensing, Imaging science, and photographic technology. He has over 30 years of experience in teaching and research at various levels of technical education. He has also vast administrative experience of serving in various capacities at technical institutions and universities. He has Worked as Investigator in major funded projects like Indo-French Research Project on Symbolic Data Analysis, DRDO Research Project on Dimensionality Reduction, and many more. He has published more than 200 papers in reputed journals & international conferences.

* * *

Abhay Kumar Agarwal is presently working as Assistant Professor in Department of Computer Science and Engineering, Kamla Nehru Institute of Technology, Sultanpur, India. He has an overall Teaching Experience of more than 17 years. He has done his Doctorate from AKTU Lucknow in 2017, M.Tech from SATI Vidisha in 2006 and B.Tech from BIET Jhansi in 1999. Besides he has also done PGDCA from Mumbai in 2001. He has published more then 45 papers in International Journals and conferences. His area of interest includes Distributed Data Warehouse and Data Mining, Machine Learning, ANN etc.

Gaurav Agarwal was born in Moradabad, India, in 1980. He is pursuing Ph.D. in Computer Science from Indian Institute of Technology (ISM), Dhanbad; M.Tech. in Computer Science & Engineering from Shobhit University, Meerut; B.E. in Computer Engineering from the North Maharastra University, Jalgaon, Maharastra, India, in 2003. Since July 2018, he has been with the Department of Computer Science and Engineering as an Assistant Professor, Raj Kumar Goel Institute of Technology, Ghaziabad, India. He has more than 14 years of academic experience. He is life member of International Association of Computer Science and Information Technology (IACSIT), Indian Society for Technical Education (ISTE) and Vijnana Bharati (VIBHA). He was the recipient of Young Investigator Award by Interscience Research Network (IRNet), Orissa, Vocational Award, Rotary International U.S.A. and The Rotary Foundation award for outstanding service for world understanding and peace. He has contributed more than thirty research papers in several International and National journals and conference proceedings of high repute. His main areas of research interest are Speech Signal Processing, Machine Learning and Genetic Algorithms.

Neelendra Badal is a Professor in Computer Science & Engineering Department at Kamla Nehru Institute of Technology (KNIT), Sultanpur, India. He received his B.Tech degree in 1997 from (BIET) Jhansi in Computer Science & Engineering. M.E. (2001) in Communication, Control and Networking from Madhav Institute of Technology and Science (MITS), Gwalior and Ph.D. (2009) in Computer Science & Engineering from Motilal Nehru National Institute of Technology (MNNIT), Allahabad. He is Chartered Engineer (CE) from Institute of Engineers (IE), India. He is Life Member of IE, IETE, ISTE and CSI-India. He has published about 30 papers in International/ National Journals, conferences and seminars. His research interests are Distributed System, Parallel Processing, GIS, Data Warehouse & Data Mining, Software Engineering and Networking.

Anjana Jain is Senior member IEEE, life member of IETE, ISTE, IE. She has completed her Ph.D. in Wireless channel characterization. She has more than 20 contributions in Reputed Journals and presented research papers in India and Abroad. Prof. Jain has been selected to participate in workshops sponsored by United Nations.

Harsh Kakadiya is currently an undergraduate student at Sardar Vallabhbhai National Institute of Technology. He is also selected for Electronics and Communication Engineering in minor degree program at Sardar Vallabhbhai National Institute of Technology.

Amritesh Kumar was born in Bihar, India, in 1986. He has received his B.Tech in Electrical and Electronics Engineering from SASTRA University, Tamilnadu, India in 2009; M.Tech. in Power Electronics and Drives from BITS PILANI, India in 2011 and Ph.D. from the from Delhi Technological University (DTU), formerly Delhi college of Engineering, Delhi, India in 2017. In June 2018 joined the Department of Electrical Engineering, NIT Silchar, India, as an Assistant Professor. He has been a recipient of VEL TECH Merit Scholarship for excellence in academics at BITS PILANI, 2011. He has been awarded under Commendable Research category at Delhi Technological University for the year 2017 and 2018 for Publication in IEEE Transactions. He has also been awarded prestigious POSOCO Power System (PPSA-2018) award under doctoral category. Dr. Amritesh Kumar has so far authored more than 16 papers in leading journals and conferences including IEEE Transactions/IET Journals and others. He is an active reviewer many leading journals line for IEEE transactions on Industrial Electronics, Informatics and Industry application. His research interests include Analysis and design of power converters/inverters and its application in renewable energy sources, Multi-level systems, Power quality issues, AC–DC microgrids and Hybrid electric vehicles.

Prabhat Kumar was born in Jharkhand, India, in 1996. He completed his B.Tech. in Electrical & Electronics Engineering from Nalanda College Of Engineering, Bihar in the year 2019. He is currently pursuing his M.Tech. in Electrical Engineering from National Institute Of Technology Silchar, Assam, India. His area of Interests include analysis and control of Power Electronics converters.

Vasundhara Mahajan is presently working as associate professor at department of electrical engineering, Sardar Vallabhbhai National Institute of Technology (SVNIT), Surat, Gujarat. She has obtained her doctoral (Ph.D.) and master degree (M. Tech.) form IIT Roorkee in 2014 and 2005 respectively. Graduated in Electrical Engineering from NIT Raipur (formerly GEC) in 1999. She worked as lecturer at Christian college of engineering and technology from Sept. 2000 to Oct. 2007. Then joined Sardar Vallabhbhai National Institute of Technology (SVNIT), Surat, Gujarat. She has published many research papers in International/national journals/conferences. She has organized 11 Short-term training programs at SVNIT. She has delivered many expert talks in various institutes/colleges. She has guided many M.Tech. and B. Tech. projects. Presently she is guiding seven doctoral thesis. Her research area is cyber security of smart grid, power system reliability, restructuring/deregulation, congestion management, energy market, power quality improvement, active power filters, FACTS, and artificial intelligence.

Mahshooq Majeed is pursuing M. Tech in Power Systems from Sardar Vallabhbhai National Institute of Technology Surat, completed B. Tech from National Institute of Technology Goa. Currently doing a major project on Machine Learning algorithms for Smart Grid.

Arun Parakh is working as Associate Professor in the Department of Electrical Engineering at Shri G.S. Institute of Technology and Science Indore (MP) India since 2003. He has done his PhD from Computer Science and Engineering, IIT Delhi, New Delhi in 2016, India. He obtained his post graduation in Computer Engineering and under graduation in Electrical Engineering from Shri G.S. Institute of Technology and Science Indore (MP) India in 2003 and 1996 respectively. His area of interests include embedded system design, high performance computing and computer architecture. He has published more than 20 research papers in various international journals and conferences. He has supervised more than 50 post graduate thesis in the area of embedded systems, also guiding several PhDs in the same area.

Janavi Popat is an undergraduate student at Sardar Vallabhbhai National Institute of Technology. She is interested in smart grid, cybersecurity, reliability, and machine learning and deep learning applications in power system planning.

Viraat Saaran was born in Modinagar, Uttar Pradesh, India. He is currently pursuing his Bachelors of Technology in Computer Science and Engineering (2017-2021).

Anurag Singh received a Bachelor of Technology in Electrical Engineering from . Allahabad in 2014, Master of Technology in Control and Instrumentation Engineering from M.M.M. University of Technology Gorakhpur in 2018. He is working as an Assistant Professor in the Department of Electrical Engineering, Veer Bahadur Singh Purvanchal University, Jaunpur (U.P.) since 2018. Main research includes in Control System, Model Order Reduction, Optimization.

Deepika Singh completed M.Tech from Motilal Nehru National Institute of Information Technology Allahabad, India. She is a Senior Research Fellow in the Department of Electronics and Communication Engineering in Indian Institute of Information Technology, Allahabad. She is working in the field of Breast Thermography for early cancer detection for the past 2 years.

Neeraj Kumar Singh received a master's degree in Electrical Power Systems from the Government Engineering College Aurangabad, Maharashtra, India. He is currently pursuing a Ph.D. degree in Electrical Engineering, Sardar Vallabhbhai National Institute of Technology (SVNIT), Surat, Gujarat. He was working as Engineer Operations at Enercon India Ltd (Wind World India Ltd). He has also contributed to the academic field as Assistant Professor at P.E.S College of Engineering, Aurangabad. His research interests include smart grid, cybersecurity, and wireless sensor networks.

Lalit Tak is pursuing his post-graduation in Power System from S.V. National Institute of Technology, Surat, Gujarat. He has completed his Bachelor of Engineering in Electrical Engineering from MBM Engineering College, Jodhpur in 2019. He has completed his diploma in Electrical Engineering from SLBS Engineering College, Jodhpur in 2015. His area of interests are switchgear and protection, cyber security, and reliability of power systems.

Padmanabh Thakur has completed his Ph.D. in Electrical Engineering from MNNIT Allahabad. Presently he is working as Professor and Head, Department of Electrical Engineering, Graphic Era (Deemed to be University), Dehradun. Also, he is holding the responsibility of Associate Editor, "IEEE Access" and Convener, IEEE SIGHT UP Section (for 2019).

Sonia Tiwari is a Consultant Radiation Oncologist in the Department of Oncology at Kamala Nehru Memorial Hospital Regional Cancer. She is an M.B.B.S., D.N.B and specialization is oncology She has also been working in pediatric oncology project at KNMH since 2013.

Atul Kumar Yadav has obtained his B.Tech in Electrical & Electronics Engineering (EEE) from Beehive College of Engineering and Technology (BCET), Dehradun, Uttarakhand, India. He received the master's degree in Power Systems from the Sardar Vallabhbhai National Institute of Technology (SVNIT), Surat, Gujarat, India. He is currently pursuing the Ph.D. degree in Electrical Engineering, Sardar Vallabhbhai National Institute of Technology (SVNIT), Surat, Gujarat. His research interests include power system reliability, Renewable Energy Sources (RES), Cyber Physical Analysis and Smart grid. He has won Student Travel Grant for research paper presentation in the 2nd International Conference on Information and Communications Technology (ICOIACT) organized by Universitas AMIKOM held on 24–25 July 2019 in Grand Zuri Malioboro Hotel, Yogyakarta, Indonesia. He got awarded with Best paper award for research paper presentation in 17th IEEE Student Conference on Research and Development (SCOReD) held at the Universiti Teknologi PETRONAS (UTP), Tronoh, Perak, Malaysia from 15th to 17th of October 2019.

Index

K

L

M

N

P

R

S

T

W

www.ingramcontent.com/pod-product-compliance
Lightning Source LLC
LaVergne TN
LVHW060305011026
849161LV00011B/2776